Marine Microplastics and the Quest for Remediation

Mohammad Oves • Anish Khan
Majed Al-Shaeri
Editors

Marine Microplastics and the Quest for Remediation

Springer

Editors
Mohammad Oves
Center of Excellence
in Environmental Studies
King Abdulaziz University
Jeddah, Saudi Arabia

Anish Khan
Center of Excellence for Advanced
Materials Research
King Abdulaziz University
Jeddah, Saudi Arabia

Majed Al-Shaeri
Department of Biological Science,
Faculty of Science
King Abdulaziz University
Jeddah, Saudi Arabia

ISBN 978-981-96-8978-1 ISBN 978-981-96-8979-8 (eBook)
https://doi.org/10.1007/978-981-96-8979-8

This Springer imprint is published by the registered company Springer Nature Singapore Pte Ltd.
The registered company address is: 152 Beach Road, #21-01/04 Gateway East, Singapore 189721, Singapore

Preface

The oceans, with their vast expanse and depth, have never failed to arouse wonder and intrigue. They provide long-lived, complex ecosystems for the earth as a whole, but not as humans define life support. For centuries, mankind have marveled at the beauty of the seas, relied on them for resources, and gained inspiration from their sustaining rhythms. Still, under its lustrous coat, this is a growing disaster, threatening more than the ecosystem of the marine world; it threatens everything we know of life. In Ocean's Invisible Scourge: Our book entitled "Marine Microplastics and the Quest for Remediation" discusses the nefarious, ubiquitous problem of global microplastic pollution.

Beginning with the unsoiled magnificence of the oceans—their quiescent equilibrium—a heartbreaking ode to the once pristine waters of our seas, this book is at once a story of our deep connection with the oceans and a call to action to protect the precious land, sea, and creatures that make the world so wondrously beautiful. Nature's finest in the oceans includes the vibrant coral reefs and the haunting melodies of humpback whales. Although we marvel at their grandeur, these ecosystems have been invaded by little plastic particles—the byproducts of humanity's ceaseless pursuit of convenience. So small and yet so spread around, microplastics have insidiously worked their way into every part of the marine environment—among the tiniest plankton and the largest predators.

But microplastics are not just an annoyance. They are an incremental catastrophe. It's overwhelming and scary data. These particles are erroneously identified by marine species, with the whole gamut, from minuscule zooplankton to mighty whales, mistaking the particles for sustenance. After ingestion, microplastics cause considerable disruption: internal blockages, release of harmful substances, and malfunctioning physiological systems. But that is not the end of the narrative. Instead, these particles traverse the food system and build up at higher trophic levels, eventually reaching humans. Our highly valued seafood is accompanied by an unintended contaminant—microscopic plastics that may serve as unintended, unknown, and possibly significant health threats.

But the devastation does not end with physical injury. Microplastics, sometimes known as Trojan horses, transmit harmful illnesses and poisons into our struggling marine ecosystems. As a result, their presence raises disease risk, disrupts reproductive cycles, and impairs aquatic populations' resilience. The oceans, which stood as an emblem of boundless vitality for so long, are now fighting with an enemy you can't quite make out because it was too busy being created with every discarded bottle, bag, and straw.

But this book is not a single record of misery, for her life was not a failure. The microplastic threat is massive but not unbeatable. In the past, humanity has shown us time and again that we are capable of finding creative, collective, and committed responses to existential threats. This will be the transition from defining the problem to finding remedies that can take care of it.

Through this text, we read stories of optimism and perseverance. So, how are researchers, engineers, legislators, and advocates trying to deal with microplastic contamination? Sophisticated filtration systems and nanotechnology-based approaches are among the new technologies being introduced to purify our oceans. At the same time, efforts are underway to arrive at sugary replacements for classic plastics, which could help prevent pollution at its source.

But technology alone cannot answer this question. The fight against microplastics will require consensus, from scientific innovation to government reform, economic encouragement, and public participation. Collective efforts, such as the commitment to a global plastic waste treaty, are an example of successfully having such conversations. Grassroots cleanup projects and promoting sustainable lifestyles by local communities are making a significant contribution.

The relevance of education and awareness needs to be educated to take on the microplastic epidemic, which this book highlights. By understanding the significant implications of every choice we make, we may start a change in what is normal behavior in our societies. It also makes you consider how you connect with plastic and how little changes, like cutting back on single-use plastic, can make a big difference.

Ocean's Invisible Scourge is primarily a clarion call for action. This forces us to rethink how we relate to seas for what they are, not limitless resources or rubbish tips, but indispensable partners in our welfare. In addition to being a source of sustenance and a way to travel, the oceans are the planet's lungs, controlling the climate and sustaining life. It is a moral imperative to safeguard them and more than an environmental necessity.

You will be both disturbed and uplifted as you read through these pages. You'll gain a complete picture of the microplastic crisis—where it comes from and its impacts. In addition, you will discover the power of human creativity and collaboration in stitching together solutions. This is not a volume about the waters; they will transcend them and will become a reflection of our ability to admit our mistakes and plot a line toward a more sustainable future.

We invite you to learn more about the intricacies of this problem and its solutions. From providing sustenance to supplying water to inspire awe and provide life, humanity has always relied on the seas. We must support and celebrate their heritage and protect their vitality for generations to come. Together, we can fight microplastic pollution and restore the oceans to their right of origin, life, and hope.

We are at the time of action. Let us realize it.

Jeddah, Saudi Arabia — Mohammad Oves
Jeddah, Saudi Arabia — Anish Khan
Jeddah, Saudi Arabia — Majed Al-Shaeri

Contents

Editors and Contributors

About the Editors

Mohammad Oves is an Associate Professor at the Center of Excellence in Environmental Studies, King Abdulaziz University, Jeddah, Saudi Arabia. With a Ph.D. in Microbiology from Aligarh Muslim University, India, his research focuses on environmental microbiology, nanotechnology, antimicrobial agents, and bioremediation. Dr. Oves has significantly contributed to the field of nanomaterials bio-fabrication, studying their applications in combating multidrug-resistant pathogens and environmental pollutants. His prolific research has resulted in over 125 peer-reviewed publications, with more than 12,500 citations and an impressive h-index of 50. Recognized among the Top 2% Scientists Worldwide, Dr. Oves has secured multiple prestigious research grants, focusing on areas such as nanomaterials for antimicrobial applications and environmental decontamination. He is a top reviewer on Web of Science for grant proposal and research articles. He has served as a visiting researcher at Tohoku University, Japan, and has collaborated extensively with institutions like King Saud University and global universities. Dr. Oves has also edited prominent books on nanotechnology, environmental sciences, and biomedicine. As a dedicated mentor, Dr. Oves supervises Ph.D. and Master's students, fostering research in microbial interactions, nanotechnology, bioenergy, and environmental sustainability. His commitment to advancing microbiology and nanoscience continues to impact both academia and industry.

Anish Khan is currently working as an Associate Professor at the Centre of Excellence for Advanced Materials Research, Faculty of Science, King Abdulaziz University, Jeddah, Saudi Arabia. He completed his Ph.D. from Aligarh Muslim University, India, in 2010 and his postdoctoral in electroanalytical chemistry at the School of Chemical Sciences, University Sains Malaysia (USM) in 2010. Over the past 3 years, he has been listed as a World 2% Influential Scientist. He has published more than 450 international publications that include research articles (around 300), edited books (more than 50), review articles (more than 20), book chapters (around 107), and two United States patents. He has a Google Scholar h-index of 51 and over 10,000 citations. He is a Member of the American Nano Society and an editorial board member of different international journals, including the *Journal of Natural Fibers* (Q1). His field of specialization includes polymer nano-composite, cation-exchangers, chemical sensors and micro-biosensors, nanotechnology, the application of nanomaterials in electroanalytical chemistry, material chemistry, ion-exchange chromatography, and electro-analytical chemistry. His research focuses on the synthesis, characterization (using different analytical techniques), and derivatization of inorganic ion-exchanger by incorporating electrically conducting polymers. This includes the preparation and characterization of hybrid nano-composite materials and their applications, polymeric inorganic cation-exchange materials, electrically conducting polymeric materials, composite material use as sensors, green chemistry by remediation of pollution, heavy metal ion-selective membrane electrode, and biosensor on neurotransmitter.

Majed Al-Shaeri is an esteemed Associate Professor of Environmental Toxicology in the Department of Biological Sciences at King Abdulaziz University, Saudi Arabia. He serves as the Director of the Environmental Protection & Sustainability Laboratory (EPS), the Toxicology Research Group, and heads the Electron Microscopy Unit. Dr. Al-Shaeri earned his Ph.D. in Environmental Toxicology and an M.Sc. in Nano-Ecotoxicology from Heriot-Watt University, Edinburgh, UK. His expertise lies in nanobiotechnology, nanotoxicology, ecotoxicology, and the synthesis and characterization of nanomaterials. His research focuses on the eco/toxicity of nanomaterials, their biomedical applications, and their environmental impact, including nanotoxicity on microbial genomics and proteomics. He has published extensively in high-impact journals, contributed to book chapters, and actively participates in international conferences and workshops. As a recognized expert, Dr. Al-Shaeri has collaborated with various industries and governmental bodies, focusing on environmental health and nanomaterial safety. He is a Fellow of the Royal Society of Biology and holds memberships in numerous scientific societies, including SETAC and the Society of Toxicology. Dr. Al-Shaeri continues to advance research in environmental and nanotoxicology, contributing significantly to human health and environmental protection.

Contributors

Syeda Tahmida Mutahara Abdal Laboratory of Environmental Health and Ecotoxicology, Department of Environmental Sciences, Jahangirnagar University, Dhaka, Bangladesh

Department of Environmental Science, Bangladesh University of Professionals (BUP), Dhaka, Bangladesh

Nasir Ahmed Al-Kinani Faculty of Sciences, Department of Biological Sciences, King Abdul Aziz University, Jeddah, Kingdom of Saudi Arabia

Majed Al-Shaeri Department of Biological Science, Faculty of Science, King Abdulaziz University, Jeddah, Kingdom of Saudi Arabia

Faculty of Sciences, Department of Biological Sciences, King Abdul Aziz University, Jeddah, Kingdom of Saudi Arabia

Juan Carlos Alvarez-Zeferino Universidad Autónoma Metropolitana, Alcaldía Azcapotzalco Ciudad de México, Mexico City, Mexico

Secretaría de la Defensa Nacional, Universidad del Ejército y Fuerza Aérea. Escuela Militar de Ingeniería, Mexico City, Mexico

Ramzi H. Amran Faculty of Marine Sciences, Department of Marine Biology, King Abdulaziz University, Jeddah, Saudi Arabia

Almaz Bereket Bokre Ministry of Agriculture, Keren, Eritrea

Amna Chaudhary Department of Fisheries and Aquaculture, Faculty of Fisheries and Wildlife, University of Veterinary and Animal Sciences, Lahore, Punjab, Pakistan

Arely Areanely Cruz-Salas Universidad Autónoma Metropolitana, Mexico City, Mexico

Universidad Autónoma Metropolitana, Alcaldía Azcapotzalco Ciudad de México, Mexico City, Mexico

Alejandro Cueva Departamento de Ciencias de la Sustentabilidad, El Colegio de la Frontera Sur, Unidad Villahermosa, Villahermosa, Tabasco, Mexico

Tesfalem W. Ghebretnsae Department of Environmental Management, Institute of Environmental Engineering, Peoples' Friendship University of Russia, Moscow, Russia

Department of Land Resources and Environment, Hamelmalo Agricultural College, Keren, Eritrea

Irfan Haidri Department of Environmental Engineering, Mining and Energy University, Gdansk University of Technology, Gdańsk, Poland

Ana Isabel Hernandez-Soriano Universidad Autónoma Metropolitana, Mexico City, Mexico

María Azahara Mesa-Jurado Laboratorio Transdisciplinario para la Sustentabilidad (LaTSu), Departamento de Ciencias de la Sustentabilidad, El Colegio de la Frontera Sur, Unidad Villahermosa, Villahermosa, Tabasco, Mexico

Anish Khan Center of Excellence for Advanced Material Research, King Abdulaziz University, Jeddah, Kingdom of Saudi Arabia

Sheila Liliana Dafne Lobato-Rocha Universidad Autónoma de Baja California, Campus Mexicali, Mexicali, BC, Mexico

George Mangse Department of Biotechnology, Faculty of Sciences, Nile University of Nigeria, Abuja, FCT, Nigeria

Carolina Martínez-Salvador Laboratorio Transdisciplinario para la Sustentabilidad (LaTSu), Departamento de Ciencias de la Sustentabilidad, El Colegio de la Frontera Sur, Unidad Villahermosa, Villahermosa, Tabasco, Mexico

Natalia Paulina Martínez-Toledo Universidad Autónoma Metropolitana, Mexico City, Mexico

Djadjiti Namla Department of Biological Science, King Abdulaziz University, Jeddah, Kingdom of Saudi Arabia

Faculty of Sciences, Department of Biological Sciences, King Abdul Aziz University, Jeddah, Kingdom of Saudi Arabia

Mohammad Orif Department of Marine Chemistry, Faculty of Marine Science, King Abdulaziz University, Jeddah, Kingdom of Saudi Arabia

Mohammad Oves Center of Excellence in Environmental Studies, King Abdulaziz University, Jeddah, Kingdom of Saudi Arabia

Beatriz Pérez-Aragón Universidad Autónoma Metropolitana, Alcaldía Azcapotzalco Ciudad de México, Mexico City, Mexico

Muhammad Qasim Department of Environmental Science, Government College University, Faisalabad, Punjab, Pakistan

School of Agricultural Technology and Food Industry, Walailak University, Nakhon Si Thammarat, Thailand

Md. Mostafizur Rahman Laboratory of Environmental Health and Ecotoxicology, Department of Environmental Sciences, Jahangirnagar University, Dhaka, Bangladesh

Department of Environmental Science, Bangladesh University of Professionals (BUP), Dhaka, Bangladesh

Tareq Rahman Department of Environmental Science, Bangladesh University of Professionals (BUP), Dhaka, Bangladesh

Department of Project Management, Charles Darwin University, Sydney, NSW, Australia

Fahad Rahul Department of Environmental Science and Disaster Management, Gopalganj Science and Technology University, Gopalganj, Bangladesh

Mohd Ahmar Rauf Barbara Ann Karmanos Cancer Institute, School of Medicine, Wayne State University, Detroit, MI, USA

Aylin Geraldine Rodríguez-Villa Universidad Autónoma Metropolitana, Mexico City, Mexico

Amamad Samman Faculty of Sciences, Department of Environmental Sciences, King Abdul Aziz University, Jeddah, Kingdom of Saudi Arabia

Sathianeson Satheesh Faculty of Marine Sciences, Department of Marine Biology, King Abdulaziz University, Jeddah, Saudi Arabia

Fotoon Sayegh Faculty of Sciences, Department of Biological Sciences, King Abdulaziz University, Jeddah, Saudi Arabia

Marine Natural Products Research Unit, King Fahad Medical Research Centre, King Abdulaziz University, Jeddah, Saudi Arabia

Faculty of Marine Science and Environments, Department of Marine Biology and Fisheries, Hodeidah University, Hodeidah, Yemen

Sadia Sikder Laboratory of Environmental Health and Ecotoxicology, Department of Environmental Sciences, Jahangirnagar University, Dhaka, Bangladesh

Department of Environmental Science, Bangladesh University of Professionals (BUP), Dhaka, Bangladesh

Department of Environmental Science and Disaster Management, Daffodil International University, Dhaka, Bangladesh

Jocelyn Tapia-Fuentes Universidad Autónoma de Baja California, Campus Mexicali, Mexicali, Mexico

Mohammad Toha Laboratory of Environmental Health and Ecotoxicology, Department of Environmental Sciences, Jahangirnagar University, Dhaka, Bangladesh

Department of Environmental Science, Bangladesh University of Professionals (BUP), Dhaka, Bangladesh

Department of Environmental Science and Disaster Management, Daffodil International University, Dhaka, Bangladesh

Qudrat Ullah Department of Environmental Science, Government College University, Faisalabad, Punjab, Pakistan

School of Agricultural Technology and Food Industry, Walailak University, Nakhon Si Thammarat, Thailand

Alethia Vázquez-Morillas Universidad Autónoma Metropolitana, Alcaldía Azcapotzalco Ciudad de México, Mexico City, Mexico

Área de investigación Tecnologías Sustentables, Departamento de Energía, División de Ciencias Básicas a Ingeniería, Universidad Autónoma Metropolitana Azcapotzalco (UAM-A), Mexico City, Mexico

Maribel Velasco-Pérez Universidad Autónoma Metropolitana, Alcaldía Azcapotzalco Ciudad de México, Mexico City, Mexico

Muhammad Waseem Department of Environmental Science, Government College University, Faisalabad, Punjab, Pakistan

The Vanishing Beauty: An Ode to Marine Majesty

1

Mohammad Oves, Djadjiti Namla, Mohd Ahmar Rauf, Anish Khan, and Mohammad Orif

Abstract

The complexity of marine systems, once perceived as endless and enduring, is diminishing due to anthropogenic climate change, overfishing, pollution, and habitat loss. Coral reefs, known as the "rainforests of the sea," are endangered due to rising ocean temperatures, resulting in severe coral bleaching that impacts many facets of food chains. As global CO_2 concentrations increase, ocean acidity rises, leading to the demise of marine organisms that utilize calcium carbonate for shell formation, and hence disrupting the entire food chain. The direct and indirect stresses from overfishing and unsustainable coastal development, par-

M. Oves (✉)
Center of Excellence in Environmental Studies, King Abdulaziz University, Jeddah, Kingdom of Saudi Arabia
e-mail: owais.micro@gmail.com

D. Namla
Department of Biological Science, King Abdulaziz University, Jeddah, Kingdom of Saudi Arabia
e-mail: dnamla@stu.kau.edu.sa

M. A. Rauf
Barbara Ann Karmanos Cancer Institute, School of Medicine, Wayne State University, Detroit, MI, USA
e-mail: ahmarrauf2@gmail.com

A. Khan
Center of Excellence for Advanced Material Research, King Abdulaziz University, Jeddah, Kingdom of Saudi Arabia
e-mail: anishkhan97@gmail.com

M. Orif
Department of Marine Chemistry, Faculty of Marine Science, King Abdulaziz University, Jeddah, Kingdom of Saudi Arabia
e-mail: mioraif@kau.edu.sa

M. Oves et al. (eds.), *Marine Microplastics and the Quest for Remediation*,
https://doi.org/10.1007/978-981-96-8979-8_1

ticularly affecting seagrasses and mangroves, heighten ocean sensitivity. Nevertheless, given such a hopeful vision, all hope is not extinguished, as numerous methods exist to construct a PowerPoint presentation that does not adversely affect the audience's general morale. Global conservation activities, including establishing Marine Protected Areas, artificial reefs, and habitat restoration programs, are beginning to produce modest outcomes. Furthermore, advancements in bioinformatics and artificial intelligence (AI) are enhancing scientific understanding and methodologies for monitoring and safeguarding marine biodiversity, creating new conservation opportunities. The United Nations has initiated efforts toward its Sustainable Development Goal 14; nevertheless, global collaboration is essential across governmental, industrial, and individual levels for effective action. Restoring marine ecosystems is critical for ethical reasons and global climate protection, human food security, and economic development. Recognizing the vitality of aquatic ecosystems and the ocean's role in sustaining terrestrial flora, we must protect and conserve the oceans appropriately to guarantee that future generations can appreciate their splendor.

Keywords

Marine · Environment · Coral reef · CO_2 concentrations · Overfishing · Diversity sustainability

1.1 Introduction

The sea can be described as a vast sea of blue, which has fascinated people for millennia, inspiring them with its great majestic beauty along with its great power. They bring light to the surface but only hide a world chock full of organisms and diverse forms of life—this complex web of oceanic and terrestrial life. The sea, from the gentle rolling of the waterline on sandy shores to the thunderous roar of the waves against cliffs, is closest to majesty and mystery. It is not just cross-cultural and cross-generational; it triggers appreciation and respect for emotions. This is a beautiful place that has a complicated structure that is gradually coming under pressure from human activity. Marine ecosystems are vital in support of global ecosystem integrity and balance (Bhatnagar and Kumar 2024). Oceans, which make up about 71% of the Earth's surface, control climate, produce half of the direct oxygen we breathe, and absorb a decent part of carbon dioxide (Hawthorne 2019). Coral reefs and seagrass beds, and mangroves provide sea creatures with their homes and feed millions of people from fishing industries reliant on these reefs (Spalding and Leal 2024). Not only do they have extreme importance and relevance from the perspective of ecology, they also have much historical, cultural, and spiritual importance to many communities across. When trying to understand precisely how life functions in the water, we quickly come to the realization that it is not immune to pollution. The balance of life within the seas and oceans is threatened as never before by such factors as shedding of plastics, over-fishing, climate change, and loss of breeding and feeding grounds (Mega and Mega 2019; Trathan et al. 2015; Barbraud et al. 2012; Everaert et al. 2020).

In this chapter, the author's intention is to explain what beauty the ocean has and what it faces. The argument begins with the breathtaking emphasis on the magnificent ocean and its elaborate analysis of the oceanic systems' biological Web of life. This point reinforces the concepts such as key terms from the existing inventory of ecological terms about resilience defined as the capacity of a system to recover from disturbance and adapt and example of how different species, including the bright coral reefs which are thriving in tropical seas and the deep-see species survived the catastrophe on the sea depths (Cremers 2018). Though, while it is rejoiced with, there is at the same time an imperative need to appreciate the difficulties threatening its future. In all its forms, pollution smothers life in the seas; the heat and acidity alter life patterns and ecosystem processes. Closely related, overfishing eradicates the most significant sea food stocks, as well as consuming the species which negatively impacts other food chains (Issifu et al. 2022). These challenges raise concern and need collective effort in understanding, addressing, and fixing the problem.

Throughout this chapter, the vulnerability of the ocean is highlighted alongside humanity's steadfast determination to adapt to change. Efforts in marine conservation carried forward under the leadership of modern scientific discovery and backed up by patriotic campaigners offer a ray of light in achieving the lost glory of oceans (Rozwadowski 2002; Moore 2011). The call to preserve the oceans has become of unprecedented importance due to the new approaches toward tackling plastic waste and also the nonrenewable agreements for lowering the emission of greenhouse gases. This chapter will also focus on education and community engagement because developing a stewardship is vital for achieving a sustainable future. This chapter honors those who came before and calls upon those of us who survived to take action. It should force the reader to think about their relationship with the ocean and embrace the sea for what it is—a granted, a shared commodity that should be preserved. Of equal value is the public understanding of the numerous benefits provided by oceans, and the challenges they face, which can help strengthen the campaign to preserve the marine environment for future generations, as inspiring as the one that has for centuries fueled imagination and research. This chapter lays down the preliminaries that provide the setting and rationale for the broad discussion on the quest for development on the one hand and conservation of the ocean, the Earth's most precious asset, on the other hand.

1.2 The Splendor of Marine Life

Marine life refers to diverse marine life systems, from colorful coral reef zones to almost boundless pelagic areas. These areas support an extraordinary number of species, each playing an appreciable role in maintaining the balance of the environment (Beaumont et al. 2008; Browman et al. 2004). The ocean's scenic beauty and species richness are necessary for the well-being of the whole world and humans.

1.2.1 Biodiversity Beneath the Waves

1.2.1.1 Overview of the Diversity of Marine Species: From Vibrant Coral Reefs to Mysterious Deep-Sea Creatures

Hidden just beneath the ocean's surface is a real world filled with fish, lobsters, and other marine life, all of which speak to the wondrous bounty of the Earth. There is a vast horizon in the sea, and the sea is full of a number of mysteries that are so marvelous that every creature wants to enjoy the beauty of sea life for a lifetime. There is an imbricated complex pattern of existence throughout the vast marine biome; while each sphere is diverse, it is entwined simultaneously (National Research Council, Division on Earth, Life Studies, Commission on Geosciences,, and Committee on Biological Diversity in Marine Systems 1995; Spalding et al. 2007). Among the most varied and productive habitats are coral reefs, which are also called marine rainforests (Reaka-Kudla 1997). These underwater gardens help provide habitat to about a quarter of marine species, and protect, feed, and raise numerous fish and invertebrate young (Naghibi et al. 2020). The complex color of coral reefs results from a symbiotic relationship between reefs and tiny photosynthetic organisms found in their tissues, the zooxanthellae that feed the brilliant coral reef ecosystem (McCoy 2010; Mishra and Singh 2024). Just as fascinating are the kelp forests, which are large underwater ecosystems abundant in giant kelp, a type of brown algae that may reach two feet/day (Bringloe et al. 2020). Figure 1.1 illustrates the intrinsic phylogenetic relationships between the different clades under the brown algae derived from 12 biomarkers. Species within each clade, as cataloged by

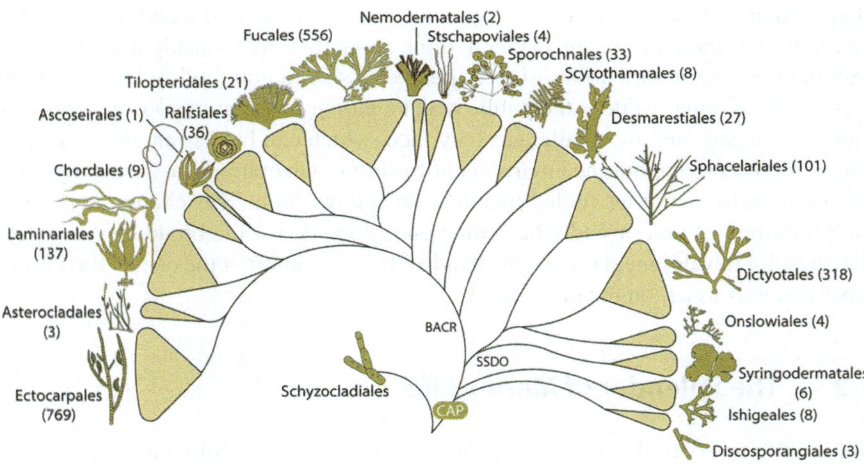

Fig. 1.1 Phylogeny of the brown algal orders based on 12 markers (18S, 5.8S, 28S, atpB, psbA, psaB, psaA, rbcL, psbC, cox1, cox3, nad1), derived from Bringloe et al. (2020). In brackets are the numbers of species within each order according to AlgaeBase. CAP stands for common ancestor of Phaeophyceae; SSDO for Sphacelariales, Syringodermatales, Dictyotales, Onslowiales clade, while BACR represents brown algal crown radiation

the AlgaeBase, highlight the biodiversity and ecological roles of the algae as important primary producers. Located in the relatively warm climate of the world, these forests support organisms from large marine mammals such as sea otters to fish and shellfish. Still, the motion of kelp using ocean currents can be compared to hypnotic motions, the charismatic lines of which express the nature of sea life. Are almost all ecosystems combating climate change by sequestering carbon and regulating global temperatures, and so are seagrass meadows, kelp forests, and mangroves? These ecosystems are not only a refuge for species within water bodies but also components of Earth in its broader environmental systems.

Another marine ecosystem that can be considered very famous now is the Great Barrier Reef, which is the most significant coral reef formation in the world. Situated in northeastern Australia, this natural delight covers more than 2300 km of coastline and includes more than 2900 reefs and 900 islands (Veron 2008). It also hosts Clownfish swimming and schooling among anemones, turtle sightings, and reef sharks, if you can spot them (Woodford 2010). Apart from being a center for population diversity, the Great Barrier Reef is an aspect of nature's design and strength. It forms the creation of evolution constructed by generations of coral polyps that have lived and formed this structure (Sheppard 2021). However, even such species are not beyond the dangers that climate change, pollution, and overfishing, among others, pose, therefore putting the existence of this marvel, as well as the many species it supports, at risk. Besides these famous cases, there is an unsurpassed catalog of life on Earth beneath the ocean waves, down in the abyss. A few kilometers beneath the ocean surface, pressure is immense, and the sun has never shone (Broad 1998) and life blooms. From the bioluminescent jellyfish to the anglerfish with its lighted fangs, and the giant tube worms that inhabit hydrothermal vents (Barbeau 2018; Wilson and Hastings 2012). These organisms have made the environment and are perfect examples of how organisms can survive under almost every condition on planet Earth. Describing the deep-sea environment as "the last frontier of our world," it is noteworthy that deep-sea ecosystems are vital for the global processes of nutrient cycling and carbon sequestration (Smith Jr et al. 2009).

Each of these marine communities, from the red-hot coral reef to the eerie blackness of the abyss, illustrates the incredible richness of life beneath the ocean's surface (Islam 2024). Yet this biotic diversity is not simply beautiful to look at; it is, in fact, fundamental for the well-being of our Earth. Marine species play various social, economic, and cultural values for existence in the ecological system, and thus support activities like fishing, tourism, and biotechnology (Alleway et al. 2019; Rotter et al. 2021). Much about life forms start here at the ocean bottom, and having a close look shows how vulnerable the environment is and how much needs to be protected. If no action is taken, such splendid phenomena as coral gardens, kelp forests, and the Great Barrier Reef might turn into the stuff of legends, and such a loss will be disastrous for any sphere of life. As a result, the tremendous number of species residing in the ocean becomes a call to guardians and cherish the inconceivable value of living in the water.

1.2.2 The Interconnectedness of Marine Ecosystems

1.2.2.1 How Ecosystems like Mangroves, Estuaries, and Coral Reefs Work Together

The ocean is a complex system of interconnected webs of other systems, which are very important in the stability of our planet (Berkowitz 2023). Some of the complex systems include mangroves, estuaries, and coral reefs, which arrive in the network. These ecosystems are not closed systems but open networks of linked and interdependent centers of sustainable resources that sustain sea life, moderate the climate, and defend coasts (Browman et al. 2004). Their mutual interactions are behaviors anatomically developed in a perfect way to perform functions that ensure the supply of food, the supply of energy, and the provision of niches to support the life cycles of numerous invertebrates and vertebrates (Prather et al. 2013). Mangroves on the tropical and subtropical coastlines are the base of many marine systems (Lee et al. 2014). These entwined roots provide what can be described as natural shields or buffers to shores, preventing erosion or damage from surges during storms. Most importantly, they serve as juvenile habitats for fish, crabs, and shrimps, with other stratified habitats, and hence being nurseries to these larger systems (De Abreu 2017). They also act as sediment and nutrient traps and deny the possibility of such sediments and nutrients being carried to the sea. Stabilized and trapped here, these nutrients form the source upon which other nearby estuaries and continuing coral reef stretches grow, thrive, and evolve. Another closely related area is estuaries, the coastal zones where river flows mix with seawater. These nutrient-charge waters are nutrient-wealthy habitats; they help act as a pathway for species that alternate between freshwater and saltwater regions. Estuaries, therefore, serve as a coupling system that links other overland systems to the ocean (Gibson et al. 2002). The yolk from fish eggs is laid in this area, where it is carried downstream to zones where organic materials are deposited. These organic materials serve as food for plankton and other aquatic creatures, such as shellfish and fish. This productivity also impacts nearby reefs; as a result, coral reefs primarily receive nutrients from estuaries (Granek et al. 2009). They also act as a barrier between the pollutants in the sea and make those in contact with the reefs less toxic.

Mangroves and estuaries, on the other hand, are closely related to coral reefs, which are the summits of marine diversity; however, their sustainability is closely related to that of the neighboring mangroves and estuaries. Mangrove and estuarine ecosystems feed the nutrient cycles that, in turn, support the complex food chains that the reef ecosystem members require (Abrantes and Sheaves 2009). Mangroves and coral reefs are also linked to each other in their roles in coastal protection; with mangroves protecting the immediate shore from the wave action, coral reefs, although not as effective in the protection of larger ocean waves, dampen the energy of the waves before they get to the shore (Mumby 2006; Guannel et al. 2016). Collectively, it provides a barrier that softens the impact of storm surges and sea level, a role important given the increasing effects of climate change.

The complexity and interconnectivity of these ecosystems can be seen even in terms of the life cycles of many marine organisms (Capra 2005). Some fish and crustaceans start life in the roots of mangroves before moving on to estuaries and later coral reefs as adults. This movement also emphasizes that the survival of one task environment depends a lot on the others. These lead to chain effects in the marine Web, like sawing down mangroves or polluting estuaries, impacting coral reef systems and the organisms that live in and around them.

All these interrelated systems are also critical for human sustenance as a source of fish, water, and protection against climate change effects. However, it is a poly equilibrium system, which means that the different ecosystems are interdependent: harm to one always creates repercussions for another—for instance, the stripping of mangrove forests results in sedimentation, which forms a cover on coral reefs, and pollution of estuaries, causing alteration of nutrient circulation in the three systems. They say that maintaining these ecosystems is a complex process that should focus not on the various individual ecosystems but instead on the interconnected system. The interaction of mangroves, estuaries, and coral reefs as beautiful examples of intricately woven works of the ocean shows a design behind the creation of the sea that supports every small part (Czerski 2023). Together, they make up a stationary system that supports life in the marine habitat and delivers numerous and highly significant services to society. Being symbiotic, they complement each other; they are also a weak link, which underlines the necessity of the combined approach to ecosystem conservation.

1.2.2.2 Importance of Marine Environments in Regulating Earth's Climate

Marine ecosystems play an important role in climate management on Earth by acting as lungs, heat controllers, and CO_2 absorbers (Lindgren et al. 2016). Oceans cover more than 70% of the Earth's surface and play a vital role in controlling the balance of the planet's climate system. The marine environment is critical to life on Earth in terms of carbon dioxide absorption, heat transmission, and water cycle (Wang et al. 2024). Another component of their relevance in management is their sequestration capacity, another significant way marine ecosystems contribute to global climate change (Oves et al. 2017; Hoegh-Guldberg and Bruno 2010). Seas collect at least 23% of the carbon dioxide in the atmosphere produced by human activities each year. Photosynthesis allows phytoplankton, tiny algae in the water, to manufacture food from carbon and other components before releasing oxygen into the atmosphere (Beer et al. 2014). It not only helps limit the quantity of carbon in the atmosphere but also serves as the foundation for the marine food chain. Carbon sequestration in deep ocean layers, such as the operation of a biological pump, removes this greenhouse gas from the atmosphere for hundreds of years thus mitigating global warming (Farrelly et al. 2013). Marine systems also help to distribute heat in space across the entire world (Trebilco et al. 2021). Most currents function as massive pipelines, transporting heat from the equator to the poles. For example, the Gulf Stream, which transports warm water from tropical latitudes to the North Atlantic, maintains the climate of Europe steady (Palter 2015). Similarly, the ACC helps to moderate climate change worldwide by mixing waters from diverse oceans, allowing heat exchange (Garcia-Soto et al. 2021). This system is critical for preserving temperature stencils, influencing the global climate and weather patterns. Also, it regulates water evaporation and precipitation. The ocean is an essential component in the water cycle (Rüpke and Gaillard 2024). It assesses how warm ocean surfaces make water evaporate and form clouds, impacting rainfall worldwide. Especially in tropical areas, farming and food chains rely on the warm moisture from the ocean waters. Most regions would witness the interruption of severe weather systems like drought or flood if the ocean were not participative in the hydrologic process. Extreme weather impacts are also moderated not only by these world-scale processes but also by maritime environments. The coastal ecosystems that predetermine the ability of coastal areas to protect the mainland and smaller coastal structures from tsunamis, which are mangroves, seagrass, and coral reef systems, may be considered organic barriers. In this way, storm effects on coastal populations are minimized, and these ecosystems act as reservoirs of wave energy. Such natural barriers will only become more critical when the sea levels rise and climate-related weather events happen more often. However, there is increasing concern about the role that the marine environment plays in regulating the Earth's climate. These ecosystems are threatened by the impact of human activities: pollution by plastics, fishing, and burning fossil fuels. Though acidification is a danger to the life of coral reefs and other organisms that form shells and are involved in the carbon cycle, warming waters are reducing carbon sequestration capacity (Adeniran-Obey et al. 2024). It says the climate cannot be regulated anymore because human beings are destroying the habitats of various species and using most of the marine resources unsustainably.

1.3 The Threats to Marine Majesty

1.3.1 Plastic Pollution and Microplastics

1.3.1.1 How Plastics Infiltrate Marine Food Chains and Ecosystems

Plastic pollution has become a common danger to marine life because it can intertwine with local food chains, threatening the balance of the entire marine population (Thushari and Senevirathna 2020). Huge quantities of plastic waste are dumped into the oceans every year, where they break down into what are called microplastics. These are particles often smaller than 5 mm, which are formed when sunlight, waves, and chemistry decompose the larger plastics. Now, they have increased the presence of microplastics on both the water surface and in the deepest parts of the ocean (La Daana et al. 2018; Peng et al. 2018). Marine living organisms think that plastics are food items, which introduces them into the food systems of the aquatic world. Tiny plankton fed at the base of oceanic food chains also regularly ingest microplastics because they look like organic matter (Malinowski et al. 2023). Microplastic pollution starts from floating in the water, consumed by small planktons, then being consumed by fish, then by even bigger fish, and then reaching humans who consume fish that have been polluted by plastic (Agbekpornu and Kevudo 2023). Often, their content includes toxic reagents that accumulate in the

Fig. 1.2 Effects of Plastics on coastal and marine biota, derived from Thushari and Senevirathna (2020): (**a**) plastics ingestion by a blueshark; (**b**) attachment on plastic debris by Goose Barnacle; (**c**) partial cover of macroplastic pollutants on Rock Oyster; and (**d**) entanglement of nestling in a synthetic plastic string

organisms and increase with each level of the food chain, including contaminants and additives soluble in seawater. The oil mixture carried by plastics also has a chemical and mechanical impact on the sea environment. These negative impacts often occur in various ways, as portrayed in Fig. 1.2, reproduced from Thushari and Senevirathna (2020). This figure illustrates some detrimental effects of plastic pollution on coastal and marine biota, highlighting a series of interactions. Another fact is that one in every four sea creatures is associated with coral, which produces disease and causes the population to be affected by plastic litter. Similarly, heaps of plastics smother mangrove and seagrass beds and their extent thus compromising their functionality as valuable ecosystems and carbon stocks. This was compounded by the fact that the constituents of the pollution known as plastics last for hundreds of years in the environment.

Emergency actions are needed because plastics are present in almost every marine environment. One way to safeguard oceans and the animals that inhabit them, including people, is to minimize the manufacturing of plastic materials, enhance ways of disposing of waste products, and support the development of the latest technologies in the removal of plastics from oceans (Gallo et al. 2020; Ogunola et al. 2018).

1.3.1.2 Case Studies Highlighting Regions Severely Affected by Microplastics

Microplastics, which are plastics smaller than 5 mm in size, are now found in environments everywhere globally. These pollutants affect many locations, and specific examples show the extent of the problem: The concentration of microplastic pollution in the Mediterranean Sea is shocking; the studies reveal up to 1.25 million particles per square kilometer of this plastic pollution (Laface 2023; Adamopoulou et al. 2021). This vulnerability results from the geographic location of the Mediterranean, which is enclosed, densely populated, and fully engaged in marine activities. These particles are ingested by marine animals, including fish, turtles, seabirds, etc., and they can harm or even kill such animals and, at the same time, introduce toxins into the food chain. Also, the area is a tourist zone, and the problem increases since plastics are left on the coast and continue to the sea. Nonetheless, the Arctic Ocean, located quite distant, is another region where microplastics play a decisive role. In September 2018, researchers identified minuscule plastic fragments in the Arctic by ocean currents and aerospace deposition from more industrially developed regions (Bergmann et al. 2022). The pollutants are ice-bound microplastics released into the marine environment every time Arctic ice melts, upsetting species and creating fragile ecosystems (Hung et al. 2022). The impact poses a serious threat to the Arctic ecosystem, particularly affecting plankton, which is the cornerstone of the food chain in the region. The cascading effects of microplastics on marine environments remain complex and far-reaching, as depicted in Fig. 1.3, reproduced from Agbekpornu et al. (Agbekpornu and Kevudo 2023). The figure displays the varied pathways via which microplastics impact marine life, from ingestion and entanglement to habitat disruption and trophic transfer. Also, the figure highlights the long-term potential consequences of this plastic pollution on human health in the food chain through the consumption of seafood. Because the general consumption of plastic is excessive and ineffective collection or recycling practices are still to be put into place, combined with large populations residing in coastal cities, microplastic pollution in Southeast Asian countries such as Thailand, Philippines, and Indonesia is significantly high (Sin et al. 2023). That is why, in some coastal regions, as many as 80% of the fish sampled contained noticeable levels of microplastic, thereby posing actual food safety and, potentially, human health risks associated with seafood consumption (Alomar et al. 2017). These case studies show how extensive the microplastic problem is and that many regions of the world have serious ecological and health problems. This indicates that much more mutual international efforts, better waste management, and increased awareness activities to reduce the utilization of plastics must be implemented to combat microplastic pollution.

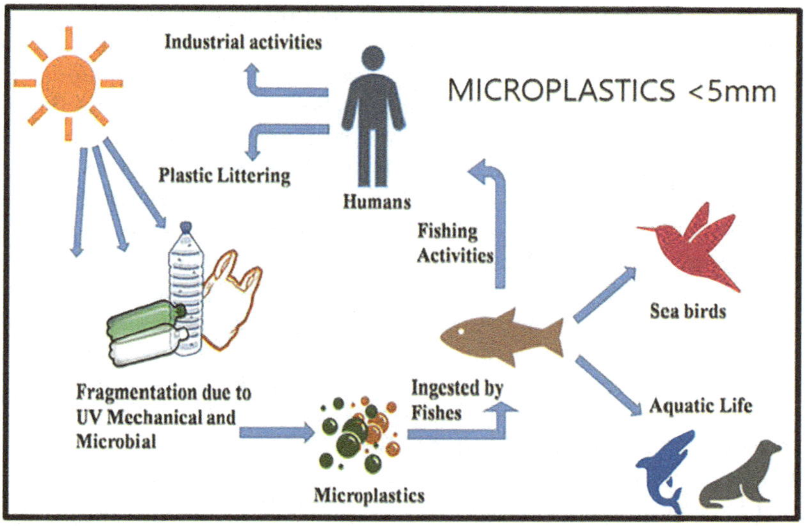

Fig. 1.3 The effect of microplastic pollution on the marine ecosystem, derived from Agbekpornu et al. (Agbekpornu and Kevudo 2023)

1.3.2 Climate Change and Ocean Warming

Climate change, which has been increasing for several decades, has seen a dramatic rise in the levels of global temperature, mainly caused by light-emitting gases such as CO_2 (Yoro and Daramola 2020). One of the most profound consequences of global warming observed in oceans is the disturbance of the already shaken seas' ecosystems by higher water temperatures (Hoegh-Guldberg et al. 2019). Coral reefs, which are fragile ecosystems in the sense that slight elevation in temperature affects them severely, are on the list. Warming of the seas impacts marine ecosystems and coral bleaching, threatening biological and social systems relevant to millions of inhabitants on the globe.

1.3.2.1 Effects of Rising Temperatures on Coral Bleaching and Marine Habitats

Another popular name for coral reefs is "marine rainforests" due to the impressive diversity of species they host. Coral reefs sustain many sea life, from sharks and sea turtles to molluskan animals and fish. Zooxanthellae are microorganisms that dwell within the tissues of coral and should be considered symbionts of the coral (Titlyanov and Titlyanova 2020). These corals have beautiful colors because algae produce their food through photosynthesis; they also obtain essential nutrients from it (Burkholder and Burkholder 1960). Consequently, corals protect the algae, allowing them to thrive. Bleaching of corals, however, happens when temperatures in the oceans rise above a certain level, generally by 1–2 °C (Goreau and Hayes 2024). Because stressed corals expel the zooxanthellae, algae are absent, so corals look white and cannot provide basic nutrients; the corals often get sick, starve, and, most often, die. Bleached corals have less stability against other changes and cannot easily stand other stresses such as water pollution and ocean acidity (Quigley and Baird 2024). An increase in seawater temperatures has been identified as the leading cause of the change in characteristics concerning the frequency and intensity of coral bleaching events. The number of large-scale bleaching events grows; examples can be observed in almost all the zones, including the Great Barrier Reef, the largest coral reef structure in the world, which has experienced several consecutive bleaching episodes in recent years. Mass coral mortality and severe reef degradation were encountered during the bleaching periods in 2016 and 2017 (Eakin et al. 2019). Since many species use coral reefs as homes, breeding grounds, and feeding locations, these corals' deaths continue to affect overall marine life. This long-term damage also reduces the ability of the coral to repair itself.

1.3.2.2 Impact on Marine Habitats

Apart from coral reefs, other communities in the marine environment are also affected by ocean warming (Abirami et al. 2021). They say that various marine animals may shift their locations and behaviors due to warmer waters. Many fish species targeted by commercial fishers favor a given temperature range for food intake, spawning, and movement. There are effects on fish populations and ecosystems, such as fish moving to higher latitudes or new deeper waters for better facilities when temperatures increase (Pörtner and Peck 2010). Some industries may be under threat, particularly those used to fishing in tropical waters where fish resources are already overstocked. Warmer ocean temperatures also lead to species change and make ocean acidification a much larger problem (Kroeker et al. 2013). About half of the dissimilar CO_2 in the sphere is solubilized in the seas, which in turn results in a decline in the pH. This is further compounded by the fact that lower pH affects the formation of calcium carbonate shells and skeletons by species such as coral, plankton, and shell fish. Ocean warming dramatically impacts the loss of important habitats such as kelp forests, mangrove forests, and seagrass meadows. These are critical systems for preserving species, coastal defense, and carbon storage. They also serve as breeding grounds for many sea organisms. These habitats are strained by increasing water temperatures, which leads to mortality and reduced performance. Thus, for instance, seagrasses and mangroves do not undergo any temperature changes and may be eradicated in areas that are gradually experiencing a change in temperature for many years.

1.3.2.3 Ocean Acidification and Its Impact on Shell-Forming Organisms

Global warming caused by the accumulation of greenhouse gases in the atmosphere is proving to bring about profound changes to the climate of the Earth. It changes the extent to which the seas are warmed, which causes ocean acidification. Ocean acidification is the tendency toward an increase in the concentration of hydrogen ions within the water mass due to the "dissolution" of CO_2 in water in the present-day atmosphere (Feely et al. 2009). The oceans absorb correspondingly greater volumes of CO_2 as their levels rise and undergo a chain of reactions that make seawater less acidic. The change across the scale poses a grave risk to aquatic wildlife since many species dependent on calcium carbonate to construct their shells and skeletons are at risk, like corals, mollusks, and the right kind of plankton (Copper 1992).

The Process of Ocean Acidification

Ocean acidification occurs in several processes, including gas carbon dioxide from the atmosphere, dissolving in the water and forming carbonic acid (Doney et al. 2009). This weak acid dissolves into hydrogen ions H^+ and bicarbonate ions HCO_3^-, reducing the pH of the seawater (McElligott and Byrne 1997). As a result, the acid content of the ocean keeps rising over time. Because saltwater possesses a higher concentration of hydrogen ions, the quantity of carbonate ions (CO_3^{2-}) that play a significant role in building and maintaining the calcium carbonate shells and skeleton in several sea organisms reduces (Hempel 2016). These organisms include major species, including clams, oysters, snails, corals, and particular types of plankton. The reduced carbonate ions and increased acidity create an environment ill-suited to organisms that build shells (Hull 2016). Over time, as carbonate ions are made less available, calcification, by which calcium carbonate is used to build shells and skeletons, is rendered much less healthy and less manageable for these species (Berman 2008). Shells of these species may crack, thin, or deform, affecting their physical framework and making them vulnerable to any next environmental factor, disease, or predator.

Impact on Shell-Forming Organisms

Shell-forming organisms are essential components of marine ecosystems. They create complex structures, such as coral reefs and shellfish beds, which support a variety of marine organisms (Harrison and Booth 2007). Such organisms act as food for a number of animals, including fish, birds, and human beings. However, ocean acidification is seriously threatening their lives and existence in the area in many ways. Figure 1.4 summarizes the diverse impacts of ocean acidification across various marine taxa, derived from results presented by Kroeker et al. (2013). In fact, the figure provides a visual overview of these negative effects. Coral reefs remain highly vulnerable due to their structural requirement for calcium carbonate (Cornwall et al. 2021). As the ocean's acid–base gets worse, corals are less able to create and sustain reefs. This leads to reduced, vulnerable coral structures that can no longer effectively cope with conditions such as storms, diseases, and bleaching events. The decline in the state of coral reefs affects the majority of the sea inhabitants, creatures that stay, feed, and breed in coral reefs. Oysters, clams, mussels and other marine animals provide examples of ocean acidification (Donkersloot 2012). These species require calcium carbonate to build shells; as carbonate ions decrease, the shells lose their solidity and dissolve. This is risky to their existence with the aid of tremendous economic consequences. Molluskan shellfish farming, an important industry in many coastal regions, is under pressure to fail the most due to reduced size and increased shellfish mortality. Among such PPCs (Planktonic Primary Producers and

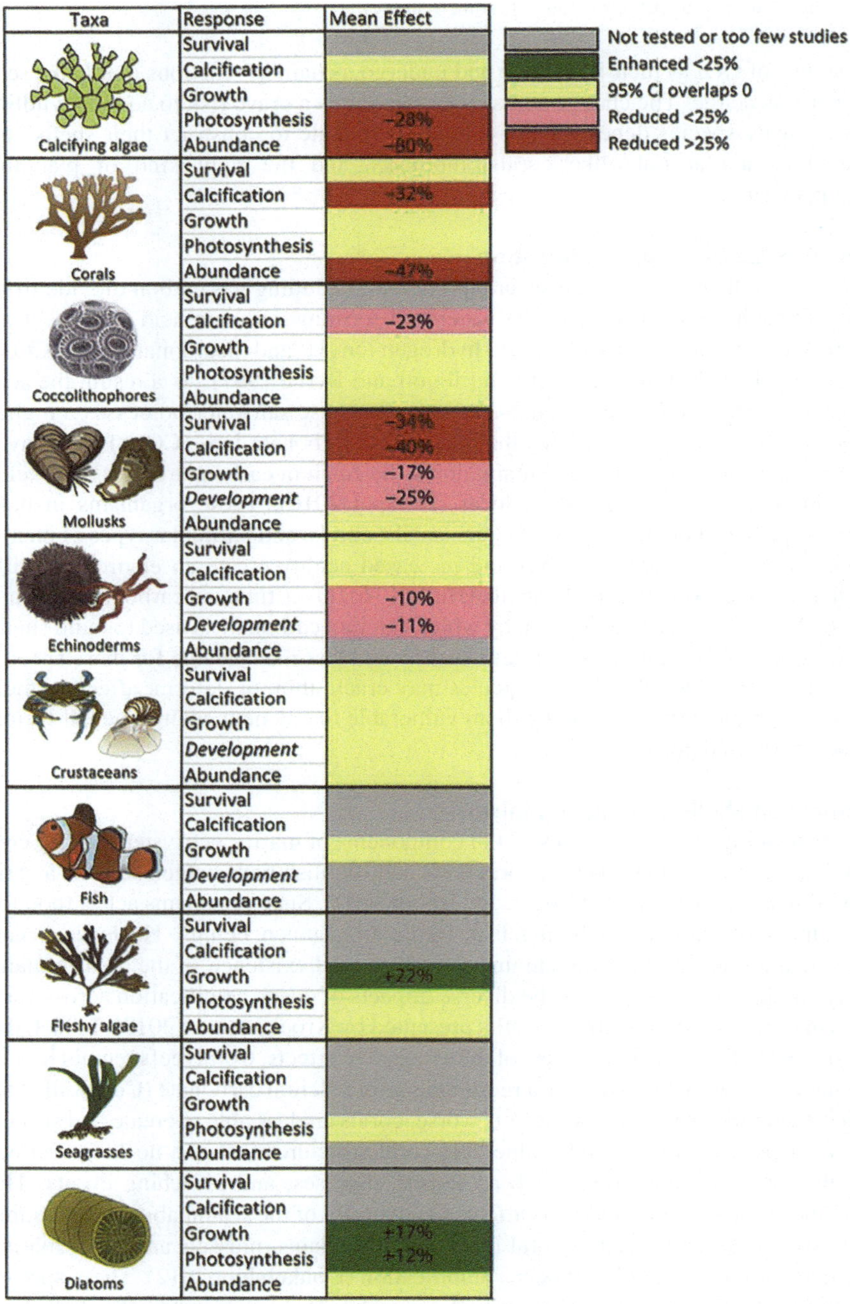

Fig. 1.4 Summary of effects of acidification among key taxonomic groups, derived from results presented by Kroeker et al. (2013). Effects are represented as either mean percent (+) increase or percent (−) decrease in each response. Percent change estimates were back transformed from the mean LnRR, and represent geometric means, which are conservative of the arithmetic means

Calcifiers), foraminifera and coccolithophores, representing a subclass of thecate plankton, are species of great importance to marine food webs (Mudie et al. 2021). These organisms belong to the lowest trophic level in the food web for many aquatic organisms. Ocean acidification affects calcium carbonate shell growth, disrupts entire ecosystems, and affects higher order trophic levels. Reduction in plankton stocks means fewer foods for fish, marine animals, and avians, affecting the world's stock of fish and food (Smayda 2019; Eby et al. 2006).

1.3.2.4 Broader Ecological and Economic Consequences

The consequences extend beyond species-level impacts. Mollusks losing their shells suffer marine food chains, and the type of differences in the species' structure alters (Gazeau et al. 2013). For fish species that feed on mollusks and plankton, this may cause a food shortage and decrease fish yields. This may affect Commercial fisheries, a central pillar in millions of people's income and staple diet. This will likely have high potential regional economic consequences for coastal countries and industries such as shellfish production, tourism, and fishing. The deterioration of healthy coral reefs might reduce the revenue from tourism businesses and the benefits of their ecosystem for storm barriers and coastlines (Goreau and Hilbertz 2005).

1.3.3 Overfishing and Habitat Destruction

1.3.3.1 The Toll of Unsustainable Fishing Practices

Both overfishing and habitat elimination represent the two significant environmental challenges facing the seas today. Evidence shows that fishing has been carried on unsustainably over the years due to increasing fish demand and inadequate measures to manage the stock's supply.

This integration threatens the welfare of sea creatures and the source of income for people whose commerce encompasses the health of oceans. Overfishing occurs when fish species are caught at rates higher than this, leading to cannibalization of specific fish populations (Telesca 2017). This offends the balance of ocean ecosystems, given that many species depend on a large and healthy stock of fish in the seas. Big fish such as tuna and cod that form crucial economic stocks are notably overfished, but there are also small but ecologically important fish like those that feed on plankton. Nothing would be as disastrous as fishing, which may lead to fish population breakdown and perhaps ecosystem imbalances (Clover 2008).

Moreover, destructive fishing practices such as bottom trawling, dynamite fishing, and cyanide fishing have significantly contributed to the destruction of the habitat (Ranjan et al. 2023). This fishing method involves the towing of huge nets along the sea bed. This action is highly detrimental to less stable environments on the sea bed, such as coral reefs, seagrass, and some deep-sea communities. These environments afford several uses to many marine species, such as shelter, breeding grounds, and feeding grounds. The extinction of these habitats means permanent damage to aquatic species and reduces the ability of these ecosystems to withstand different pressures. It causes many impacts; the fish population is reduced, and fundamental aspects of marine life are compromised. There is much to learn about the specific role of fisheries management, fishing methods, and conservation methods in protecting marine life, which forms the subject of this chapter, to ensure the oceans remain a source of energy and food in the future. Doing nothing will continue to expose the world's food supply and wildlife population to further threats by overfishing and the destruction of habitats.

1.3.3.2 Loss of Habitats Due to Coastal Development and Trawling

Both overfishing and destruction of habitat remain major threats to marine organisms, where coastal development and destructive fishing practices, for example, trawling, are primary threats to the necessary habitats. Land reclamation by cities, tourists, and industries worldwide changes the coasts in ways that, more often than not, negatively impact the structure of critical coastal habitats such as mangroves, seagrass beds, and coral reefs (Sahavacharin et al. 2022). The ecosystems provide shelter, breeding grounds for some marine animals, and feeding grounds for some organisms. It pointed out that the formation of coastlines leads to the removal of habitats, which erodes the biological diversity and considerably erodes the marine ecosystems' ability to withstand environmental pressures. Of all the fishing techniques deployed, bottom trawling is the worst, which causes a significant amount of damage to the aquatic environment. Bottom trawlers use big nets horizontally on the seabed, which are detrimental to vulnerable structures like coral and benthic species (Steadman et al. 2021). It also poses a lot of damage to habitats and affects the species relying on these habitats in one way or the other. The perpetual loss of critical habitats erodes the wolf populations and also affects the problem of overfishing of stocks. This destruction of these essential marine habitats, combined with overfishing, creates a vicious cycle that overburdens the aquatic resources and weakens the solidity of the oceans. Reducing habitat loss, therefore, requires improvement in management practices, proper Coastal Development, and the stoppage of damaging fishing practices to protect the sea species.

1.3.4 Oil Spills and Chemical Contamination

1.3.4.1 Long-Term Impacts of Oil Spills on Marine Ecosystems

Dave et al.'s paper reveals that oil spillage and cases of chemical pollution rank as some of the worst natural disasters that hit marine habitats (Dave and Ghaly 2011). Such events, arising from tanker mishaps, offshore drilling calamities, or pipeline breakages, emit considerable levels of oil and dangerous materials into the aquatic environment. The first impacts of accidents can often be observed; however, the long-term effects on the marine biome are rather extensive and potentially impact species abundance, food chain, and fishing economies.

The direct impacts of oil spills are most often devastating (Zhang et al. 2019). Oil accumulations cover large portions of the ocean, which saturate the sea water and spread on the coast. Avian life, fish, and marine mammals come in direct contact with oil and drown or freeze to death in some cases, or die of suffocation. If the seabirds are oculared with oil, they will lose the ability to insulate, and they will die from hypothermia or drowning. Similarly, seals and otters die from the effects of oil on their skin; their ability to control temperatures is also affected. However, the consequences of oil spills are often longer standing and less noticeable. The compounds in the oil are toxic to sea life. When they enter the food chain, the toxins accumulate within species as time progresses and eventually cause chronic diseases and low reproduction rates. Marine life, including fish, shellfish, and plankton, can accumulate toxins, which, with time, transfers up the food chain, affecting larger fish and aquaculture fed by people. Some chemical constituents, such as the polycyclic aromatic hydrocarbons (PAHs) in oil, are toxic and cause long-term damage to the DNA of all the living forms in seas (Mallah et al. 2022).

Oil pollution has an ill effect on marine lives have a terrible effect, especially on coral reefs, which are inherently sensitive ecosystems. Oil can smother the corals, prevent incoming sunlight, and upset the balance of the tropical sea. Furthermore, it will continue to impact the reproduction cycles of several marine species, worsening the effect on biodiversity. The disastrous oil blowout of the Deepwater Horizon rig in the Gulf of Mexico in 2010 caused a significant loss of coral, and the marine life that depends upon it continues to have an impact on fish stocks and the coastal milieu (Sumaila et al. 2012). If an oil spill occurs, it takes years or even decades to restore the affected ecosystems. While numerous species may recover fairly fast, the overall environment might take much longer before it returns to normal. Oil biodegradation in the environment is relatively slow, and in various cases, oil residues remain in the bottom deposit for many years, continually affecting the benthos (Bacosa et al. 2022). Besides the adverse environmental effects, the oil spill dramatically impacts the economy. Fisheries, tourism, and numerous coastal businesses are severely affected. Workers lose employment, and there is less economic activity in those zones. The recovery may take many years for these businesses, which will only deepen the impact on affected local regions.

1.4 The Ripple Effect: Impacts Beyond the Ocean

1.4.1 How Marine Degradation Affects Global Food Security and Economies

Marine degradation, with concerns including overfishing, impacts on habitat structures, oil slicks, and pollution, affects several fields that surface far more than what is observable in marine areas. The marine environment fundamentally affects world food security and economics; any change has substantial potential impacts. They are the source of food, as over 3.1 billion people depend on fish and other marine animals as their staple source of protein (Tacon and Metian 2018; Dongyu 2024). The problems with overfishing lead to the collapse of wild fish stocks due to unsustainable fishing techniques, coastal development, and pollution that qualitatively degrade marine ecosystems and significantly reduce the production of fish and other marine life (Hilborn 2011). This threatens food security for all coastal towns and populations worldwide, especially considering that a population consumes seafood in their diet. Many people depend on fisheries as their primary source of livelihood, including fishers, and are facing a problem of reduced stock, which reduces income and employment. The tourism industry, especially for the location-based service (LBS) that heavily depends on Helix One's vibrant coral reef stretches of unspoiled coastline, results in high economic costs by losing or eliminating critical ecosystems (Donkor 2022). Oil spillage or pollution along the coastal area leads to low tourism, increasing economic loss. Besides, it indirectly affects world business and trade since marine deterioration is responsible for the loss of trade. Sound Ocean empowers the commercial stream and provides essential amenities and products like salt and marine products. These ecosystems may degrade thus leading to a rise in transportation costs and resource availability constraints on world supplies of food and other amenities. Deterioration of the marine environment challenges food security and economic well-being, and environmental conservation laws are needed to address the effects of degradation on ecosystems and human welfare.

1.4.2 Cultural and Spiritual Significance of Oceans to Coastal Communities

Seas still hold significant symbolic connotations to the peoples of the world residing along the sea's shores (Gillis 2012; McNiven 2004). To many, the sea is not only an object on which to survive but also an object that defines the most essential aspects of personal and social life. According to the traditions of some cultures, the ocean is a spiritual world that connects people with their ancestors and the whole world (Dickie 2005; Taylor 2007). It is often viewed as a giver, which gives people food, support, and spiritual advice. Coastal societies associate the sea with general existence and include it in their custom, ceremonies, and festivals. Many native cultures, specifically those of the Pacific Islands and African shoreline, respect the ocean and marine life; commonly practiced rituals offer prayers of prosperity in fishing, safety in travels, and protection from disasters (Fay 2018). In the case of the ocean, it is considered a giant force and personified as a creator and destroyer. People pray to it and sing to it. Moreover, the sea is also a symbol of endurance and unity. It has been essential in constructing constructive cultural paradigms and the archetypical performing function in oral traditions and artistic displays. For coastal communities, the ocean is a crucial source of food and, therefore, fosters and enhances an appreciation of spirits with the Earth, and, hence, a level of care for the environment.

1.4.3 Impacts on Global Climate Systems and Human Health

Its impacts are not limited to the marine ecosystem; they include disturbing effects on climate systems and human health. Oceans play crucial roles in regulating the climate of our planet and maintaining balance in ecosystems (Bigg et al. 2003). In these habitats, disruptions of maritime regions precipitate disruptions of these processes, triggering responses worldwide. Ocean pollution affects global climate systems since carbon sinks are interfered with (Lu et al. 2018). The seas act as a primary buffer reservoir, and effectively, the seas currently absorb a large portion of the total carbon dioxide (CO_2) emitted into the atmosphere. Indeed, phytoplankton grows massively in productive ocean provinces, expatriates CO_2, and converts it to organic carbon. When the habitats of organisms such as coral reefs and plankton are destroyed, degraded, or changed by pollution, acidification, and overfishing, the ocean's ability to absorb more CO_2 is reduced. This aggravates climate change and accelerates global warming, and we experience extreme and frequent climate changes, high floods, and other climatic changes. Apart from climate systems, the degradation of oceans poses a serious threat to human health. Sulfides, gases, plastics, heavy metals, and hazardous compounds concentrate in the seas and penetrate the food chain (Shah 2021). Consumption of contaminated seafood leads to human interaction with poisonous substances likely to cause several diseases, disorders, and conditions affecting the nervous system, reproduction, and cancers, respectively (Mohammad et al. 2018; National Academies of Sciences, Engineering, and Medicine 2024). However, there has also been a steady increase in the vulnerability of populations in coastal areas who rely on healthy marine ecosystems for food sources or income due to marine degradation. Moreover, a high concentration of CO_2 in water causes ocean acidification and affects key marine species, such as shellfish and plankton. Reducing these species may result in a limited number of species, which means world food security is vulnerable. Their reduced abundance and species diversity then reduce overall ecosystem stability and recovery from new pressures, such as climate change. Marine degradation erodes crucial ocean functions in providing climate mitigation through carbon capture, food supply, and enhancing climate change. It also threatens people's health by polluting food sources and eradicating income sources. The solution to ocean degradation requires international cooperation to protect wildlife, reduce pollutants, and address climate change. Conserving oceans should be considered important in maintaining the climate in the world and human health, to ensure that all would get to live better in the future.

1.5 Conservation Efforts: Restoring the Majesty

Marine biology is giving new means to protect and restore significant marine habitats. In the modern world, actual tiles and stones are replaced with concrete, steel, and other environmentally friendly alternatives, and artificial constructions resemble natural reefs. Such reefs act as shelter, breeding grounds, and food for many marine life, helping to rehabilitate areas damaged by human activities. Marine Protected Areas (MPAs) are crucial conservation tools because they put stakes in and demarcate areas forbidden to several activities like fishing and mining for recovery. Marine Protected Areas (MPAs) play a vital role in acting as a refuge for marine species by providing an environment in which endangered species can regain their strength. Coral reef restoration, mangrove planting, and seagrass bed rehabilitation programs aim at reconstructive measures aimed at putting back the ecosystem. The restoration efforts highlighted in the study have potential regarding the revival of biological diversity and enhancing the resistance of marine systems to climate change. In combination, these developments help to protect the sea life, reduce the impact of fishing and pollution, and maintain the complicated balance of the marine environment.

1.5.1 The Role of Bioinformatics and AI in Monitoring and Protecting Marine Biodiversity

As for the future, bioinformatics and artificial intelligence (AI) techniques are gradually becoming increasingly crucial to monitoring and conserving marine biological resources. Bioinformatics assists scientists in analyzing large genomic databases to detect species, genetic variations and the health of marine populations (Oleksiak and Rajora 2020). On the other hand, an extensive amount of data about the environment is processed using AI, which may consist of satellite photos, ocean conditions, and underwater sounds (Lou et al. 2023). Machine learning can understand patterns and shifts in aquatic environments, predict changes in the type of flora and fauna, and identify possible signs of other environmental vices like pollution or unlawful fishing practices. These technologies help in timely and effective conservation drives since they present up-to-date information to policymakers, and conservationists enable them to take the proper measures to protect marine species. Marine bioinformatics and AI are increasing the capability to observe worldwide aquatic ecosystems and ensure that conservation efforts are data-led and directed accordingly.

1.5.2 International Agreements and Initiatives

1.5.2.1 Policies like the UN's Sustainable Development Goal 14 (Life Below Water)

Sustainable Development Goal SDG14: Life below water, as it aims at promoting the conservation and sustainable use of the ocean, seas, and marine resources for sustainable development (Friess et al. 2019). This objective outlines the necessity of protecting ocean life, reducing pollution, ensuring the sustainability of the Andes region, and preserving coastal resources and environments. SDG 14, given its sweep and importance, sets a baseline to address significant marine conservation challenges like habitat degradation, overfishing, and ocean acidification (Atieno and Hendriks 2024). SDG 14 calls for cooperation among countries and the passage of policies that can help restore aquatic habitats, guiding governments, nongovernmental organizations (NGO), and communities to achieve considerable progress in caring for the seas. It also enhances the setting up of maritime protected areas and reduces adverse practices like furthering unlawful fishing and mitigating the impact of climate upon seas.

1.5.2.2 Success Stories of Conservation Efforts from Around the World

Nowadays, various conservation approaches have been realized as significant in conserving and restoring marine habitats. Through the enforcement of strict laws, promotion of research, and participation of the people in and around the area, the Great Barrier Reef Marine Park Authority in Australia has been instrumental in conserving the most extensive coral reef system (Day 2016). Several case studies on rehabilitating mangroves, particularly in the Philippines and Indonesia, show considerable progress in increasing the population of species, repopulation of coastal structures, and improving carbon sequestration (Macintosh and Ashton 2002). In the Caribbean, positive outcomes have been produced by the "ReefBase", which has restored the coral reefs through coral farming and transplantation (Riegl et al. 2009). These initiatives show that there is global awareness and pact as well as creativity on how marine habitats can recover and gain the capability to bounce back into productive ecosystems through functional management.

1.5.3 Grassroots Movements and Citizen Science

1.5.3.1 Community-Driven Ocean Cleanups and Awareness Campaigns

There is increasing and significant participation of all customary populations in organizing themselves for cleanups and awareness campaigns. People, including citizens, environmental organizations, and volunteer groups, are carrying out cleanups of beaches and rivers and conserving the coastal regions because of the increasing problem of marine pollution, especially plastics (Thushari and Senevirathna 2020). Activities like "The Ocean Conservancy's International Coastal Cleanup" call volunteers worldwide to clean up the coasts and rivers (Jorgensen et al. 2021). The above exercises also help get rid of waste and increase an understanding of how pollution affects the seas. Locally driven projects foster ownership of environments, creating awareness of sustainable activities for improved environmental management, such as reducing plastics and promoting environmentally friendly products (Mihai et al. 2021). Moreover, such controversial programs often involve local schools, companies, and governments and make their network beneficial for strengthening the message of ocean conservation and a responsible attitude toward the environment.

1.5.3.2 Engaging Individuals in Marine Preservation Through Technology and Education

In the areas of marine conservation, technology and education are fundamental prerequisites for fostering the capacity of people to engage in conservation. Marine citizen science programs leverage technology to enable individuals to conduct marine biology monitoring and research. Some applications like "Marine Debris Tracker" allow volunteers to gather info about the kind and quantity of debris found, which becomes valuable data for scientists and environmentalists (Burgess et al. 2021). Similarly, applications that track the state of coral reefs or applications that evaluate water conditions enable users to make changes, including feeding the research community with problems like ocean acidification or temperature rise. Institutional projects, web-based and communal, introduce people to themes about the marine environments, the challenges faced, and ways of making changes. There are special programs offered in schools, museums, and even the caring of environmental nonprofit groups about the practices of sustainability, preserving nature, as well as learning more about the role of oceans in regulating climate on the planet. These efforts immerse education within technology to engage the public in playing their part in marine conservation and empower individuals to take an active role and make a difference in conserving the oceans.

1.6 A Call to Action: Safeguarding the Ocean's Beauty

Unfortunately, the oceans are threatened today, representing a primary need for our world's sustainability. Marine lives are at risk due to overfishing, pollution, habitat loss, and climate change. Seas regulate temperatures, support life, and provide food to over 3 billion people worldwide. Nonetheless, their pressures are increasing while the chance to intercede is narrowing. Different players, including governments, industry, and citizens, must embrace synergy and act on policies and strategies to protect the ocean's glory to meet future generations' needs.

1.6.1 Urging Governments, Industries, and Individuals to Take Immediate Action

This is because the government comes in handy when it comes to implementing laws on the maintenance of the oceans. International practices similar to the United Nations' Sustainable Development Goal 14 (Life Below Water) provide guidance regarding the protection of the oceans; however, states should step up their engagement in implementing and adhering to the laws. Actions to address plastic pollution, establish Marine Protected Areas (MPAs), and implement better management control of fisheries are needed (Da Costa et al. 2020). At the same time, industries need to become accountable for the imprint they have left on their environment. Shippers, the plastics and fishing industries, should learn to do things sustainably. This chapter argues that the private sector's engagement in promoting marine conservation activities is crucial in enhancing intervention and ensuring sustainable access to aquatic resources. People also have quite a significant role to play. Many changes that the individuals effect are as follows: abstaining from the use of plastics and promoting eco-friendly seafood. Positive changes to the population's behavior require significant actions that include public education. It can involve cleaning up beaches, supporting organizations that protect marine life and environment, and demanding that governments create better environmental policies that will help to protect aquatic life. If individuals think about what they can do to change, it all becomes a part of something that sends out a powerful message for change.

1.6.2 Emphasizing the Need for Global Collaboration in Tackling Marine Issues

Numerous and complex threats to our seas require an integrated and international effort. Global warming, climate change, marine acidification, over-exploitation of fish stocks, and pollution are problems that are not recognized by national jurisdictions. These difficulties cannot be addressed by a single nation, industry, or organization through its efforts. Cooperation on an international level is also a crucial factor in the successful solution of maritime issues (Solesvik 2022). International treaties like the Paris Agreement and the United Nations Convention on the Law of the Sea create cooperation structures; however, better commitments and other improvement measures are critical to protecting the ocean. Joint endeavors, for instance, large MPA formation in various nations, will ensure that the marine systems have the space they require for growth. Understanding and research activities, and the spread of information, have been made core aspects of this kind of relationship. Marine systems are complex, and explaining how such systems work is essential, especially for managing the systems. The global community of governments, NGOs, scientists, and industries need to work in partnership with regard to monitoring marine health status, sharing and extrapolating pollution rates, tracking endangered species, and determining ways and means of assessing the effectiveness of the set conservation measures. Furthermore, a primary focus on funding for marine-related research and conservation initiatives is needed. They are the oceans as a common concern of humanity, and only in concept and acting united can we dare to dream about conserving and restoring them.

1.6.3 Vision of a Restored Ocean Ecosystem for Future Generations

The goal of a revitalized ocean is possible only if all state governments, industries, marine scientists, and citizens are committed to making it a reality. Envision a realm where marine biodiversity thrives: there are diversified communities in coral reefs, fish stocks are strong and copious, and marine habitats are restored to the optimal state. A healthy sea also sustains the Earth's climatic balance by limiting the extent of global warming by absorbing carbon monoxide (Ayyam et al. 2019). Coastal towns benefit from gains accrued from sustainable access and utilization of the fishery resource base, well-developed tourism industries, and stable and healthy coastal ecosystems, which minimize impacts from storm surges and flooding (McIvor et al. 2012).

In this view, humanity recognizes the existence and importance of oceans for humanity and human welfare, and they coexist with the oceans. Long-termism remains the norm in sectors, while all consider ocean conservation a responsibility. Geographic marine protected areas, or Marine Protected Areas (MPAs), have become familiar worldwide and have acted as havens for several aquatic creatures and other natural systems. Apart from enjoying an existing beautiful sight, its beauty is being restored and preserved through collective unity efforts toward reducing pollution and protecting and increasing the number of species that inhabit the ocean. Real-time tracking of oceanic health is achieved through various technologies, such as AI and bioinformatics, ensuring that threats are well identified and negated in good time. Great is the significance of this vision, which depicts procreating waters full of life, whereby the flowing generations will be bequeathed to a prosperous, balanced, and thriving planet. Every effort to rehabilitate the oceans will improve marine life's presence while improving human health, food resources, and economic systems. We have the information, the resources, and the responsibility to act now.

The time of comfort is now over. The call to action is clear: the first step. Governments should implement stricter policies, industries should employ friendly measures for the environment, and people should purposely decide on their actions. Through cooperation and commitment from people from different countries, we can revive the ocean's beauty and provide the necessary level of preservation for future generations. Modern oceans and the future of our Earth can be either worse or better depending on what we do now. It is time to join our efforts to protect the oceans' magnificence and give everyone a chance.

1.7 Conclusion

This infinite ocean, which harbors life in its bosom, is the matrix of the life force. The multiple diverse ecosystems portray the creative work of nature, which is essential to the Earth's well-being. However, the possibility of a climate change situation raises a question mark over this vibrant sector. Marine life, especially endearing coral reefs, which are deemed as the cities of the oceans, is currently suffering from bleaching, an indication of an increase in water temperatures. Along these parameters, seagrass meadows and mangroves— vital for marine life and coastal resilience— are under pressure, the betterment of which directly declines with every increase in global temperature. This crisis is worsened by ocean acidification caused by excessive emission of CO_2. Organisms that secrete shells play a critical role in the seas. Still, they struggle with heightened acidity levels, risking marine life and the stability of millions whose income comes from the oceans. The ocean is threatened by its existence and is affected by the legacy effects of oil leaks and contamination, in addition to human abuse and exploitation. However, there is hope in these constraints, as seen below. Human beings alone can change the course of the history of the oceans as we know them. It falls under our ability to decrease greenhouse gas emissions, embrace sustainability, and encourage the engagement of the world community. The ocean is so strong because it has had to withstand a variety over millennia. However, it now challenges us to accept this strength via our actions. This moment is crucial. However, it is pertinent to note that the ocean mentioned goes beyond the degree of being a resource; it is, in fact, a heritage—a factory of life that deserves protection. Preserving that beauty is of the essence for future generations, speaking about the shared responsibility for this blue sphere.

References

Abirami B, Radhakrishnan M, Kumaran S, Wilson A (2021) Impacts of global warming on marine microbial communities. Sci Total Environ 791:147905

Abrantes K, Sheaves M (2009) Food web structure in a near-pristine mangrove area of the Australian wet tropics. Estuar Coast Shelf Sci 82(4):597–607

Adamopoulou A, Zeri C, Garaventa F, Gambardella C, Ioakeimidis C, Pitta E (2021) Distribution patterns of floating microplastics in open and coastal waters of the eastern Mediterranean Sea (Ionian, Aegean, and Levantine seas). Front Mar Sci 8:699000

Adeniran-Obey SO, Isibor PO, Imoobe TO (2024) Marine water acidification and coral bleaching. In: Arctic marine ecotoxicology. Springer, Cham, pp 403–420

Agbekpornu P, Kevudo I (2023) The risks of microplastic pollution in the aquatic ecosystem. In: Advances and challenges in microplastics. IntechOpen, London

Alleway HK, Gillies CL, Bishop MJ, Gentry RR, Theuerkauf SJ, Jones R (2019) The ecosystem services of marine aquaculture: valuing benefits to people and nature. Bioscience 69(1):59–68

Alomar C, Sureda A, Capó X, Guijarro B, Tejada S, Deudero S (2017) Microplastic ingestion by Mullus surmuletus Linnaeus, 1758 fish and its potential for causing oxidative stress. Environ Res 159:135–142

Atieno P, Hendriks SL (2024) Sustainable development goal 14: life below water. In: Handbook on public policy and food security. Edward Elgar Publishing, Northampton, pp 333–343

Ayyam V, Palanivel S, Chandrakasan S (2019) Coastal ecosystems of the tropics-adaptive management. Springer, Singapore

Bacosa HP, Ancla SMB, Arcadio CGLA, Dalogdog JRA, Ellos DMC, Hayag HDA, Adhikari PL (2022) From surface water to the deep sea: a review on factors affecting the biodegradation of spilled oil in marine environment. J Marine Sci Eng 10(3):426

Barbeau MM (2018) The luminous sea (Doctoral dissertation, Memorial University of Newfoundland)

Barbraud C, Rolland V, Jenouvrier S, Nevoux M, Delord K, Weimerskirch H (2012) Effects of climate change and fisheries bycatch on Southern Ocean seabirds: a review. Mar Ecol Prog Ser 454:285–307

Beaumont NJ, Austen MC, Mangi SC, Townsend M (2008) Economic valuation for the conservation of marine biodiversity. Mar Pollut Bull 56(3):386–396

Beer S, Björk M, Beardall J (2014) Photosynthesis in the marine environment. Wiley, Hoboken

Bergmann M, Collard F, Fabres J, Gabrielsen GW, Provencher JF, Rochman CM, Tekman MB (2022) Plastic pollution in the Arctic. Nat Rev Earth Environ 3(5):323–337

Berkowitz H (2023) From organizations as systems of ocean destruction to organizations as systems of ocean thriving. Bus Soc Rev 128(1):71–94

Berman A (2008) Biomineralization of calcium carbonate. The interplay with biosubstrates. Biomineralization 4:167–205

Bhatnagar DG, Kumar S (2024) Ecological conservation and sustainable management of marine environments. In: Nature-based solutions in achieving sustainable development goals: harmonizing nature and progress. Springer, Cham, pp 233–261

Bigg GR, Jickells TD, Liss PS, Osborn TJ (2003) The role of the oceans in climate. Int J Climatol 23(10):1127–1159

Bringloe TT, Starko S, Wade RM, Vieira C, Kawai H, De Clerck O, Verbruggen H (2020) Phylogeny and evolution of the brown algae. Crit Rev Plant Sci 39(4):281–321

Broad WJ (1998) The universe below: discovering the secrets of the deep sea. Simon and Schuster
Browman HI, Stergiou KI, Cury PM, Hilborn R, Jennings S, Lotze HK, Mace PM (2004) Perspectives on ecosystem-based approaches to the management of marine resources. Mar Ecol Prog Ser 274:269–303
Burgess H, Herring C, Lippiatt S, Lowe S, Uhrin AV (2021) NOAA marine debris monitoring and assessment project shoreline survey guide
Burkholder PR, Burkholder LM (1960) Photosynthesis in some Alcyonacean corals. Am J Bot 47(10):866–872
Capra F (2005) Complexity and life. Theory Cult Soc 22(5):33–44
Clover C (2008) The end of the line: how overfishing is changing the world and what we eat. University of California Press, Oakland
Copper P (1992) Paleoscene 14. Organisms and carbonate substrates in marine environments. Geosci Can 19:3767
Cornwall CE, Comeau S, Kornder NA, Perry CT, van Hooidonk R, DeCarlo TM, Lowe RJ (2021) Global declines in coral reef calcium carbonate production under ocean acidification and warming. Proc Natl Acad Sci 118(21):e2015265118
Cremers L (2018) The protection of coral reefs (Doctoral dissertation, Ghent University)
Czerski H (2023) Blue machine: how the ocean shapes our world. Random House, New York
Da Costa JP, Mouneyrac C, Costa M, Duarte AC, Rocha-Santos T (2020) The role of legislation, regulatory initiatives and guidelines on the control of plastic pollution. Front Environ Sci 8:104
Dave DAEG, Ghaly AE (2011) Remediation technologies for marine oil spills: a critical review and comparative analysis. Am J Environ Sci 7(5):423
Day J (2016) The great barrier reef marine park: the grandfather of modern Marine Protected Areas (MPAs). In: Big, bold and blue: lessons from Australia's marine protected areas. Csiro, Canberra, pp 65–97
De Abreu D (2017) Evaluation of suitable nursery areas for penaeid shrimps in shallow water systems in southern Mozambique
Dickie RA (2005) Indigenous traditions and sacred ecology in the Pacific Islands. UW J Undergrad Res VIII:1–9
Doney SC, Fabry VJ, Feely RA, Kleypas JA (2009) Ocean acidification: the other CO2 problem. Annu Rev Mar Sci 1(1):169–192
Dongyu Q (2024) 2024 the state of world fisheries and aquaculture-blue transformation in action. In: The state of world fisheries and aquaculture. FAO, Rome, pp 1–232
Donkersloot R (2012) Ocean acidification and Alaska fisheries: views and voices of Alaska's fishermen, marine industries and coastal residents. Alaska Marine Conservation Council, Coastal Voices on Ocean Acidification Project
Donkor FK (2022) Role of international law in effective governance of the marine environment. In: Life below water. Springer, Cham, pp 882–894
Eakin CM, Sweatman HP, Brainard RE (2019) The 2014–2017 global-scale coral bleaching event: insights and impacts. Coral Reefs 38(4):539–545
Eby LA, Roach WJ, Crowder LB, Stanford JA (2006) Effects of stocking-up freshwater food webs. Trends Ecol Evol 21(10):576–584
Everaert G, De Rijcke M, Lonneville B, Janssen CR, Backhaus T, Mees J, Vandegehuchte MB (2020) Risks of floating microplastic in the global ocean. Environ Pollut 267:115499
Farrelly DJ, Everard CD, Fagan CC, McDonnell KP (2013) Carbon sequestration and the role of biological carbon mitigation: a review. Renew Sust Energ Rev 21:712–727
Fay L (2018) Human connections with the ocean represented in African and Japanese oral narratives: ecopsychological perspectives. California Institute of Integral Studies, San Francisco
Feely RA, Doney SC, Cooley SR (2009) Ocean acidification: present conditions and future changes in a high-CO_2 world. Oceanography 22(4):36–47
Friess DA, Aung TT, Huxham M, Lovelock C, Mukherjee N, Sasmito S (2019) SDG 14: life below water–impacts on mangroves. Sustainable Dev Goals 445:445–481
Gallo F, Fossi C, Weber R, Santillo D, Sousa J, Ingram I, Romano D (2020) Marine litter plastics and microplastics and their toxic chemicals components: the need for urgent preventive

measures. In: Analysis of nanoplastics and microplastics in food. CRC Press, Boca Raton, pp 159–179

Garcia-Soto C, Cheng L, Caesar L, Schmidtko S, Jewett EB, Cheripka A, Abraham JP (2021) An overview of ocean climate change indicators: sea surface temperature, ocean heat content, ocean pH, dissolved oxygen concentration, arctic sea ice extent, thickness and volume, sea level and strength of the AMOC (Atlantic meridional overturning circulation). Front Mar Sci 8:642372

Gazeau F, Parker LM, Comeau S, Gattuso JP, O'Connor WA, Martin S, Ross PM (2013) Impacts of ocean acidification on marine shelled molluscs. Mar Biol 160:2207–2245

Gibson RN, Barnes M, Atkinson RJA (2002) Impact of changes in flow of freshwater on estuarine and open coastal habitats and the associated organisms. Oceanogr Mar Biol 40:233

Gillis JR (2012) The human shore: seacoasts in history. University of Chicago Press, Chicago

Goreau TJ, Hayes RL (2024) 2023 record marine heat waves: coral reef bleaching HotSpot maps reveal global sea surface temperature extremes, coral mortality, and ocean circulation changes. Oxford open. Climate Change 4(1):kgae005

Goreau TJ, Hilbertz W (2005) Marine ecosystem restoration: costs and benefits for coral reefs. World Resour Rev 17(3):375–409

Granek EF, Compton JE, Phillips DL (2009) Mangrove-exported nutrient incorporation by sessile coral reef invertebrates. Ecosystems 12:462–472

Guannel G, Arkema K, Ruggiero P, Verutes G (2016) The power of three: coral reefs, seagrasses and mangroves protect coastal regions and increase their resilience. PLoS One 11(7):e0158094

Harrison PL, Booth DJ (2007) Coral reefs: naturally dynamic and increasingly disturbed ecosystems. In: Marine ecology. Oxford University Press, Oxford, pp 316–377

Hawthorne N (2019) Earth's oceans—Atlantic, Pacific, Arctic, Indian, and southern—are the storehouse of earth's saline water. Oceans cover about 71 percent of earth's surface. Average depth of earth's oceans is about 3,800 m with the greatest ocean depth recorded at 11,036 m in the Mariana trench. At the present time, the oceans contain a volume of about 1.35 billion cubic kilometers (96.5 per-cent of earth's total water supply), but the volume fluctuates with the growth and melting of glacial ice. In: The handbook of nature. Comstock Publishing Company, Ithaca, p 176

Hempel S (2016) The functional response of the Antarctic bivalve Laternula elliptica to ocean warming and acidification (Doctoral dissertation, Open Access Te Herenga Waka-Victoria University of Wellington)

Hilborn R (2011) Overfishing: what everyone needs to know. Oxford University Press, Oxford

Hoegh-Guldberg O, Bruno JF (2010) The impact of climate change on the world's marine ecosystems. Science 328(5985):1523–1528

Hoegh-Guldberg O, Jacob D, Taylor M, Guillén Bolaños T, Bindi M, Brown S, Zhou G (2019) The human imperative of stabilizing global climate change at 1.5 C. Science 365(6459):eaaw6974

Hull EV (2016) Ocean acidification: legal and policy responses to address climate change's evil twin. Washington J Environ Law Policy 6:349

Hung H, Halsall C, Ball H, Bidleman T, Dachs J, De Silva A, Wilson S (2022) Climate change influence on the levels and trends of persistent organic pollutants (POPs) and chemicals of emerging Arctic concern (CEACs) in the Arctic physical environment–a review. Environ Sci: Processes Impacts 24(10):1577–1615

Islam MZ (2024) Ocean wildlife and megafauna protection. In: The blue book: smart sustainable coastal cities and blue growth strategies for marine and maritime environments. Springer, Cham, pp 15–33

Issifu I, Alava JJ, Lam VW, Sumaila UR (2022) Impact of ocean warming, overfishing and mercury on European fisheries: a risk assessment and policy solution framework. Front Mar Sci 8:770805

Jorgensen B, Krasny M, Baztan J (2021) Volunteer beach cleanups: civic environmental stewardship combating global plastic pollution. Sustain Sci 16(1):153–167

Kroeker KJ, Kordas RL, Crim R, Hendriks IE, Ramajo L, Singh GS, Gattuso JP (2013) Impacts of ocean acidification on marine organisms: quantifying sensitivities and interaction with warming. Glob Chang Biol 19(6):1884–1896

La Daana KK, Gårdfeldt K, Lyashevska O, Hassellöv M, Thompson RC, O'Connor I (2018) Microplastics in sub-surface waters of the Arctic Central Basin. Mar Pollut Bull 130:8–18

Laface F (2023) Integrated approach for the assessment of microplastics pollution and its impacts in under-explored areas of the southern Italy (Central Mediterranean Sea)

Lee SY, Primavera JH, Dahdouh-Guebas F, McKee K, Bosire JO, Cannicci S, Record S (2014) Ecological role and services of tropical mangrove ecosystems: a reassessment. Glob Ecol Biogeogr 23(7):726–743

Lindgren JF, Salo K, Brynolf S, Andersson K, Svensson E, Zetterdahl M, Magnusson M (2016) The natural environment and human impacts. In: Shipping and the environment: improving environmental performance in marine transportation. Springer, Cham, pp 29–74

Lou R, Lv Z, Dang S, Su T, Li X (2023) Application of machine learning in ocean data. Multimedia Syst 29(3):1815–1824

Lu Y, Yuan J, Lu X, Su C, Zhang Y, Wang C, Sweijd N (2018) Major threats of pollution and climate change to global coastal ecosystems and enhanced management for sustainability. Environ Pollut 239:670–680

Macintosh DJ, Ashton EC (2002) A review of mangrove biodiversity conservation and management. University of Aarhus, Aarhus

Malinowski CR, Searle CL, Schaber J, Höök TO (2023) Microplastics impact simple aquatic food web dynamics through reduced zooplankton feeding and potentially releasing algae from consumer control. Sci Total Environ 904:166691

Mallah MA, Changxing L, Mallah MA, Noreen S, Liu Y, Saeed M, Zhang Q (2022) Polycyclic aromatic hydrocarbon and its effects on human health: an overview. Chemosphere 296:133948

McCoy M (2010) Reef. In: Reef and rainforest. CSIRO Publishing, Clayton, pp 18–104

McElligott S, Byrne RH (1997) Interaction of B and in seawater: formation of B. Aquat Geochem 3(4):345–356

McIvor AL, Spencer T, Möller I, Spalding M (2012) Storm surge reduction by mangroves. The Nature Conservancy and Wetlands International, Wageningen

McNiven I (2004) Saltwater people: spiritscapes, maritime rituals and the archaeology of Australian indigenous seascapes. World Archaeol 35(3):329–349

Mega VP, Mega VP (2019) Threatened urban and ocean biodiversity: the imperative of resilience. In: Eco-responsible cities and the global ocean: geostrategic shifts and the sustainability trilemma. Springer, Cham, pp 43–84

Mihai FC, Gündoğdu S, Markley LA, Olivelli A, Khan FR, Gwinnett C, Molinos-Senante M (2021) Plastic pollution, waste management issues, and circular economy opportunities in rural communities. Sustain For 14(1):20

Mishra SB, Singh S (2024) Coral reef: a source of value added compounds beneath the water. In: Immune-boosting nutraceuticals for better human health. Apple Academic Press, Palm Bay, pp 305–330

Mohammad AM, Chowdhury T, Biswas B, Absar N (2018) Food poisoning and intoxication: a global leading concern for human health. In: Food safety and preservation. Academic Press, New York, pp 307–352

Moore C (2011) Plastic Ocean: how a sea captain's chance discovery launched a determined quest to save the oceans. Penguin, London

Mudie PJ, Marret F, Gurdebeke PR, Hartman JD, Reid PC (2021) Marine dinocysts, acritarchs and less well-known NPP: tintinnids, ostracod and foraminiferal linings, copepod and worm remains. Geol Soc 511:159–232

Mumby PJ (2006) Connectivity of reef fish between mangroves and coral reefs: algorithms for the design of marine reserves at seascape scales. Biol Conserv 128(2):215–222

Naghibi Z, Stagner JA, Carriveau R, Ting DS (2020) Underwater vegetation, aquatic cultivation. In: Environmental management of air, water, agriculture, and energy. CRC Press, Boca Raton, pp 33–52

National Academies of Sciences, Engineering, and Medicine (2024) Exposure to contaminants associated with consumption of seafood. In: The role of seafood consumption in child growth and development. National Academies Press, Washington

National Research Council, Division on Earth, Life Studies, Commission on Geosciences, & Committee on Biological Diversity in Marine Systems (1995) Understanding marine biodiversity. National Academies Press, Washington

Ogunola OS, Onada OA, Falaye AE (2018) Mitigation measures to avert the impacts of plastics and microplastics in the marine environment (a review). Environ Sci Pollut Res 25:9293–9310

Oleksiak MF, Rajora OP (2020) Marine population genomics: challenges and opportunities. In: Population genomics: marine organisms. Springer, Cham, pp 3–35

Oves M, Hussain FM, Ismail IMI, Felemban NM, Qari HA (2017) Microbiological carbon sequestration: a novel solution for atmospheric carbon–carbon sequestration through biological approach. In: Handbook of research on inventive bioremediation techniques. IGI Global, Hershey, pp 108–133

Palter JB (2015) The role of the Gulf stream in European climate. Annu Rev Mar Sci 7(1):113–137

Peng X, Chen M, Chen S, Dasgupta S, Xu H, Ta K, Bai S (2018) Microplastics contaminate the deepest part of the world's ocean. Geochem Perspect Lett 9(1):1–5

Pörtner HO, Peck MA (2010) Climate change effects on fishes and fisheries: towards a cause-and-effect understanding. J Fish Biol 77(8):1745–1779

Prather CM, Pelini SL, Laws A, Rivest E, Woltz M, Bloch CP, Joern A (2013) Invertebrates, ecosystem services and climate change. Biol Rev 88(2):327–348

Quigley KM, Baird AH (2024) Future climate warming threatens coral reef function on world heritage reefs. Glob Chang Biol 30(7):e17407

Ranjan D, Verma P, Kshatri AS, Patel A, Gupta V, Chaudhary V, Yadav B (2023) Destructive fishing practices and their impacts on fisheries. In: Latest trends in fisheries and aquatic animal health, vol 3. AkiNik Publications, New Delhi

Reaka-Kudla ML (1997) The global biodiversity of coral reefs: a comparison with rain forests. Biodiversity II: understanding and protecting our biological resources, 2. Joseph Henry Press, Washington, p 551

Riegl B, Bruckner A, Coles SL, Renaud P, Dodge RE (2009) Coral reefs: threats and conservation in an era of global change. Ann N Y Acad Sci 1162(1):136–186

Rotter A, Barbier M, Bertoni F, Bones AM, Cancela ML, Carlsson J, Vasquez MI (2021) The essentials of marine biotechnology. Front Mar Sci 8:629629

Rozwadowski HM (2002) Sea knows no boundaries (cl). University of Washington Press, Seattle

Rüpke L, Gaillard F (2024) The geological history of water: from earth's accretion to the modern deep water cycle. Elements 20(4):253–258

Sahavacharin A, Sompongchaiyakul P, Thaitakoo D (2022) The effects of land-based change on coastal ecosystems. Landsc Ecol Eng 18(3):351–366

Shah SB (2021) Heavy metals in the marine environment—an overview. In: Heavy metals in Scleractinian corals. Springer, Cham, pp 1–26

Sheppard C (2021) Coral reefs: a natural history. Princeton University Press, Princeton

Sin LT, Balakrishnan V, Bee ST, Bee SL (2023) A review of the current state of microplastic pollution in south Asian countries. Sustain For 15(8):6813

Smayda T (2019) Global epidemic of noxious phytoplankton blooms and food chain consequences in large ecosystems. In: Food chains, yields, models, and management of large marine ecosoystems. Routledge, London, pp 275–307

Smith KL Jr, Ruhl HA, Bett BJ, Billett DSM, Lampitt RS, Kaufmann RS (2009) Climate, carbon cycling, and deep-ocean ecosystems. Proc Natl Acad Sci 106(46):19211–19218

Solesvik MZ (2022) International cooperation on technology innovation development: a case study from the maritime industry. In: Research handbook on innovation in international business. Edward Elgar Publishing, Northampton, pp 264–274

Spalding MD, Leal M (2024) The state of the world's mangroves 2024

Spalding MD, Fox HE, Allen GR, Davidson N, Ferdaña ZA, Finlayson MAX, Robertson J (2007) Marine ecoregions of the world: a bioregionalization of coastal and shelf areas. Bioscience 57(7):573–583

Steadman D, Thomas JB, Villanueva VR, Lewis F, Pauly D, Palomares ML, Collinson T (2021) New perspectives on an old fishing practice: scale, context and impacts of bottom trawling. Our Shared Seas, Report, 44

Sumaila UR, Cisneros-Montemayor AM, Dyck A, Huang L, Cheung W, Jacquet J, Pauly D (2012) Impact of the Deepwater horizon well blowout on the economics of US gulf fisheries. Can J Fish Aquat Sci 69(3):499–510

Tacon AG, Metian M (2018) Food matters: fish, income, and food supply—a comparative analysis. Rev Fish Sci Aquac 26(1):15–28

Taylor B (2007) Surfing into spirituality and a new, aquatic nature religion. J Am Acad Relig 75(4):923–951

Telesca JE (2017) Accounting for loss in fish stocks: a word on life as biological asset. Environ Soc 8(1):144–160

Thushari GGN, Senevirathna JDM (2020) Plastic pollution in the marine environment. Heliyon 6(8):e04709

Titlyanov EA, Titlyanova TV (2020) Symbiotic relationships between microalgal zooxanthellae and reef-building coral polyps in the process of autotrophic and heterotrophic nutrition. Russ J Mar Biol 46:307–318

Trathan PN, García-Borboroglu P, Boersma D, Bost CA, Crawford RJ, Crossin GT, Wienecke B (2015) Pollution, habitat loss, fishing, and climate change as critical threats to penguins. Conserv Biol 29(1):31–41

Trebilco R, Fleming A, Hobday AJ, Melbourne-Thomas J, Meyer A, McDonald J, Pecl GT (2021) Warming world, changing ocean: mitigation and adaptation to support resilient marine systems. Rev Fish Biol Fish 32:39–63

Veron JEN (2008) A reef in time: the great barrier reef from beginning to end. Harvard University Press, Cambridge

Wang B, Hua L, Mei H, Wu X, Kang Y, Zhao N (2024) Impact of climate change on the dynamic processes of marine environment and feedback mechanisms: an overview. Arch Comput Methods Eng 31:3377–3408

Wilson T, Hastings JW (2012) Bioluminescence: living lights, lights for living. Harvard University Press, Cambridge

Woodford J (2010) The great barrier reef. Macmillan Publishers, New York

Yoro KO, Daramola MO (2020) CO2 emission sources, greenhouse gases, and the global warming effect. In: Advances in carbon capture. Woodhead Publishing, Sawston, pp 3–28

Zhang B, Matchinski EJ, Chen B, Ye X, Jing L, Lee K (2019) Marine oil spills—oil pollution, sources and effects. In: World seas: an environmental evaluation. Academic Press, New York, pp 391–406

2 Marine Microplastic Insulation and Characterization Techniques

Djadjiti Namla, Mohammad Oves, Majed Al-Shaeri, and George Mangse

Abstract

Microplastics (MPs), viz. nanoscale size plastic particles less than 5 mm in size, have emerged as a significant concern in the marine environment due to their ubiquitous presence and detrimental impact on aquatic ecosystems. It is estimated that there are approximately 51 trillion MPs in the seas, which is 500 times more than the number of stars in our galaxy. Marine organisms ingest these particles, leading to their accumulation and potential entry into the human diet through seafood consumption. Their detection and characterization remain challenging due to the diverse range of physical and chemical characteristics, along with their varying composition and concentration. Additionally, the rate at which fragmentation occurs in natural settings of MPs remains uncertain, even though they are typically generated from the breakdown of larger plastic items or product wear. This chapter focuses on current methods used for the quantitative and qualitative detection of MPs including visual inspection, spectroscopic analysis, and chromatography. It provides an in-depth examination of insulation methods, which encompass a spectrum of sampling, extraction, concentration, and purification techniques tailored to isolate MPs from complex marine matrices. These methods are specifically tailored to isolate MPs from the complex matrices of

D. Namla (✉) · M. Al-Shaeri
Department of Biological Science, Faculty of Science, King Abdulaziz University, Jeddah, Kingdom of Saudi Arabia
e-mail: dnamla@stu.kau.edu.sa; malshaeri@kau.edu.sa

M. Oves
Center of Excellence in Environmental Studies, King Abdulaziz University, Jeddah, Kingdom of Saudi Arabia

G. Mangse
Department of Biotechnology, Faculty of Sciences, Nile University of Nigeria, Abuja, FCT, Nigeria

M. Oves et al. (eds.), *Marine Microplastics and the Quest for Remediation*,
https://doi.org/10.1007/978-981-96-8979-8_2

marine environments, aiming for an explicit understanding of MPs and mitigating the impacts of these pervasive pollutants.

Keywords

Microplastics · Marine pollution · Detection methods · Spectroscopic analysis · Fragmentation dynamics

2.1 Introduction

Microplastics (MPs) have gained significant attention, with studies revealing alarming statistics in the last decade due to their pervasive effect on ecosystems. They get into marine water bodies mostly via urban runoff, fragmentation of larger plastics, and industrial discharges. These plastic types are often ingested, accumulated, and amplified in different marine biota, including fish, whales, plankton, aquatic invertebrates, and corals, posing a significant threat that could later get into the food chain to posing problems to human health (Ziani et al. 2023; Bhardwaj et al. 2024; Paul et al. 2024). Recent studies have estimated that humans consume approximately 39,000 to 52,000 microplastic (MP) particles per year (Cox et al., 2019). This statistic was obtained from screening different processed or ready-to-eat food items exposed to microplastics. In addition to the amount of inhaled suspended MPs in the air, ingested from food items, another significant volume of MP particles gets into the human body yearly from drinking tap water, making approximately 87,000 MP particles that get into the human body system yearly. These statistics reveal the emerging concern over microplastic pollution (Hossain et al. 2023; Sun et al. 2024). Shreds of evidence from recent studies have shown that MPs could pose a serious threat to human health by inducing a wide range of health aberrations, including gastrointestinal inflammation, oxidative stress, and endocrine disruption, which could culminate into long-term health detriments (Bora et al. 2024; Kadac-Czapska et al. 2024; Calzoni et al. 2024). Besides, results of preliminary studies have dictated the presence of trace amounts of MPs in critical areas of the human body, such as the placenta and lungs (Jenner et al. 2022; Sharma et al. 2024). These classes of plastics cause an enormous threat to marine biodiversity via biomagnification in tissues due to bioaccumulation upon ingestion. Their exceptional surface-to-volume ratio facilitates the bioaccumulation in the living systems. The scope of microplastic endangerments on the ecosystem has been shown to be extended through trophic transfer, as larger predators gobble up smaller organisms that have ingested varying amounts of these pollutants in their ambient environment alongside other food particles. Besides, MPs, due to their high surface-to-volume ratio, serve as carriers of other harmful chemical species, including heavy metals, pesticides, persistent organic pollutants (POPs), and hydrophobic pollutants that empty into water bodies from nearby industries, agriculture sites, or urban runoff (Saeedi 2023; Weis and Alava 2023). Although marine ecosystems have been referred to as the epicentre of MPs accumulation as a result of their ubiquitous presence in these environments, these micro-sized materials have permeated into almost every part of the earth, including the deep ocean trenches and the high mountains. These plastics get into all types of water bodies, including oceans, seas, lakes, and rivers, posing significant threats to both the biotic and the abiotic components present in such ecosystems (Kaliszewicz et al. 2023; Bhardwaj et al. 2024). Natural transport mechanisms, such as river systems, wind patterns, and ocean currents that aid the movement of water, nutrients, sediments, and materials across different parts of the Earth's surface, facilitate the widespread propagation of MPs, enabling their reach even to remote areas, such as the Arctic. This supports recent reports of trace amounts of MPs detected in marine sediments and ice cores. Atmospheric determinants, such as turbulence, convection lift, drafts, and wind speed, which affect flux mechanisms and source–sink dynamics of pollution in both terrestrial and oceanic environments could aid the transport of microplastics to distant sites across the ocean surfaces (Ward et al. 2020; Evangelou et al. 2024). Plastic pollution continues to exert debilitating economic detriments on sectors, such as tourism, fishing, and water treatment plants, placing economies around the world under pressure due to lost income from distorted ecosystems and rising cleaning costs. Considering these, developing countries with weak waste management systems disproportionately bear the burden of MP pollution. These have warranted the surge in research aiming to estimate the pollution scale of MPs and elucidate their negative impacts on marine ecosystems and human health. Initiatives at the national, regional, and international levels are essential to reduce plastic waste and lessen its environmental impact. Enacting extended producer responsibility programs and placing bans on single-use plastics, as well as innovative technologies like bioplastics and improved recycling techniques, are crucial in mitigating plastic waste hazards (Kibria et al. 2023; Bidashimwa et al. 2023).

2.2 Definition, Source, and Type of Microplastics

Microplastics (MPs) are generally defined as minute-sized plastic particles ranging between 1μm and 5 mm. They are ubiquitous and diversified in terms of chemical composition, dimensions, and density. Additionally, plastic fragments can be as small as < 1μm in size—and are termed as nanoplastics (NPs). Recently, synthetic materials including polyamide, poly (butylene succinate), polycaprolactone, polyethylene, polyhydroxyalkanoates (PHAs), polyhydroxy butyrate, polylactic acid, polypropylene, polystyrene, polyvinyl alcohol, and polyvinyl chloride have been identified as plastic pollutants in the marine waters, under different forms and sizes (Ren and Ni 2023). Primarily, these plastic materials are intentionally made into microbeads for diverse uses, including synthetic fibers fragmented from clothing or personal care products. Secondary, saltwater waves, and sunlight under mechanical forces, induce the disintegration of larger plastics into microplastics. Between 80 and 90% of MPs present in the marine water systems are classified as land-based—and include plastic bottles, textiles, personal care products, construction materials, and plastic incinerators. The remaining percentage is discharged into these

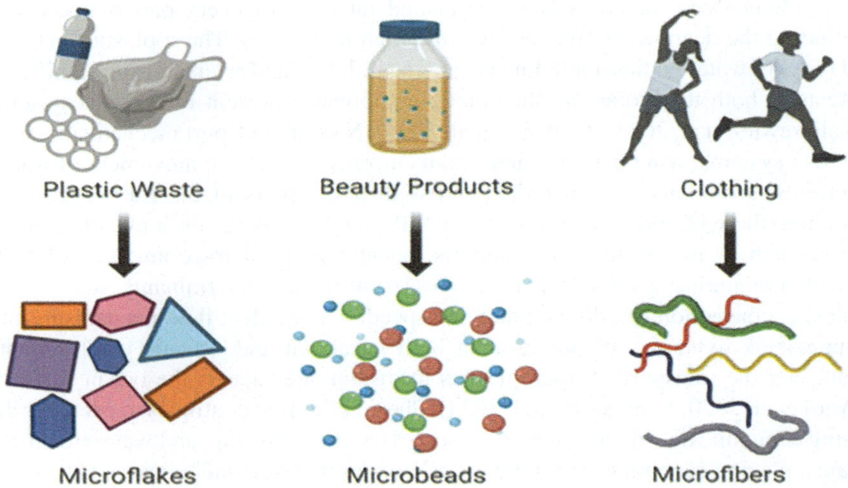

Fig. 2.1 Forms of microplastics based on their primary sources

ecosystems from fishing gear, marine vessels, and plastic litter on beaches (Ziani et al. 2023; Râpă et al. 2024). However, the primary sources of microplastic pollution in marine waters include the laundering of synthetic clothes—which account for 35% of contamination in these environments (Fig. 2.1). Additionally, 28% of MP pollution is induced by the abrasion of tires during driving, while 2% of the pollution is intentional, involving the addition of microbeads in personal care products for cosmetics (Bikiaris et al. 2024).

2.2.1 Primary Vs Secondary Microplastics

Primary MPs are intentionally deliberately made and introduced into various commercial goods and consumables, including insecticides, cosmetics, pharmaceuticals, personal care products, and detergents. Secondary MPs are inadvertently produced from the degradation of larger polymers through biological, chemical, and physical processes in marine ecosystems. Additionally, these plastic debris are grouped as polystyrene, polyethylene, polypropylene, and other plastic types based on their chemical composition, as depicted in Fig. 2.2 (Hettiarachchi and Meegoda 2023; Kibria et al. 2023; Osman et al. 2023).

2.2.2 Microplastics in Different Environments

The long-range transport, sources of emission, and meteorological conditions are known to influence the atmospheric MPs, just like other airborne pollutants. It has been stated that the MPs present in ambient air are associated with varying

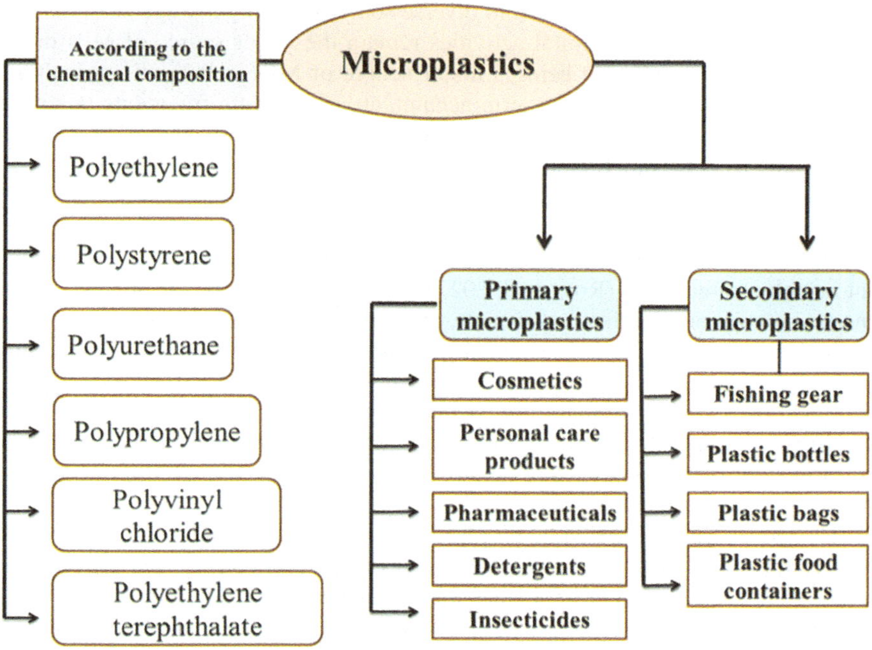

Fig. 2.2 Classifications of microplastics. Adopted from Osman et al. (2023)

concentrations based on location. As a matter of fact, higher amounts of plastic particles are generally recorded in urban areas. However, there have been some conflicting findings about the latter statement. Few studies have shown that some rural areas possess elevated MP concentrations than the metropolitan areas—and this is probably linked to plant filtration (Campanale et al. 2020; Mihai et al. 2022; Abad López et al. 2023). The pervasive presence of MPs in wastewater, surface, and ground waters raises concerns about their contamination of drinking water—with limited findings confirming their presence in bottled and tap water. Indeed, significant efforts and studies are ongoing to further enlighten the presence of MP beads and fibers in drinking water, attracting attention from the media and scientific communities. This necessitates a standardized study of characteristics and methodologies to assess the negative impacts of human exposure to MPs through drinking water. Most recently, MPs were detected in the stools of pregnant women in the Makassar city region of South Sulawesi, Indonesia, because of different levels of seafood consumption (Gambino et al. 2022; Conti et al. 2024; Hasanah et al. 2024). Initially, MP studies focused on their pollution of aquatic environments, but a certain interest has been given to their impacts on the terrestrial ecosystem because previous research has confirmed soil to be their sinking grounds. Their effects on soil ecology have been ascertained in several studies. These minute plastics disrupt the soil in many ways, such as affecting the biotic components of the environment, decreasing the water-holding capacity of the soil, also by carrying other pollutants.

Although MPs get into the soil from diverse sources, mainly sewage sludge, landfills, and fertilizers, agricultural activities remain the major source of MPs in the soil. With plastic mulching being a major source of MPs in the soil, agricultural residuals contribute to the long-term accumulation of plastic fragments in soil as environmental contaminants. Atmospheric deposition, wastewater irrigation, and industrial waste are other sources of MP's pollution of the soil (Raza et al. 2022; Siddiqua et al. 2022; Liu et al. 2023; Yarahmadi et al. 2024). Microplastics have been shown to be pervasive in aquatic environments, emanating from unregulated urban runoff, fragmented pieces of larger plastics, and industrial effluents emptying into nearby waterbodies (Ross et al. 2023). These minute fragments of plastics permeate into various marine ecosystems, ranging from surface waters to deep-sea sediments, and are being internalized by marine life directly or via the food chain, as evident in plankton and whales, respectively (Yu and Singh 2023). Their small size facilitates their easy ingestion by marine life, and results in bioaccumulation and disruption in body physiology (Fig. 2.3). MPs have also been shown to play key roles in stabilizing and shuttling heavy metals and POPs, aggravating their detrimental impacts on the ecosystem (Miller et al. 2020).

The unwarranted intrusion of microplastics into the food chain continues to raise serious human health concerns as fishes and other aquatic life forms increasingly ingest MPs and relay them to humans that consume them (Alberghini et al. 2022). MPs travel through the food web via the trophic transfer, potentially accruing even at aberrantly higher concentrations. In humans and higher organisms, MPs have been implicated to cause endocrine disruption, oxidative stress, and gastrointestinal

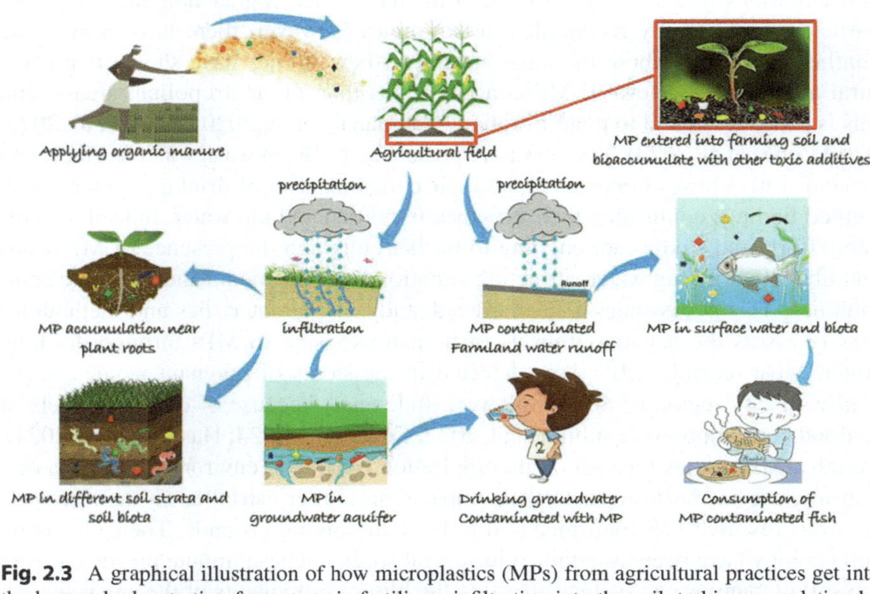

Fig. 2.3 A graphical illustration of how microplastics (MPs) from agricultural practices get into the human body, starting from organic fertilizers infiltration into the soil, to bioaccumulation by soil organisms, and entry into groundwater aquifers (Rana et al. 2023)

inflammation (Liu et al. 2023). In fact, the recent detection of MPs in drinking water and agricultural products at varying levels depicts their ubiquity and potential risk profile and warrants urgent intervention and tailored research on their possible mitigation (Brancaleone et al. 2024).

2.3 Challenges Associated with Microplastics Analysis

Microplastics are indeed fragments from larger plastics such as polyethylene, polystyrene, and more. However, the rate at which the fragmentation into minute size occurs within the natural settings of water bodies remains uncertain. Hence, both the exposure and potential risks associated with MP pollution hamper the study of these plastics (Ziani et al. 2023). The diminutiveness in size and heterogeneity in form of microplastics make their isolation and quantification in various environmental matrices difficult (Qian et al. 2024). Besides, the use of traditional sampling paradigms has been associated with a handful of technical bottlenecks, such as their ineffectiveness in capturing the full scope of microplastics, in terms of sizes and shapes, and this often results in incomplete assessments (Lamichhane et al. 2023). Additionally, the mixing and blending of other recalcitrant pollutants with the microplastic polymers further complicates their definitive determination in the ecosystems. The technology is also very expensive to run, in terms of facilities procurement and management (Rani et al. 2023).

2.3.1 Sampling and Sample Preparation

Due to their nanoscale size, membrane techniques are mostly employed to filter MPs from the marine water during analysis. The choice of membrane type depends on the analysis and treatment involved because it could be targeting the filtration of other materials, such as degradable and inert. Also, the pore size of these membranes is considered during the selection of suitable membranes, as the pore size has a significant influence on the effectiveness of the filter, the retained fraction of MPs, and the amount of residual matrix. The filtration setup has also been reported to influence the number of plastic particles retained, including their shape and size. For instance, when dealing with plastic fibers, multiple-layer filters are the most suitable. Although the selection of a better filter membrane can be overlooked, a poor membrane is likely to lead to contamination. Similarly, the pore size and decontamination measures significantly affect the utilization of the sieve as a filter membrane (Tan and Rodrigue 2019; Agboola et al. 2021; Cevallos-Mendoza et al. 2022).

Dielectrophoretic filtration (using an electric field to filter plastic particles based on their dielectric signatures differences), chromatographic techniques including hydrodynamic chromatography (using a mobile phase to separate particles by sizes of solid beads, in a small amount of the specific sample contained in a column), and field flow fraction (applying either electric or hydraulic field perpendicularly to the direction of plastic particles contained in a narrow channel, resulting in the accumulation of certain particles in one of the walls) have emerged as novel techniques to separate MPs of different size and chemical compositions (Adelugba and Emenike 2024; Prata et al. 2024).

Elsewhere, by considering their optical properties, MPs can be characterized, sorted, and counted using the flow cytometry technique. This analytical method enables the high throughput of these plastic particles. Certain studies have used flow cytometry in a synergic approach with fluorescent dyes in pre-concentration steps for the effective detection and characterization of microplastics. These nanoscale plastics have also been filtered from the marine water samples via mechanical fractioning. However, it is even more expedient to centrifuge the less dense and minute plastic particles at a higher speed and at an extended duration. And the use of a glass tube in the centrifugation process is preferable to bypass the plastic contamination (Robinson et al. 2023; Salvia et al. 2023; Sacco et al. 2023).

2.3.2 Lack of Standardized Methods

One of the main impediments to the thorough evaluation and elucidation of the characteristic distribution and environmental impacts of microplastics (MPs) is the dearth in standardized procedures. Arriving at a definitive and validated conclusion on the ecological status of these pollutants is extremely difficult when study data is inconsistent owing to disparities in sampling strategies, isolation processes, and screening methods. Misalignment in the description and classification of polymers due to their varied forms, shapes, and sizes make widen the variations in results from various studies. For example, some researchers screen for plastics at the nanoscale, while others concentrate on bigger microplastic pieces. Furthermore, inconsistency in the procedures used in isolating MPs from their intricate ecological matrices represents a stiff blockade to arriving at a comprehensive and reproducible MPs investigation. The dearth in definitive standards makes it even more difficult to establish effective baselines, predict long-term trends, and implement effective mitigation modalities, and this highlights the dire need for methodological harmonization and international alliance (Toussaint et al. 2019; Ghosh et al. 2023; Ziani et al. 2023; Du et al. 2024; Hapich et al. 2024; Vdovchenko and Resmini 2024).

2.3.3 Diversity in Microplastic Forms

The variety of forms of microplastics (MPs) makes their identification in the surroundings and the biological systems difficult. In general, MPs originate from a variety of sources and on this premise, they are classified as primary microplastics, which include plastic pellets occurring at industrial sites, and secondary microplastics, which are derived from the breakdown of bigger plastics. Microplastics often exist in a variety of sizes and shapes, such as fibers, beads, and fragments, and each one has a unique level of resistance and challenges in terms of difficulty to isolate and quantify, warranting different screening regimens. While traditional methods could be used to detect some MPs, others, such as those that exist on the nano scale, require the use of sophisticated screening techniques. In the context of their chemical makeup, MPs vary as well and comprise various blends of polymers ranging from polyethylene, polypropylene, polystyrene to polyvinyl chloride, making them portray a broad variety of chemical and physical characteristics. Chemical heterogeneity of microplastics significantly impacts their interaction in the environment. Fluctuations in the texture and porosity of MPs also affect their participation as carriers of pollutants in the ecosystem. To properly assess the prevalence and environmental and human health impact of MPs, it is imperative to deploy inclusive analytical techniques and tailored research initiatives (Borriello et al. 2023; Sarkar et al. 2023).

2.3.4 Complex Fragmentation Process

Microplastics usually surface in marine ecosystems when larger pieces of plastic debris are degraded. The degradation process is slow and highly complex and depends on several determining factors, including luminance, temperature, and oxygen content. The intrinsic properties of the degrading plastic item, such as the molecular weight distribution and additive composition, all play a combinatorial role in initiating the plastic fragmentation process. To create accurate models for forecasting the future environmental condition of the oceans and seas, it is crucial to obtain estimates of the disintegration rates of plastic materials occurring in the marine environment. Classical chemical kinetics, however, may fall short in proffering definite determination of the fragmentation rates since the plastic item often remains unbroken during weathering and is only accompanied by polymer chain breakage and oxygen coupling via oxidation. Only until enough weathering has occurred do tiny pieces begin to break off when modest mechanical tension is applied, marking the beginning of fragmentation. A larger percentage of the original plastic material is converted to microplastics as weathering continues. Prior research has examined the weathering of polyethylene in both actual and simulated marine and coastal settings, assessing plastics deterioration using chemical and mechanical stress methods. Changes in mechanical characteristics have been seen in several studies, albeit the findings are frequently incongruous (Ali et al. 2024).

2.3.5 Limitations of Current Detection Technologies

Effective detection and monitoring techniques are necessary to investigate the ecological and environmental health effects of micro- and nano-sized plastics. One of the hottest topics in study right now is the extraction, separation, and identification of plastic particles of various sizes— particularly nanoscale particles—from intricate biological and environmental materials. Uniform criteria for the extraction, qualitative, and quantitative analysis of microplastics from various sample matrices are lacking, despite the abundance of techniques and procedures. To gather comparable data, it is imperative that effective and widespread extraction, identification, and quantification techniques be developed (Schymanski et al. 2021). In addition, current screening methods commonly used in micro- and nano-sized plastics research rely on spectroscopic data from Fourier-transform infrared (FTIR) and Raman spectroscopy. Although these techniques are highly sensitive, they are time-intensive and expensive. Besides, the sample preparation requirements for these screening platforms are also cumbersome. Resolution limitations resonating from commonly accessible spectroscopic techniques and technical limitations arising from a dearth of trained staff with a certain level of technical expertise, particularly in developing countries, hamper qualitative microplastics research (Huang et al. 2023).

2.4 Microplastic Isolation and Concentration Techniques

2.4.1 Introduction to Isolation and Concentration Techniques

Purification and extraction serve as the foundation for the identification and characterization of the MPs, which are then identified and quantified, with quality-control procedures incorporated at each stage. The most important stage is extracting MPs from any environmental matrix, which depends on the density, size, and form of the MPs and the kind of matrix. Thus, extraction procedures are being adopted depending on the kind of matrix being examined. The role of concentration techniques in the MPs' analysis cannot be overemphasized. They are usually employed to level up the recovered titer volume and improve the detectability of MPs in the sample. Different concentration strategies are being utilized for MPs isolation, depending on the suitability and complexity of the sample matrix. Furthermore, factors such as financing, demography, laboratory resources, and matrix type, several modifications and improvements are being developed by different researchers (Wu et al. 2020; Magalhães et al. 2022; Thornton Hampton et al. 2023).

2.4.2 Density Separation

The density-based separation method functions on the basis that polymers with different densities exhibit different settling velocities in a fluid medium. Polymers can be selectively separated according to their settling behavior by changing the medium's density. It should be noted that although the settling equation offers a theoretical framework for comprehending the procedure, several variables may affect the actual separating efficiency. These determinants include the size distribution of the particles, the characteristics of the polymers involved, and the layout of the separation apparatus. In practice, effective density separation of polymers is frequently thought to need experimental validation and optimization. Density separation is used to isolate microplastics from environmental samples. Dense salt solutions, such as water (H_2O), sodium bromide (NaBr), sodium chloride (NaCl), and sodium hexametaphosphate (SHMP), are usually employed in isolating MPs using density separation. Eventually, the denser components sink to the bottom of the solution, while materials with lesser density float on the surface. Thereafter, the floating MPs are recovered and subjected to further determinations. Density separation techniques demonstrate high efficiency; however, several drawbacks have been identified. Their low MP recovery rates, especially when it involves densities of close parity with the other organic materials and debris in the matrices. In addition, the cost of purchasing dense solutions and the possibilities of altering the intrinsic properties of MPs, and the potential for contaminating the mixture through repeated processing (Wu et al. 2020; Schütze et al. 2022; Jessieleena et al. 2023; Rani et al. 2023; Crutchett and Bornt 2024).

2.4.3 Filtration

Filtration is a direct analytical procedure that is used in the isolation of MPs from water samples. Filters with minute pore sizes of approximately 0.45μm are deployed to retrieve MPs from its aqueous solution. This technique is typically suitable for capturing MPs of specific sizes. However, limitations, such as its inability to capture plastics of minute size, such as those on the nanoscale, and clogging of filters with organic debris, could primarily reduce its efficiency. The process is also time-consuming and labor-intensive, especially when it involves large volumes of samples (Dey et al. 2024; Schymanski et al. 2021; Vohl et al. 2024).

2.4.4 Chemical Digestion

The use of hydrolyzing agents, such as potassium hydroxide (KOH), hydrogen peroxide (H_2O_2), or Fenton's reagent to break down organic matter in environmental samples in order to isolate MPs enjoys wide utility and is usually referred to as chemical digestion. These reagents degrade organic materials sparing MPs, hence enhancing their effective determinations. However, some limitations associated with chemical digestion include the possibility of altering MPs during harsh chemical reaction regimens, which may affect their shape or polymer structure. Besides, some reagents are hazardous, costly, or less efficient against certain organic pollutants, warranting multiple treatment procedures (Kazimierowicz et al. 2022; Wang et al. 2024; Lu et al. 2024).

2.4.5 Sedimentation

The sedimentation method is used to separate MPs from particles with larger weights, which settle at the bottom under gravity. This technique is often deployed to isolate soil or sediment samples. Generally, MPs, which are intrinsically less dense, remain suspended or float and are separated. The sedimentation method is simple and less-expensive; however, several draw backs to this technique have been reported. The technique requires much time to separate a given sample, making it less suitable for isolating MPs with densities at par with other constituents in the sample. Besides, the agitation process that characterizes sedimentation techniques could result in mixing of various sample constituents, which could affect the accuracy of the technique (Vojnović et al. 2024).

2.4.6 Magnetic Separation

Magnetic separation is a novel isolation technique that leverages the magnetic properties of the various component materials in solution. In this method, magnets are used to attract magnetizable constituents and isolate them from nonmagnetic components. In recent times, magnetic separation techniques have shown promise in isolating MPs; however, its application is limited to samples with specific magnetic attributes or MPs bonded to magnetizable additives. Besides, the method is not suitable when working with MPs occurring in complex environmental matrices with minimal magnetic composition (Rani et al. 2023; Wu et al. 2020;Vohl et al. 2024).

2.5 Microplastic Characterization Techniques

2.5.1 Microscopy

Microscopy is usually the initial step in microplastic analysis and is often accompanied by spectroscopy to ascertain the material type. The relevance of employing microscopy largely lies in its propensity to afford count, color, size, and morphological attributes about the sample and to inform the researcher of the toxicity profile and source apportionment. Based on cost-effectiveness, stereoscope microscopy has been shown to be less expensive and a simpler technique for MPs analysis. However, the use of stereoscope microscopes to screen MPs is usually subjective, owing to MPs' diversity in size, shape, and form compared with other nanoscale materials, and this often results in misclassification. Staining dyes play a crucial role in the microscopic screening of MPs as they enable an effective objective criterion when selected for plastics, even without other stainable particles. The commonly used staining dye in MP studies is Nile Red. Under a UV light at a wavelength of 250 nm, this dye could produce a red-to-yellow fluorescence under the blue light for stained MPs. Surprisingly, studies conducted using bottled water and sediment samples have reported that the staining dye approach reduces the number of false or abnormal positives to 100% and 90%, respectively, as plastic features. Elsewhere, Nile Red can be used as a preselection method prior to any further chemical screening, to improve the sample throughput in synergy with a fluorescent microscope for the analysis of >1 µm MPs.

2.5.2 Chemical

Macroscopic screening and staining dyes alone remain ineffective in studying MPs, as they only yield speculative outputs. Thus, for more information about the chemical composition or the source of these suspected plastics, specific chemical characterizations are required. Vibrational spectroscopy (VS) constitutes the main foundation of these chemical techniques. Currently, the frequently used chemical analytical approaches include Fourier transform infrared spectroscopy (FTIR) and Raman microscopy. Technically, FTIR exploits the principles of infrared spectroscopy to detect alterations in the dipole moment. Also, the FTIR technique is predicated on the measurement of infrared radiation that is transmitted, absorbed, and reflected. In contrast, vibrational spectroscopy relies on the detection of inelastically scattered light caused by alterations in the molecule's polarization. However, even though the latter techniques seem effective for MP characterization, their time-consuming and cost explain their infrequent use in research settings—and researchers may only access them via scientific collaborations. Additionally, there are certain pigments or contaminants, such as carbon black and titanium oxide, that could interfere with MPs, limiting the emission, absorption, and reflectance of the infrared radiation of these plastics.

Additionally, there are restrictions when it comes to utilizing the VS in forensic investigations such as environmental interferences, sample complexity, and instrumental limitations. Therefore, to bypass such challenges, when identifying certain trace materials such as the low concentration drugs, explosives, or even organic compounds like CO_2, the obtained peaks are often subjected to baseline correction due to the affinity of chemical bonds to specific spectral peaks. However, open online databases, such as SiMPle and OpenSpecy are available and could be utilized for polymer characterization, especially when the plastic item involved is known (Seeley and Lynch 2023).

2.5.3 Scanning Electron Microscopy

Scanning Electron Microscopy (SEM) is a valuable analytical method frequently used to describe the size, form, and chemical makeup of microplastics. Objects with dimensions ranging from millimeters to nanometers are imaged and measured using this technology. Because the deBroglie wavelength of the electron beam used to probe a material is far smaller than optical wavelengths, the optical diffraction constraint is removed, allowing the SEM method to achieve nanoscale spatial resolution. SEMs may create pictures of samples with a resolution of only a few nanometers because of the electron beams they utilize. SEM, therefore, aids in the observation of minute particles that an optical microscope could overlook (Huang et al. 2023; Lamichhane et al. 2023; Sarkar et al. 2023). Microscopy and electron microscopy (such as transmission electron microscopy [TEM], scanning electron microscopy [SEM], and atomic force microscopy [AFM]) are common techniques for analyzing MPs. Other techniques include spectrometry (such as Fourier transform infrared spectroscopy (FTIR) and Raman spectroscopy), mass spectrometry (such as pyrolysis-gas chromatography coupled to mass spectrometry [Pyr-GC/MS] and chemical depolymerization-liquid chromatography—tandem mass spectrometry [CD-LC-MS/MS]). The amount and form of MPs can be determined using microscopy and electron microscopy, but not their composition. The resolution of an electron microscope is less than 1 nm, while that of a standard optical microscope is around 250 nm. Micro-FTIR and micro-Raman spectrometry methods have the benefit of allowing the composition of individual MPs to be determined without damaging the material. However, comparing the results among different research by employing microscopy and spectroscopic techniques becomes problematic due to the uneven form and variable sizes of plastic items (Alqaheem and Alomair 2020; Wang et al. 2024).

2.5.4 Mass Spectrometry Techniques

Mass spectrometry (MS) is a highly advanced technique used to identify the structure and chemical characteristics of materials, as well as for molecular screening and quantification of the material. Numerous allied scientific disciplines, including chemistry, biochemistry, pharmacology, medicine, and many more, have been transformed by MS methodology. The emergence of MS as an indispensable analytical platform for X-raying the chemical composition, fragmentation rates, and associated contaminants of microplastics in marine ecosystems has afforded a profound understanding of their environmental impact. The pyrolysis-gas chromatography-mass spectrometry (Pyr-GC/MS) operates by degrading microplastics, heating them into volatile chemical species. Thereafter, these compounds are extracted and identified based on their chemical composition and additive contents. The high sensitivity of this method makes it suitable for characterizing microplastics and aids in tracking their degradation within the marine ecosystems. The thermal desorption-gas chromatography-mass spectrometry (TD-GC-MS) platform is utilized in screening volatile organic compounds and additives emanating from microplastics. This technique enables researchers to gain a thorough understanding of how toxic substances leach from littered plastics items into marine habitats (Seeley and Lynch 2023). The matrix-assisted laser desorption/ionization time-of-flight mass spectrometry (MALDI-TOF-MS) enables the rapid screening of polymers and their molecular weight distributions. The fact that the sample preparation requirements of this method are relatively lesser and could perform iterative screening of different substances in a common assay distinguishes it as an efficient method for studying large-scale marine dispersed microplastics. In addition, the Orbitrap mass spectrometry and Fourier-transform ion cyclotron resonance mass spectrometry (FT-ICR MS) are usually deployed for analysis that warrants high resolution. This analytical technique is suitable for handling complex mixtures, such as fragmentation by-products bound to microplastic surfaces and other persistent organic pollutants. The high resolution of this analytical tool makes it stand out for the determination of the chemical crosstalk that exists between microplastics and pollutants in the marine ecosystems (Li et al. 2024).

Secondary ion mass spectrometry (SIMS) is another advanced analytical platform that affords researchers the ability to analyze the surface composition of microplastics, unearthing a comprehensive elemental mapping and unravelling the mechanisms of pollutant adherence and interactions with microplastic particles (Feng et al. 2021). Recently, researchers used liquid chromatography-mass spectrometry (LC-MS) as complementary tools with the techniques to enable the analysis of nonvolatile compounds such as dyes, plasticizers, and soluble organic pollutants (Magalhães et al. 2022). Although mass spectrometry techniques afford detailed insights into the chemical complexity of microplastics and their ecological implications in marine ecosystems, however, bottlenecks such as sample contamination and matrix effects in quantitative analysis hamper this technique (Sarkar et al. 2023).

2.5.5 Other Techniques

A variety of other analytical techniques are employed in addition to mass spectrometry and microscopy to characterize microplastics. Transmission electron microscopy, or TEM, is another technique for analyzing microplastics. Although the visual identification of microplastics using TEM may vary depending on the analyzer, cross-validating data from several users using a tried-and-true procedure helps reduce inaccuracies. The elemental composition of the plastic material is determined by additional investigation using energy-dispersive X-ray spectroscopy (EDS). Antioxidants or additive components like Ca, Na, Al, Mg, and Si are commonly used as markers to detect microplastics when utilizing EDS. One of the most important methods for studying materials is thermal analysis, which examines the functional connection of this change. Thermal analysis is used to evaluate microplastics in environmental samples after they have been heated. As the temperature rises, the microplastics absorb a lot of heat, which finally transforms the polymers from a solid to a liquid or gas. An endothermic peak then appears at that temperature. Because different kinds of polymers have different thermal stabilities, standard thermograms of polymers can be used to study the type and content of microplastics and their additions (Campanale et al. 2020; Woo et al. 2021; Li et al. 2022; Huang et al. 2023). Differential scanning calorimetry (DSC) facilitates material analysis by analyzing the thermal properties of polymers. Nowadays, DSC is frequently used to detect polyethylene and other significant microplastics that are regularly detected in the environment. Differential scanning calorimetry is typically employed with other methods since DSC has limited uses. Another method for thermal analysis is thermogravimetric analysis (TGA). This method confirms the sample's qualitative and quantitative data by assessing the sample's dependence on temperature and time, as well as by measuring the sample's weight loss when heated at a certain rate under preset ambient and temperature conditions. TGA and DSC have previously been used to study microplastics in wastewater (Blanco and Siracusa 2021; Huang et al. 2023).

2.6 Emerging Trends and Future Directions

Microplastics result from the breakdown of different types of polymers including the degradation products of plastic containers, bottles, foils, and several other consumer items. Their small size, portability, and characteristic innocuity have made them to be considered relatively harmless to humans and aquatic life for a few decades (Cai et al. 2023). Several reports have highlighted the pervasiveness and recalcitrant nature of microplastics in various aspects of the environment, including rivers, precipitation, soil, and wildlife (Amobonye et al. 2021; Ziani et al. 2023). Of note, traces of microplastics have been detected in the human organ systems, including the respiratory organs (Jenner et al. 2022), digestive organs (Zhang et al. 2021), blood tissue (Leslie et al. 2022; Leonard et al. 2024), and reproductive organs (Codrington et al. 2024). Recently, a report submitted by the Washington Post opined that the occurrence of trace quantities of microplastics may be in the ambient air emanating from plastic packaging, clothing fibers, car tire particles, AstroTurf, and spheres permeating from cosmetic exfoliating agents. According to WHO (2019), "microplastics in drinking water don't seem to pose a health risk at current levels based on the limited information." Evidence-based studies showed that current platforms where microplastics sprout elicit insignificant levels of risk to humans. However, a rising body of research on microplastics has indicated that there are serious problems due to the increasing amounts in the environment. The recent surge in investigations on microplastics underscores heightened concerns the scientific community shares regarding the grievous and snowballing impacts of microplastics on the ecosystem and warrants more tailored investigations into their occurrence, distribution, and effects on human health and the environment. Numerous studies have linked or discovered tenuous connections between the presence of these microscopic plastic particles and health problems like cancer, heart disease, and other chronic illnesses, despite the lack of conclusive evidence that their presence could have a negative impact on human health (Campanale et al. 2020; Ghosh et al. 2023; Bhardwaj et al. 2024). These studies contain a lot of confounding factors that the authors were unable to control for, making it difficult, if not impossible, to draw conclusions about the cause of the disease. Although some research on rodents has indicated that elevated blood levels of microplastics may have neurotoxic consequences, it is challenging to apply these findings to people, particularly since possible environmental exposures are hundreds of times lower than those that the animals experienced (Li et al. 2022; Liu et al. 2023). To accurately determine what health risks, if any, are posed at current or anticipated concentrations of microplastics, a thorough risk assessment would be required. Depending on where they come from, microplastics may contain and release harmful substances, and they may also induce some degree of inflammation on their own (Jeong et al. 2016; Kadac-Czapska et al. 2024). Since cellular irritation may be a significant mechanism, the toxicity of microplastics is most likely influenced by their kind, size, shape, and, of course, concentration. The immune system (macrophages) could absorb microplastics that are not immediately expelled from the body. The type of plastic will probably determine its capacity to break down the substance.

2.6.1 Advances in Microplastic Detection

Because of its widespread presence and detrimental effects on ecosystems and human health, microplastic (MP) pollution has become a major environmental problem on a global scale. As a result, MP evaluation in ambient and experimental samples has grown in importance and frequency (Lamichhane et al. 2023). MPs are often characterized chemically and physically using spectroscopy and microscopy. However, these analytical techniques frequently require costly equipment or time-consuming sample pretreatments. Emerging research in microplastics are geared toward the design and development of more portable and cheaper microplastic detection platforms that incorporates advanced technologies, such as machine learning and deep learning functionalities, to enhance high-throughput screening and dynamic determinations of MPs in the environments, particularly, in the oceanic ecosystem (Blevins et al. 2021). Incorporating these technologies in MPs detection processes holds auspicious prospects in revolutionizing the identification and stratification of MPs, abrogating extraneous sample preparation requirements (Torres Moral et al. 2022). Li et al. (2024) recently developed a high-throughput platform for assessing microplastic using polarization holographic imaging. In their study, they developed a microplastic detection system that concurrently captures trends of holographic interference and states of polarization, enabling multimodal data collection and rapid MPs detection. The system design allows for the recording of light waves' characteristics and leverages various birefringence features to categorize the material composition and structures of MPs. In addition, the device operated on automated real-time summations and morphological estimations of component materials, including microplastic sheets and additional natural constituents. By developing an enhanced flow cytometry-based microplastic and nanoplastic screening platform, Li et al. (2023) presented an essential leverage on some technical bottleneck for the effective assessment and quantification of the risk profile of microplastic and nanoplastic pollution. Primpke et al. (2017) presented an automated technique for analyzing microplastics using focal plane array (FPA) FTIR microscopy and image analysis to shorten the time demand for complete data analyses. Their study showed marked potential of automation in improving data quality and time requirements in interpreting complex FTIR-imaging data.

2.6.2 Improved Sampling Methods

Drones, passive samplers, and high-volume air samplers are all cutting-edge methods used in microplastic sampling, and each has unique benefits. Large volumes of air are drawn in by the high-volume air samplers, which capture microplastics for further examination through filters. Over time, passive samplers are also used to collect microplastics from the ambient air. Their dependence on natural currents causes them to gradually acquire microplastics. Specialized sampling equipment may be installed on drones and unmanned aircraft to gather air samples at various elevations, offering information on the vertical distribution of microplastics. They effectively traverse wide regions, gathering samples and aerial photos to evaluate the spread of microplastics. Compared to conventional sample techniques, these technologies have benefits including greater spatial coverage, economic effectiveness, and less environmental disturbance, which improve our comprehension of microplastic contamination in many ecosystems. The most accurate and repeatable way for gathering plastic particles is active sampling, which is more user-friendly than passive sampling. Visual counting is the most often used physical analysis methodology. The most common method for identifying polymers gathered by visual sorting is spectroscopy (FTIR; Raman). Nevertheless, accurate chemical information can also be obtained by destructive analytical methods (thermal degradation) (Amobonye et al. 2021; Cerasa et al. 2021; Razeghi et al. 2021; Pasquier et al. 2022; Dong et al. 2023; Huang et al. 2023).

2.6.3 Incorporating Nanoplastics and Secondary Microplastics

Plastic materials finding their way into the natural ecosystem display a varying degree of recalcitrancy, eventually disintegrate into minute particles due to the actions of different environmental factors. Microplastics, comprising plastic materials ranging within the size of 1–5 mm, and NPs, with a size less than 1 mm, pose danger as pollutants to all life forms on the planet Earth (Ali et al. 2024). They are readily ingested by living organisms, resulting in their biomagnification and bioaccumulation. Both intentionally (primary nanoplastic) and inadvertently (secondary nanoplastic)-synthesized nanoplastic production have occurred in the environment (Trevisan et al. 2022). Regardless of whether they are top-down milled or bottom-up manufactured, primary nanoplastics comprise pristine non-surface-modified material. Unlike primary nanoplastics, secondary nanoplastics are composed of a complex mixture of polymers (Dube and Okuthe 2023). Their physicochemical characteristics and interactions with the environment are markedly altered in the face of environmental weathering. Moreover, secondary nanoplastic particles are a blend of plastics with varying origins in terms of their additive content and polymer composition. Among the commonly used analytical approaches for the identification and quantification of microplastics, there are SEM, Fourier transform infrared spectroscopy, Raman spectroscopy, flow cytometry, mass spectrometry, matrix-assisted laser desorption/ionization, pyr olysis gas chromatography–mass spectrometry, and liquid chromatography–tandem mass spectrometry (Hale et al. 2022). Nevertheless, limitations have been noted, including the inability to identify MP/NP precisely and the challenge of establishing beyond a reasonable doubt that the polymers found came from the sample rather than contamination during sample collection, processing, and analysis (Mendonça et al. 2022).

2.6.4 Integration of Multidisciplinary Approaches

The integration of multidisciplinary strategies is key in surmounting the encumbrances of microplastic pollution. Even though a great deal of study has been done to comprehend the problem of microplastic contamination of different aquatic ecosystems, there is still a significant research deficit in areas such as their ecotoxicology, destiny, dispersion, and techniques of detection in aquatic settings. Furthermore, nothing is known about the combined harmful effects of microplastics and other environmental contaminants, which calls for further investigation (Alimi et al. 2021; Picó and Barceló 2023). Ecotoxicology plays a pivotal role in understanding the dynamics and impact of microplastics on ecosystems, including their distorting effect on the physiology, reproductive capacity, and behavioral pattern of living organisms in their various niches. Ecotoxicology is also valuable in studying important environmental determinants, such as trophic transfer and material bioaccumulation, and the identification of sensitive bioindicators of the ecosystem's biosafety. Besides, ecotoxicological studies are used to estimate the risk profile, toxicity thresholds, and state of exposure of pollutants (Khan et al. 2022; Vandenberg et al. 2023; Dey et al. 2024). The deployment of advanced chemical techniques complements these efforts, enhancing efficient identification and characterization of microplastics with high precision and sensitivity. Analytical platforms, such as Raman Spectroscopy and Fourier Transform Infrared Spectroscopy (FTIR) are utilized in polymer identification. In addition, data from thermal analysis, such as Pyrolysis-GC/MS and mass spectrometry, are explored to shed more light on surface characteristics and elemental or polymer composition of the microplastics. The use of imaging techniques, such as Scanning Electron Microscopy (SEM) also assist in unearthing the underpinnings of microplastic distribution, while additives and adsorbed contaminants, which may be contributory to secondary toxicity are detected using chemical analysis (Whiting et al. 2022; Sarkar et al. 2023; Seeley and Lynch 2023).

2.6.5 Sustainable Solutions and Alternatives

Effective and sustainable interventions and remediation of microplastics occurrence and impacts in the ecosystem warrant an amalgamation of innovative materials, systemic modifications, and global alliances. Recently, the use of biodegradable plastics as an alternative to conventional petrochemical-based polymers has been advocated and encouraged due to their ease in breaking down into natural substances through the actions of decomposing microbes. For instance, polylactic acid (PLA), which is used in making consumer products such as disposable tableware, cutlery, and electronics, has been harnessed from renewable biomass, such as sugarcane and cornstarch. Bacteria-derived polyhydroxyalkanoates (PHA) are a tuneable, biodegradable, and biocompatible material used in tissue engineering and other medical applications. However, several bottlenecks hamper their extensive adoption and utilization, such as the industrial composting requirements for adequate degradation, the detrimental effects of contamination in recycling streams, and dependence on fossil-based energy sources during production (Ayilara et al. 2020; Rezvani Ghomi et al. 2021; Pulingam et al. 2022; Che Ab Aziz et al. 2023; Lamichhane et al. 2023). The adoption of other eco-friendly alternatives, such as natural fiber products and plant-based packaging, is also contributory in the fight to mitigate microplastic distribution. Textiles fabricated from natural products such as wool, hemp, or cotton can be used to substitute synthetic fiber-based textiles. In addition, seaweeds, mushrooms, and algae could be harnessed to produce sustainable packaging materials. The adoption of reusable and recyclable materials, such as glass, metal, or durable plastics in place of plastics, could contribute to long-term waste mitigation (Chia et al. 2020; Periyasamy 2023).

References

Abad López AP, Trilleras J, Arana VA, Garcia-Alzate LS, Grande-Tovar CD (2023) Atmospheric microplastics: exposure, toxicity, and detrimental health effects. RSC Adv 13(11):7468–7489. https://doi.org/10.1039/d2ra07098g

Adelugba A, Emenike C (2024) Comparative review of instrumental techniques and methods for the analysis of microplastics in agricultural matrices. Microplastics 3(1):1–21. https://doi.org/10.3390/microplastics3010001

Agboola O, Fayomi OSI, Ayodeji A, Ayeni AO, Alagbe EE, Sanni SE, Okoro EE, Moropeng L, Sadiku R, Kupolati KW, Oni BA (2021) A review on polymer nanocomposites and their effective applications in membranes and adsorbents for water treatment and gas separation. Membranes 11(2):139. https://doi.org/10.3390/membranes11020139

Alberghini L, Truant A, Santonicola S, Colavita G, Giaccone V. Microplastics in Fish and Fishery Products and Risks for Human Health: A Review. Int J Environ Res Public Health. 2022;20(1):789. Published 2022 Dec 31. https://doi.org/10.3390/ijerph20010789

Ali SS, Elsamahy T, Al-Tohamy R, Sun J (2024) A critical review of microplastics in aquatic ecosystems: degradation mechanisms and removing strategies. Environ Sci Ecotechnol 21:100427. https://doi.org/10.1016/j.ese.2024.100427

Alimi OS, Fadare OO, Okoffo ED (2021) Microplastics in African ecosystems: current knowledge, abundance, associated contaminants, techniques, and research needs. Sci Total Environ 755(Pt 1):142422. https://doi.org/10.1016/j.scitotenv.2020.142422

Alqaheem Y, Alomair AA (2020) Microscopy and spectroscopy techniques for characterization of polymeric membranes. Membranes 10(2):33. https://doi.org/10.3390/membranes10020033

Amobonye A, Bhagwat P, Raveendran S, Singh S, Pillai S (2021) Environmental impacts of microplastics and nanoplastics: a current overview. Front Microbiol 12:768297. https://doi.org/10.3389/fmicb.2021.768297

Ayilara MS, Olanrewaju OS, Babalola OO, Odeyemi O (2020) Waste management through composting: challenges and potentials. Sustain For 12(11):4456. https://doi.org/10.3390/su12114456

Bhardwaj LK, Rath P, Yadav P, Gupta U (2024) Microplastic contamination, an emerging threat to the freshwater environment: a systematic review. Environ Syst Res 13:8. https://doi.org/10.1186/s40068-024-00338-7

Bidashimwa D, Hoke T, Huynh TB, Narkpitaks N, Priyonugroho K, Ha TT, Burns A, Weissman A (2023) Plastic pollution: how can the global health community fight the growing problem? BMJ Glob Health 8(3):e012140. https://doi.org/10.1136/bmjgh-2023-012140

Bikiaris N, Nikolaidis NF, Barmpalexis P (2024) Microplastics (MPs) in cosmetics: a review on their presence in personal-care, cosmetic, and cleaning products (PCCPs) and sustainable alternatives from biobased and biodegradable polymers. Cosmetics 11(5):145. https://doi.org/10.3390/cosmetics11050145

Blanco I, Siracusa V (2021) The use of thermal techniques in the characterization of bio-sourced polymers. Materials 14(7):1686. https://doi.org/10.3390/ma14071686

Blevins MG, Allen HL, Colson BC, Cook A-M, Greenbaum AZ, Hemami SS, Hollmann J, Kim E, LaRocca AA, Markoski KA, Miraglia P, Mott VL, Robberson WM, Santos JA, Sprachman MM, Swierk P, Tate S, Witinski MF, Kratchman LB, Michel APM (2021) Field-portable microplastic sensing in aqueous environments: a perspective on emerging techniques. Sensors 21(10):3532. https://doi.org/10.3390/s21103532

Bora SS, Gogoi R, Sharma MR, Anshu B, Deka P, Bora J, Naorem RS, Das J, Teli AB (2024) Microplastics and human health: unveiling the gut microbiome disruption and chronic disease risks. Front Cell Infect Microbiol 14:1492759. https://doi.org/10.3389/fcimb.2024.1492759

Borriello L, Scivicco M, Cacciola NA, Esposito F, Severino L, Cirillo T (2023) Microplastics, a global issue: human exposure through environmental and dietary sources. Food Secur 12(18):3396. https://doi.org/10.3390/foods12183396

Brancaleone E, Mattei D, Fuscoletti V, Lucentini L, Favero G, Cecchini G, Frugis A, Gioia V, Lazzazzara M (2024) Microplastic in drinking water: a pilot study. Microplastics 3(1):31–45. https://doi.org/10.3390/microplastics3010003

Cai Z, Li M, Zhu Z, Wang X, Huang Y, Li T, Gong H, Yan M (2023) Biological degradation of plastics and microplastics: A recent perspective on associated mechanisms and influencing factors. Microorganisms 11(7):1661. https://doi.org/10.3390/microorganisms11071661

Calzoni E, Montegiove N, Cesaretti A, Bertoldi A, Cusumano G, Gigliotti G, Emiliani C (2024) Microplastic and extracellular vesicle interactions: recent studies on human health and environment risks. Biophysica 4(4):724–746. https://doi.org/10.3390/biophysica4040047

Campanale C, Massarelli C, Savino I, Locaputo V, Uricchio VF (2020) A detailed review study on potential effects of microplastics and additives of concern on human health. Int J Environ Res Public Health 17(4):1212. https://doi.org/10.3390/ijerph17041212

Cerasa M, Teodori S, Pietrelli L (2021) Searching Nanoplastics: from sampling to sample processing. Polymers 13(21):3658. https://doi.org/10.3390/polym13213658

Cevallos-Mendoza J, Amorim CG, Rodríguez-Díaz JM, Montenegro M, d. C. B. S. M. (2022) Removal of contaminants from water by membrane filtration: A review. Membranes 12(6):570. https://doi.org/10.3390/membranes12060570

Che Ab Aziz NI, Zakaria Y, Md Muslim NZ, Nik Hassan NF (2023) Emerging and advanced Technologies in Biodegradable Plastics for sustainability. In: Ali GAM, Makhlouf ASH (eds) Handbook of biodegradable materials. Springer, Cham. https://doi.org/10.1007/978-3-031-09710-2_21

Chia WY, Ying Tang DY, Khoo KS, Kay Lup AN, Chew KW (2020) Nature's fight against plastic pollution: algae for plastic biodegradation and bioplastics production. Environ Sci Ecotechnol 4:100065. https://doi.org/10.1016/j.ese.2020.100065

Codrington J, Varnum AA, Hildebrandt L, Pröfrock D, Bidhan J, Khodamoradi K, Höhme AL, Held M, Evans A, Velasquez D, Yarborough CC, Ghane-Motlagh B, Agarwal A, Achua J, Pozzi E, Mesquita F, Petrella F, Miller D, Ramasamy R (2024) Detection of microplastics in the human penis. Int J Impot Res. https://doi.org/10.1038/s41443-024-00930-6

Conti I, Brenna C, Passaro A, Neri LM (2024) Bioaccumulation rate of non-biodegradable polystyrene microplastics in human epithelial cell lines. Int J Mol Sci 25(20):11101. https://doi.org/10.3390/ijms252011101

Cox KD, Covernton GA, Davies HL, Dower JF, Juanes F, & Dudas SE (2019). Human consumption of microplastics. Environmental Science & Technology, 53(12):7068–7074. https://doi.org/10.1021/acs.est.9b01517

Crutchett TW, Bornt KR (2024) A simple overflow density separation method that recovers >95% of dense microplastics from sediment. Methods X 12:102638. https://doi.org/10.1016/j.mex.2024.102638

Dey P, Bradley TM, Boymelgreen A (2024) Trophic transfer and bioaccumulation of nanoplastics in Coryphaena hippurus (mahi-mahi) and effect of depuration. PLoS One 19(11):e0314191. https://doi.org/10.1371/journal.pone.0314191

Dong H, Wang X, Xu L, Ding J, Wania F (2023) A flow-through passive sampler for microplastics in air. Environ Sci Technol 57(6):2362–2370. https://doi.org/10.1021/acs.est.2c07016

Du J, Zhan L, Zhang G, Zhou Q, Wu W (2024) Antibiotic sorption onto MPs in terrestrial environment: a critical review of the transport, bioaccumulation, ecotoxicological effects and prospects. Drug Chem Toxicol. https://doi.org/10.1080/01480545.2024.2433075

Dube E, Okuthe GE (2023) Plastics and micro/Nano-plastics (MNPs) in the environment: occurrence, impact, and toxicity. Int J Environ Res Public Health 20(17):6667. https://doi.org/10.3390/ijerph20176667

Evangelou I, Tatsii D, Bucci S, Stohl A (2024) Atmospheric resuspension of microplastics from bare soil regions. Environ Sci Technol 58(22):9741–9749. https://doi.org/10.1021/acs.est.4c01252

Gambino I, Bagordo F, Grassi T, Panico A, De Donno A (2022) Occurrence of microplastics in tap and bottled water: current knowledge. Int J Environ Res Public Health 19(9):5283. https://doi.org/10.3390/ijerph19095283

Ghosh S, Sinha JK, Ghosh S, Vashisth K, Han S, Bhaskar R (2023) Microplastics as an emerging threat to the global environment and human health. Sustain For 15(14):10821. https://doi.org/10.3390/su151410821

Hale RC, Seeley ME, King AE, Yu LH (2022) Analytical chemistry of plastic debris: sampling, methods, and instrumentation. In: Bank, M.S. (ed) Microplastic in the environment: pattern and process. Environmental contamination remediation and management. Springer, Cham. https://doi.org/10.1007/978-3-030-78627-4_2

Hapich H, Cowger W, Gray AB (2024) Microplastics and trash cleaning and harmonization (MaTCH): semantic data ingestion and harmonization using artificial intelligence. Environ Sci Technol 58(46):20502–20512. https://doi.org/10.1021/acs.est.4c02406

Hasanah U, Amqam H, Septami AE, Chalid M, Aris AZ (2024) Plasticizing pregnancy: microplastics identified in expectant mothers' feces. Environ Health Insights 18:11786302241235810. https://doi.org/10.1177/11786302241235810

Hettiarachchi H, Meegoda JN (2023) Microplastic pollution prevention: the need for robust policy interventions to close the loopholes in current waste management practices. Int J Environ Res Public Health 20(14):6434. https://doi.org/10.3390/ijerph20146434

Hossain MB, Yu J, Banik P, Noman MA, Nur A-AU, Haque MR, Rahman MM, Albeshr MF, Arai T (2023) First evidence of microplastics and their characterization in bottled drinking water from a developing country. Front Environ Sci 11:1232931. https://doi.org/10.3389/fenvs.2023.1232931

Huang Z, Hu B, Wang H (2023) Analytical methods for microplastics in the environment: a review. Environ Chem Lett 21(1):383–401. https://doi.org/10.1007/s10311-022-01525-7

Jenner LC, Rotchell JM, Bennett RT, Cowen M, Tentzeris V, Sadofsky LR (2022) Detection of microplastics in human lung tissue using μFTIR spectroscopy. Sci Total Environ 831:154907

Jessieleena A, Rathinavelu S, Velmaiel KE, John AA, Nambi IM (2023) Residential houses - a major point source of microplastic pollution: insights on the various sources, their transport, transformation, and toxicity behaviour. Environ Sci Pollut Res Int 30(26):67919–67940. https://doi.org/10.1007/s11356-023-26918-1

Kadac-Czapska K, Ośko J, Knez E, Grembecka M (2024) Microplastics and oxidative stress-current problems and prospects. Antioxidants 13(5):579. https://doi.org/10.3390/antiox13050579

Kaliszewicz A, Panteleeva N, Karaban K, Runka T, Winczek M, Beck E, Poniatowska A, Olejniczak I, Boniecki P, Golovanova EV, Romanowski J (2023) First evidence of microplastic occurrence in the marine and freshwater environments in a remote polar region of the Kola peninsula and a correlation with human presence. Biology 12(2):259. https://doi.org/10.3390/biology12020259

Kazimierowicz J, Dębowski M, Zieliński M (2022) Effect of pharmaceutical sludge pre-treatment with Fenton/Fenton-like reagents on toxicity and anaerobic digestion efficiency. Int J Environ Res Public Health 20(1):271. https://doi.org/10.3390/ijerph20010271

Khan FR, Catarino AI, Clark NJ (2022) The ecotoxicological consequences of microplastics and co-contaminants in aquatic organisms: a mini review. Emerg Top Life Sci 6(4):339–348. https://doi.org/10.1042/ETLS20220014

Kibria MG, Masuk NI, Safayet R, Nguyen HQ, Mourshed M (2023) Plastic waste: challenges and opportunities to mitigate pollution and effective management. International J Environ Res 17(1):20. https://doi.org/10.1007/s41742-023-00507-z

Lamichhane G, Acharya A, Marahatha R, Modi B, Paudel R, Adhikari A, Raut BK, Aryal S, Parajuli N (2023) Microplastics in environment: global concern, challenges, and controlling measures. Int J Environ Sci Technol 20(4):4673–4694. https://doi.org/10.1007/s13762-022-04261-1

Leonardo, Alberghini Alessandro, Truant Serena, Santonicola Giampaolo, Colavita Valerio, Giaccone (2023) Microplastics in Fish and Fishery Products and Risks for Human Health: A Review International Journal of Environmental Research and Public Health 20(1):789. https://doi.org/10.3390/ijerph20010789

Leonard SVL, Liddle CR, Atherall CA, Chapman E, Watkins M, Calaminus SDJ, Rotchell JM (2024) Microplastics in human blood: polymer types, concentrations and characterisation using μFTIR. Environ Int 188:108751. https://doi.org/10.1016/j.envint.2024.108751

Leslie HA, van Velzen MJM, Brandsma SH, Vethaak AD, Garcia-Vallejo JJ, Lamoree MH (2022) Discovery and quantification of plastic particle pollution in human blood. Environ Int 163:107199. https://doi.org/10.1016/j.envint.2022.107199

Li C, Jiang B, Guo J, Sun C, Shi C, Huang S, Liu W, Wu C, Zhang Y (2022) Aging process of microplastics in the aquatic environments: aging pathway, characteristic change, compound effect, and environmentally persistent free radicals formation. Water 14(21):3515. https://doi.org/10.3390/w14213515

Li J, Huang F, Zhang G, Zhang Z, Zhang X (2023) Separation and flow cytometry analysis of microplastics and nanoplastics. Front Chem 11:1201734. https://doi.org/10.3389/fchem.2023.1201734

Li Y, Zhu Y, Huang J, Ho YW, Fang JK, Lam EY (2024) High-throughput microplastic assessment using polarization holographic imaging. Sci Rep 14(1):2355. https://doi.org/10.1038/s41598-024-52762-5

Liu X, Lin H, Xu S, Yan Y, Yu R, Hu G (2023) Occurrence, distribution, and characteristics of microplastics in agricultural soil around a solid waste treatment center in Southeast China. J Soils Sediments 23(2):936–946. https://doi.org/10.1007/s11368-022-03341-6

Lu H, Wang X, Cong Q, Chen X, Li Q, Li X, Zhong S, Deng H, Yan B (2024) Research progress on the degradation of organic pollutants in water by activated persulfate using biochar-loaded nano zero-valent iron. Molecules 29(5):1130. https://doi.org/10.3390/molecules29051130

Magalhães S, Alves L, Romano A, Medronho B, Rasteiro M (2022) Extraction and characterization of microplastics from Portuguese industrial effluents. Polymers 14(14):2902. https://doi.org/10.3390/polym14142902

Mendonça JDS, Guimarães RCA, Zorgetto-Pinheiro VA, Fernandes CDP, Marcelino G, Bogo D, Freitas KC, Hiane PA, de Pádua Melo ES, Vilela MLB, Nascimento VAD (2022) Natural antioxidant evaluation: a review of detection methods. Molecules 27(11):3563. https://doi.org/10.3390/molecules27113563

Mihai F-C, Gündoğdu S, Markley LA, Olivelli A, Khan FR, Gwinnett C, Gutberlet J, Reyna-Bensusan N, Llanquileo-Melgarejo P, Meidiana C, Elagroudy S, Ishchenko V, Penney S, Lenkiewicz Z, Molinos-Senante M (2022) Plastic pollution, waste management issues, and circular economy opportunities in rural communities. Sustain For 14(1):20. https://doi.org/10.3390/su14010020

Miller ME, Hamann M, Kroon FJ (2020) Bioaccumulation and biomagnification of microplastics in marine organisms: a review and meta-analysis of current data. PLoS One 15(10):e0240792. https://doi.org/10.1371/journal.pone.0240792

Osman AI, Hosny M, Eltaweil AS, Omar S, Elgarahy AM, Farghali M, Yap PS, Wu YS, Nagandran S, Batumalaie K, Gopinath SCB, John OD, Sekar M, Saikia T, Karunanithi P, Hatta MHM, Akinyede KA (2023) Microplastic sources, formation, toxicity and remediation: a review. Environ Chem Lett. https://doi.org/10.1007/s10311-023-01593-3

Pasquier G, Doyen P, Carlesi N, Amara R (2022) An innovative approach for microplastic sampling in all surface water bodies using an aquatic drone. Heliyon 8(11):e11662. https://doi.org/10.1016/j.heliyon.2022.e11662

Paul S, Nath S, Bhattacharjee S et al (2024) Unveiling the effects of microplastics pollution on marine fauna. Blue Biotechnol 1:6. https://doi.org/10.1186/s44315-024-00006-6

Periyasamy AP (2023) Environmentally friendly approach to the reduction of microplastics during domestic washing: prospects for machine vision in microplastics reduction. Toxics 11(7):575. https://doi.org/10.3390/toxics11070575

Picó Y, Barceló D (2023) Microplastics and other emerging contaminants in the environment after COVID-19 pandemic: the need of global reconnaissance studies. Curr Opin Environ Sci Health 33:100468. https://doi.org/10.1016/j.coesh.2023.100468

Prata JC, Padrão J, Khan MT, Walker TR (2024) Do's and don'ts of microplastic research: a comprehensive guide. Water Emerg Contam Nanoplastics 3:8. https://doi.org/10.20517/wecn.2023.61

Primpke S, Lorenz C, Rascher-Friesenhausen R, Gerdts G (2017) An automated approach for microplastics analysis using focal plane array (FPA) FTIR microscopy and image analysis. Anal Methods 9(9):1499–1511. https://doi.org/10.1039/c6ay02476a

Pulingam T, Appaturi JN, Parumasivam T, Ahmad A, Sudesh K (2022) Biomedical applications of polyhydroxyalkanoate in tissue engineering. Polymers 14(11):2141. https://doi.org/10.3390/polym14112141

Qian N, Gao X, Lang X, Deng H, Bratu TM, Chen Q, Stapleton P, Yan B, Min W (2024) Rapid single-particle chemical imaging of nanoplastics by SRS microscopy. Proc Natl Acad Sci USA 121(3):e2300582121. https://doi.org/10.1073/pnas.2300582121

Rana MM, Haque MR, Tasnim SS, Rahman MM (2023) The potential contribution of microplastic pollution by organic fertilizers in agricultural soils of Bangladesh: quantification, characterization, and risk appraisals. Front Environ Sci 11:1205603. https://doi.org/10.3389/fenvs.2023.1205603

Rani M, Ducoli S, Depero LE, Prica M, Tubić A, Ademovic Z, Morrison L, Federici S (2023) A complete guide to extraction methods of microplastics from complex environmental matrices. Molecules 28(15):5710. https://doi.org/10.3390/molecules28155710

Râpă M, Cârstea EM, Şăulean AA, Popa CL, Matei E, Predescu AM, Predescu C, Donţu SI, Dincă AG (2024) An overview of the current trends in marine plastic litter management for a sustainable development. Recycling 9(2):30. https://doi.org/10.3390/recycling9020030

Raza M, Lee JY, Cha J (2022) Microplastics in soil and freshwater: understanding sources, distribution, potential impacts, and regulations for management. Sci Prog 105(3):368504221126676. https://doi.org/10.1177/00368504221126676

Razeghi N, Hamidian AH, Wu C et al (2021) Microplastic sampling techniques in freshwaters and sediments: a review. Environ Chem Lett 19:4225–4252. https://doi.org/10.1007/s10311-021-01227-6

Ren SY, Ni HG (2023) Biodeterioration of microplastics by bacteria isolated from mangrove sediment. Toxics 11(5):432. https://doi.org/10.3390/toxics11050432

Rezvani Ghomi E, Khosravi F, Saedi Ardahaei A, Dai Y, Neisiany RE, Foroughi F, Wu M, Das O, Ramakrishna S (2021) The life cycle assessment for Polylactic acid (PLA) to make it a low-carbon material. Polymers 13(11):1854. https://doi.org/10.3390/polym13111854

Robinson JP, Ostafe R, Iyengar SN, Rajwa B, Fischer R (2023) Flow Cytometry 12(14):1875. https://doi.org/10.3390/cells12141875

Ross MS, Loutan A, Groeneveld T, Molenaar D, Kroetch K, Bujaczek T, Kolter S, Moon S, Huynh A, Khayam R, Franczak BC, Camm E, Arnold VI, Ruecker NJ (2023) Estimated discharge of microplastics via urban stormwater during individual rain events. Front Environ Sci 11:1090267. https://doi.org/10.3389/fenvs.2023.1090267

Sacco NA, Zoppas FM, Devard A, González Muñoz M, García G, Marchesini FA (2023) Recent advances in microplastics removal from water with special attention given to photocatalytic degradation: review of scientific research. Microplastics 2(3):278–303. https://doi.org/10.3390/microplastics2030023

Saeedi M (2023) How microplastics interact with food chain: a short overview of fate and impacts. J Food Sci Technol 61(3):1–11. https://doi.org/10.1007/s13197-023-05720-4

Salvia R, Rico LG, Bradford JA, Ward MD, Olszowy MW, Martínez C, Madrid-Aris ÁD, Grífols JR, Ancochea Á, Gomez-Muñoz L, Vives-Pi M, Martínez-Cáceres E, Fernández MA, Sorigue M, Petriz J (2023) Fast-screening flow cytometry method for detecting nanoplastics in human peripheral blood. Methods X 10:102057. https://doi.org/10.1016/j.mex.2023.102057

Sarkar S, Diab H, Thompson J (2023) Microplastic pollution: chemical characterization and impact on wildlife. Int J Environ Res Public Health 20(3):1745. https://doi.org/10.3390/ijerph20031745

Schütze B, Thomas D, Kraft M, Brunotte J, Kreuzig R (2022) Comparison of different salt solutions for density separation of conventional and biodegradable microplastic from solid sample matrices. Environ Sci Pollut Res Int 29(54):81452–81467. https://doi.org/10.1007/s11356-022-21474-6

Schymanski D, Oßmann BE, Benismail N, Boukerma K, Dallmann G, von der Esch E, Fischer D, Fischer F, Gilliland D, Glas K, Hofmann T, Käppler A, Lacorte S, Marco J, Rakwe ME, Weisser J, Witzig C, Zumbülte N, Ivleva NP (2021) Analysis of microplastics in drinking water and other clean water samples with micro-Raman and micro-infrared spectroscopy: minimum requirements and best practice guidelines. Anal Bioanal Chem 413(24):5969–5994. https://doi.org/10.1007/s00216-021-03498-y

Seeley ME, Lynch JM (2023) Previous successes and untapped potential of pyrolysis-GC/MS for the analysis of plastic pollution. Anal Bioanal Chem 415(15):2873–2890. https://doi.org/10.1007/s00216-023-04671-1

Sharma RK, Kumari U, Kumar S (2024) Impact of microplastics on pregnancy and fetal development: a systematic review. Cureus 16(5):e60712. https://doi.org/10.7759/cureus.60712

Siddiqua A, Hahladakis JN, Al-Attiya WAKA (2022) An overview of the environmental pollution and health effects associated with waste landfilling and open dumping. Environ Sci Pollut Res Int 29(39):58514–58536. https://doi.org/10.1007/s11356-022-21578-z

Sun J, Sui M, Wang T, Teng X, Sun J, Chen M (2024) Detection and quantification of various microplastics in human endometrium based on laser direct infrared spectroscopy. Sci Total Environ 906:167760

Tan X, Rodrigue D (2019) A review on porous polymeric membrane preparation. Part II: production techniques with polyethylene, polydimethylsiloxane, polypropylene, polyimide, and polytetrafluoroethylene. Polymers 11(8):1310. https://doi.org/10.3390/polym11081310

Thornton Hampton LM, De Frond H, Gesulga K, Kotar S, Lao W, Matuch C, Weisberg SB, Wong CS, Brander S, Christansen S, Cook CR, Du F, Ghosal S, Gray AB, Hankett J, Helm PA, Ho KT, Kefela T, Lattin G, Lusher A, Rochman CM (2023) The influence of complex matrices on method performance in extracting and monitoring for microplastics. Chemosphere 334:138875

Torres Moral T, Sanchez-Niubo A, Monistrol-Mula A, Gerardi C, Banzi R, Garcia P, Demotes-Mainard J, Haro JM, The Permit Group (2022) Methods for stratification and validation cohorts: a scoping review. J Personal Med 12(5):688. https://doi.org/10.3390/jpm12050688

Toussaint B, Raffael B, Angers-Loustau A, Gilliland D, Kestens V, Petrillo M, Van den Eede G (2019) Review of micro- and nanoplastic contamination in the food chain. Food Addit Contamin Part A 36(5):639–673. https://doi.org/10.1080/19440049.2019.1583381

Trevisan R, Ranasinghe P, Jayasundara N, Di Giulio RT (2022) Nanoplastics in aquatic environments: impacts on aquatic species and interactions with environmental factors and pollutants. Toxics 10(6):326. https://doi.org/10.3390/toxics10060326

Vandenberg LN, Rayasam SDG, Axelrad DA, Bennett DH, Brown P, Carignan CC, Chartres N, Diamond ML, Joglekar R, Shamasunder B, Shrader-Frechette K, Subra WA, Zarker K, Woodruff TJ (2023) Addressing systemic problems with exposure assessments to protect the public's health. Environ Health 21(Suppl 1):121. https://doi.org/10.1186/s12940-022-00917-0

Vdovchenko A, Resmini M (2024) Mapping microplastics in humans: analysis of polymer types, and shapes in food and drinking water—a systematic review. Int J Mol Sci 25(13):7074. https://doi.org/10.3390/ijms25137074

Vohl S, Kristl M, Stergar J (2024) Harnessing magnetic nanoparticles for the effective removal of micro- and nanoplastics: a critical review. Nanomaterials 14(14):1179. https://doi.org/10.3390/nano14141179

Vojnović B, Mihovilović P, Dimitrov N (2024) Approaches for sampling and sample preparation for microplastic analysis in laundry effluents. Sustain For 16(8):3401. https://doi.org/10.3390/su16083401

Wang B, Hu S, Teng Y et al (2024) Current advance of nanotechnology in diagnosis and treatment for malignant tumors. Sig Transduct Target Ther 9:200. https://doi.org/10.1038/s41392-024-01889-y

Ward ND, Megonigal JP, Bond-Lamberty B et al (2020) Representing the function and sensitivity of coastal interfaces in earth system models. Nat Commun 11:2458. https://doi.org/10.1038/s41467-020-16236-2

Weis JS, Alava JJ (2023) (Micro)plastics are toxic pollutants. Toxics 11(11):935. https://doi.org/10.3390/toxics11110935

Whiting QT, O'Connor KF, Potter PM, Al-Abed SR (2022) A high-throughput, automated technique for microplastics detection, quantification, and characterization in surface waters using laser direct infrared spectroscopy. Anal Bioanal Chem 414(29-30):8353–8364. https://doi.org/10.1007/s00216-022-04371-2

Woo H, Seo K, Choi Y, Kim J, Tanaka M, Lee K, Choi J (2021) Methods of analyzing microsized plastics in the environment. Appl Sci 11(22):10640. https://doi.org/10.3390/app112210640

World Health Organization. Microplastics in Drinking-Water. 20 Aug. 2019. www.who.int/news/item/20-08-2019-microplastics-in-drinking-water.

World Health Organization (2023) Microplastics in drinking-water, https://www.who.int/news/item/20-08-2019-microplastics-in-drinking-water (2019).

Wu M, Yang C, Du C, Liu H (2020) Microplastics in waters and soils: occurrence, analytical methods and ecotoxicological effects. Ecotoxicol Environ Saf 202:110910. https://doi.org/10.1016/j.ecoenv.2020.110910

Yarahmadi A, Heidari S, Sepahvand P, Afkhami H, Kheradjoo H (2024) Microplastics and environmental effects: investigating the effects of microplastics on aquatic habitats and their impact on human health. Front Public Health 12:1411389. https://doi.org/10.3389/fpubh.2024.1411389

Yu R-S, Singh S (2023) Microplastic pollution: threats and impacts on global marine ecosystems. Sustain For 15(17):13252. https://doi.org/10.3390/su151713252

Zhang N, Li YB, He HR, Zhang JF, Ma GS (2021) You are what you eat: microplastics in the feces of young men living in Beijing. Sci Total Environ 767:144345

Ziani K, Ioniță-Mîndrican CB, Mititelu M, Neacşu SM, Negrei C, Moroşan E, Drăgănescu D, Preda OT (2023) Microplastics: a real global threat for environment and food safety: a state of the art review. Nutrients 15(3):617. https://doi.org/10.3390/nu15030617

A Toxic Journey: Microplastics as Vectors for Contaminants 3

Tesfalem W. Ghebretnsae and Almaz Bereket Bokre

Abstract

Sufficient and latest information on the journey of the toxicity venture of microplastics (MPs) are given in many literatures. MPs play vital roles as a vector for chemical contaminants because of having unique surface properties such as a huge specific surface area, pore size distribution, and elevated hydrophobic behavior along with their durable characteristics. Thus, these characteristics help MPs to have the capacity to take up organic and inorganic chemical contaminants on their surfaces by the forces of adsorption. Therefore, significant concentrations of many harmful hydrophobic organic chemicals, heavy metals, and antibiotics are becoming increasingly evident on MPs' surfaces and biological agents. Consequently, MPs' characteristics to become vectors help them to disperse chemical contaminants worldwide from polar regions to tropics in terrestrial, aquatic, and atmospheric ecosystems and food webs. Like climate change, these toxic chemicals are testing the health and survival of humans and all organisms on earth. In the twenty-first century, microplastic contamination and climate change crises have become the most urgent, unbearable burdens of the earth and organisms living in it. Previous studies and narratives merely explained the relationship among microplastics, aquatic life, and climate change. But nowadays, undesirable consequences of microplastic pollution and their derived microplas-

T. W. Ghebretnsae (✉)
Department of Environmental Management, Institute of Environmental Engineering, Peoples' Friendship University of Russia, Moscow, Russia

Department of Land Resources and Environment, Hamelmalo Agricultural College, Keren, Eritrea
e-mail: tesfaweld333@gmail.com

A. B. Bokre
Ministry of Agriculture, Keren, Eritrea

M. Oves et al. (eds.), *Marine Microplastics and the Quest for Remediation*,
https://doi.org/10.1007/978-981-96-8979-8_3

tic and nanoplastic fragments are becoming self-evident on air, aquatic, and terrestrial ecosystems, thereby on the climate change. As in the case of marine life, microplastics have paramount importance in hindering carbon storage and nitrogen cycling. A soil polluted by microplastics affect climate change in two ways. The microplastics have tangible potential to directly accelerate climate change through the emission of greenhouse gases (GHGs) from the soil into the atmosphere. The second way happens indirectly, when microplastics induce trees and soil organisms to respire more. As the availability of microplastics in the soil increases, more toxic chemicals will accumulate on the surfaces of microplastics near roots' water and nutrient absorption area, consequently, many trees will be unable to absorb carbon dioxide. In both ways, more GHGs are released into the atmosphere because of microplastics. If microplastics are exposed to ultraviolet radiation and physical and chemical weather processes, they will disperse in the environment, and many GHGs get a way to accumulate into the atmosphere. In all ecosystems, as their exposure time increases, their specific surface area will increase to absorb heavy metals and other contaminants, and, by and large, they will pose irreversible damage to ecology. This chapter is intended to give some insights on the key roles of microplastics as vectors of contaminants in the atmosphere, hydrosphere, and pedosphere in the context of global warming.

Keywords

Vectors of contaminants · MP pollution · Climate change · GHGs

3.1 Introduction

Sufficient and latest information on the journey of toxicity of microplastics (MPs) are given in many literatures. MPs play vital roles in dispersing detrimental inorganic, organic, and biological pollutants from region to region across all global ecological systems. MPs have a good capacity to adsorb chemical contaminants and also the potential to form layers over living organisms. To put it simply, MPs act as paths of contaminants from the environment into living organisms, the so-called vector-effect (Santos-Echeandía et al. 2021).

Plastic materials originate from by-products of crude oil refinery, cellulose (natural matter), salt, coal, and natural gas with the help of the polymerization process. They are discarded into the environment without proper management after serving several purposes. After being exposed to anthropogenic and natural degrading forces, they become the potential source for ecological contamination. Plastic degradation is accompanied by environmental weathering. The result of plastic degradation is different-sized microplastics; it is the breaking of long-chain polymers into smaller ones. Actually, the plastic particles are known as MPs if each of their diameter is between 0.011 and 5 mm (Thompson et al. 2004). Plastic fragments smaller than microplastics are called nanoplastics (with an upper threshold of 0.020 mm).

Generally, two major microplastic classes have been identified: main microplastic and derivative microplastics. Main MPs, or sometimes named as primary MPs, include microfibers, microbeads, plastic glitter, and pellets. They have been produced in micro size during their fabrication, while derivative MPs are new forms of particulates as the result of fragmentation and degradation of plastic materials such as plastic bags, containers, bottles, tires, and nets. In terms of chemical composition, it is possible to mention six types of polymers of microplastic (which have active roles in MPs contamination). These are polyvinyl chloride (PVC), polyethylene terephthalate (PET), polystyrene (PS), polypropylene (PP), low-density polyethylene (LDPE), and high-density polyethylene (HDPE). Environmental weathering mechanisms of plastic materials include both abiotic and biotic degradation. Abiotic degradation employs heat, chemical reactions, and UV-radiation, while biotic degradation (e.g., composting processes) is caused by enzymatic breakdown and biodisintegration in which plastics are disrupted into finer and finer particles or microplastics.

With increasing human population and modernization, plastic waste has become part of human lifestyles. Sharma et al. (2023) pointed out historical, ongoing, and future levels of plastic production. In 2050, plastic production will hold up to 13% of the global total carbon budget. At least 80×10^6 quintals of plastic waste are known to be deposited in oceans each year, and this represents more or less about 85% of all marine litter; and this quantity is still expected to triple by 2040. In terms of global plastic market size, production of plastic materials has increased to 593.00 billion US dollars in 2021 and, collectively, will even climb up steadily at a yearly growth rate of 3.7% up to 2030. These figures help to understand the development of plastic toxicity on our planet, how far plastic wastes are injected into the food chain, and their probable effects on the health of the society.

Nowadays, three great global challenges, such as the climate change crisis, plastic pollution, and antibiotic resistance, which are intricately related and interconnected through various pathways, impact biosphere, pedosphere, hydrosphere, and atmosphere of the earth and, above all, the health of the ecosystem and survival of the planet's creatures. It is becoming understandable that the global increase of temperature causes breaking of plastic materials via solar radiation, acid, temperature, mechanical forces, and biotic agents (Li et al. 2024). Plastic material degradation results in microplastics, and microplastics are the main vectors of toxic chemical contaminants (Tang 2023). Microplastics again also intensify the impact of climatic crises by emitting GHGs during production and disrupting carbon pumping aquatic and terrestrial organisms.

This chapter will explain how MPs, being vectors, contribute to environmental pollution by interacting with water, air, and soil. It will also try to address the links that exist between microplastic pollution and climate change. Hence, this chapter is planned as follows: it starts by ecotoxicological impact of microplastics, the next section deals with MPs as vectors of chemical contaminants, and then proceeds to the impacts of MPs on climate change and vice versa.

3.2 Ecotoxicological Impact of Microplastics

Many reports have realized the effects of MPs on the issues of human and animal health, ecosystem functioning, aquatic environment, terrestrial ecosystems, soils, and overall life. MPs can easily enter humans, animals, and plants when MPs, along with chemical contaminants, are inhaled from the air and consumed with food and drink. They (MPs) are observed to affect not only psychological well-being but also the function of all body systems and organs. For example, Kanungo et al. (2023) demonstrated how the ecotoxicity of microplastic impair the health of nerves, eggs, immune, reproduction, endocrine, and other systems of aquatic organisms and humans. Non-degradable characteristics, slow recycling efforts, the use of single-use plastic products, and inadequate global plastic management, combined with high plastic consumption, lead to significant plastic waste accumulation. This results in widespread plastic pollution and, ultimately, severe impacts on ecological health and human well-being (Ribeiro et al. 2019; Sharma et al. 2023). Conclusively, microplastics proved to be the sources and vectors of mutagenic, carcinogenic, and teratogenic chemicals.

The impact of plastic pollution on environmental wellness was first noted since the 1970s (Tumwasigye et al. 2023), although the manufacturing of plastic materials for trade goes before the 1950s (Sharma et al. 2023). Problems associated with plastics pollution have become worse as many chemicals are incorporated into plastics in the form of additives and resins. Durability of plastics is improved by adding additives to plastics (Wiesinger et al. 2021) though many of them are partly made up of persistent organic pollutants (POPs). POPs are known to be dangerous to the health of humans and animals because of their genotoxic behavior (Sharma et al. 2023). Resins are also added along with additives to boost the efficiency of plastics. Murphy (2001) listed some of the additives such as silica to make plastics stronger, lead heat stabilizers in order to process plastic at high temperature, plasticizers mainly phthalates (esters used in personal care products like soap, perfume, hairsprays, etc.) and terephthalates (esters used in making containers during food packing) to improve flexibility, chlorine and bromine as flame retardants, and ultraviolet stabilizing agents (chemicals use to absorb ultraviolet radiations, e.g., hydroxyphenyl triazine in thermoplastics) to protect exposed plastics against photodegradation.

Uncontrolled microplastics have abundant environmental crisis, and plant and animal health problems. For animals, microplastics can cause physical damage, inflammation, and other health problems, which can reduce growth if they are ingested. Increased global surface air temperature (global warming) leach out harmful chemicals (POPs and heavy metals), which are a threat to the ecosystem and human well-being, into the food chain (Tang 2023). The chemicals also have a high potential to contaminate aquatic ecosystem negatively, and impact the terrestrial ecosystem and biodiversity. Increased leachates of tires affected the mutual relationship between microbes and algae negatively (O'Brien et al. 2022). The growth of microalgae and the ocean carbon cycle are adversely affected by microplastics. Soil organisms are also becoming victimized by the ecotoxicological effects of the combined action of pollution due to microplastics and climate variabilities. Li et al. (2024) asserted the impact of plastic pollution on soil carbon-based stock. In their works, microbially available SOC content was found to decrease by 18.9%, but mineral-associated SOC content was found to increase by 52.5%. As the volume of plastic materials in soil increases, its effects on soil physical properties, microbial activities, stability, soil moisture loss, and soil evaporation become more significant. The growth of earthworms is inhibited as the MP concentration in soil increases. Polyethylene terephthalate (PET) toxicity on plants becomes terrible if it joins soil components. The contaminants, which are derived from MPs, inhibit the plant's growth by reducing biomass, leaf number, and photosynthesis performance.

3.3 Roles of MPs as Vectors of Pollutants

MPs are well evident transmitters of persistent organic pollutants (POPs), specially hydrophobic organic chemicals (HOCs) such as polychlorinated diphenyls (PDBs), dichloro diphenyl trichloroethane (DDT), polybrominated diphenyl ethers (PBDEs), brominated diphenyl ether (BDE), hexachlorocyclohexane (HCHs), polycyclic aromatic hydrocarbons (PAHs), hazardous metals such as Cr, Pb, Cd, Cu, Hg, and Co, pharmaceuticals (ibuprofen, ibuprofen, diclofenac, and naproxen), pathogens, antibiotic infectious microbes, and inheritable factors (Fig. 3.1). The actual potentiality of MPs as vectors and transporters of harmful chemicals in aqua-sphere and terrestrial ecology emanates from their physical surface characteristics (their tiny sizes, increased hydrophobicity, huge surface areas, sponginess, or porosity), chemical composition, functional groups, and carbon chains (Tumwasigye et al. 2023). Luo et al. (2022) mentioned the main forces, which help MPs to interface with contaminants, as van der Waals (intermolecular forces due to uncharged atoms/molecules or due to polarization), hydrogen bonding (when hydrogen reacts with electronegative atoms, e.g., oxygen in water), halogen bonding (electrostatic type of attraction), hydrophobic forces (water repellent forces), electrostatic attractive and repulsive forces, pi-pi interaction, or dispersion effect. Thus, MPs adsorb and carry inorganic and organic contaminants by the aforementioned forces, and these associated forces create ideal conditions for the sorption of all harmful chemicals in marine, air, and

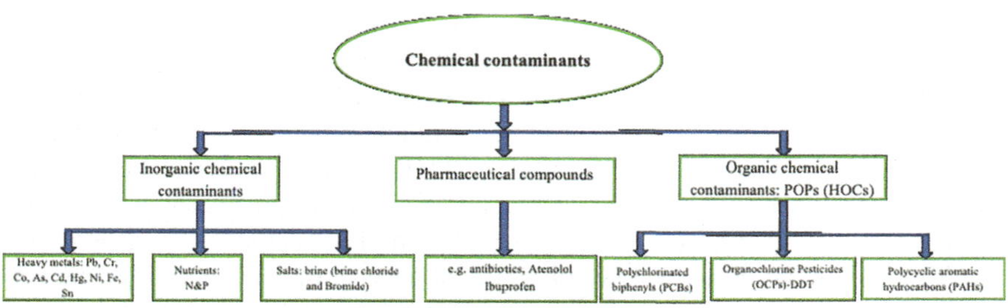

Fig. 3.1 Common chemical contaminants in MPs pollution

soil ecosystem (Lee et al. 2014). Environmental factors like temperature and salinity also have significant impacts on MPs' capacity to act as vectors and spreaders of pollutants.

3.3.1 Inorganic Chemical Contaminants

Inorganic chemical contaminants include nutrients, salts, and toxic metals (particularly heavy metals). The negative impacts of nutrients (carbon, nitrogen, and phosphorus) on the health of aquatic ecosystem can be aggravated by plastic detritus when the colonized microorganisms on the plastic particles facilitate the nutrient cycles and leach. There is a dearth of information on the relationship between MPs and salts. However, chlorination treatment was identified to have a major role in enabling the formation of MP clusters (polyethylene and thermoplastic polyurethane) and their elevated interaction with chromium (Miao et al. 2023).

Microplastics are well-known adsorbers and vectors of potentially toxic inorganic chemical contaminants, particularly those of heavy metals. The concentration of heavy metals on the surface of MPs can become a threat to the health of the ecosystem, people, and all living things. Different studies use different metal concentration evaluation techniques during the detection, measurement, and analysis of metal concentration on the surface of MPs (Kutralam-Muniasamy et al. 2021; Melo-Agustín et al. 2022; Tumwasigye et al. 2023). Transmission electron microscopy (TEM), Fourier-Transform Infrared Spectroscopy (FTIR), and energy-dispersive spectroscopy (EDS) are applied directly, while inductively coupled plasma mass spectrometry (ICP-MS) is somehow used indirectly. To study the concentration of metals on the external part of micro-sized plastic materials, Vedolin et al. (2018) and Turner et al. (2019) used Infrared and X-ray fluorescence analysis and inductively coupled plasma optical emission spectroscopy, respectively. Guo and Wang (2019) also used a novel film-pore mass transfer (FPMT) model to analyze concentration status of Sr^{2+} on PET, PE, and PVC. Currently, the nanoscale secondary ion mass spectrometry (NanoSIMS) method is used to study the distribution of Cu, Cl, and biofilm on MP surfaces in the aquatic ecosystem (Zhou et al. 2022).

MPs have enormous surface-to-volume ratios. The layer of these chemicals on the surface of MPs is much thicker than their surrounding water. These very crucial characteristics of MPs help them to spread adsorbed heavy metals to different levels of food webs and aquatic ecosystems. Heavy metals represent the larger proportion of inorganic chemical contaminants that associate with microplastic sphere. Sorption of toxic metals like lead, zinc, chromium, nickel, cobalt, cadmium, and copper was evidenced on the surfaces of PET, HDPE, PVC, LDPE, and PP (Rochman et al. 2014; Godoy et al. 2019). PE and PVC fragments show substantial sorption affinity with zinc, lead, and chromium, but PET has the least sorption affinity with these metals. Many studies on different parts of the planet divulge the concentration of heavy metals on the surface of MPs. For example, Turner et al. (2019) measured the concentration of Lead (Pb), Bromine (Br), Antimony (Sb), and

Cadmium (Cd) on the surface of MPs in Western Europe as 14, 0.8, 1.3, 20, and 4.4 μg g^{-1}, respectively. Santos-Echeandía et al. (2021) also determined mercury (Hg) concentration (5–8216 nmol kg^{-1}), sourced from plastic items, collected from the Spanish Mediterranean coast. Massos and Turner (2017) also measured the adsorbed heavy metals (Cd and Pb) on the surfaces of MPs in south-western England. Brennecke et al. (2016) determined the concentration of Cu (1.67, 1100.97, 1320 ppm) and Zn (0.53, 160.27, 102.09 ppm) in seawater and on PS and PVC, respectively. Vedolin et al. (2018) were able to determine the concentration of Fe (22.78 mg kg^{-1}) and Al (45.27 mg kg^{-1}), and they also observed traces of Cr, Cu, Mn, Sn, and Ti on the surface of MPs.

3.3.2 Organic Chemical Contaminants (OCCs)

OCCs, carbon- and hydrogen-based chemicals, are anthropogenic pollutants that bring about undesirable or harmful effects on air, water, soil, all living things, and the whole environment. They may be sourced from tissues of living organisms, insecticides, weedkillers, soil erosion, runoff, inappropriate use of chemicals in industries, agriculture, and pharmaceutical products. Their concentration on the surface of microplastics can be determined with the help of gas chromatography-mass spectrometry (GC-MS). But, before applying the techniques for measuring the concentrations of adsorbed organic pollutants, extraction techniques such as washing, maceration (putting MPs in solvents), Soxhlet (extracting device), and ultrasound-assisted and solvent-type extractions must be used. Nowadays, many studies have been conducted to quantify the adsorption of OCCs on the surfaces of microplastics. Accordingly, many reports have shown us the abundance of organic chemical pollutants on the surface of MPs and aquatic ecosystems. For example, Mai et al. (2018) measured the concentration of PAHs (3400–119,000 ng. g^{-1}) on PE and PS. Pozo et al. (2019) reported the concentration of POPs (e.g., PBDEs = 10–133, PCBs = 3–60, and DDTs = 0.1–7 ng g^{-1}) on HDPE and PP. Gorman et al. (2019) evaluated the concentration of PAH (1454–6000 ng g^{-1}) and PCBs (0.8–104.6 ng g^{-1}) on MPs. Llorca et al. (2018) also estimated the adsorption ability of PS, PE, and PET to be within the range of 20–60%. The adsorption coefficients of ibuprofen (0.034 L kg^{-1}) and sulfamethoxazole (0.025 L kg^{-1}) on PET, HDPE, PVC, and LDPE were determined by Magadini et al. (2020). Li et al. (2018) were able to measure the sorption coefficient between antibiotics and MPs (PVC, PP, PE, PS) within the range of 7.1–804 L kg^{-1}. Systematic studies were conducted to observe closely the transfer of POP chemicals from PS and PE to fishes (Tivefalth and Carney Almroth 2018) and the bioaccumulations of POPs (e.g., PBDE and BDE) in fish (Gorman et al. 2019). Yeo et al. (2020) investigated and measured the quantity of organic chemical contaminants (PCBs range from 0.04 ng g^{-1} to 124 ng g^{-1} and PBDEs were found within the range of 0–2158 ng g^{-1}) in microplastics and aquatic animals in the Pacific Ocean (around Tokyo Bay).

Tumwasigye et al. (2023) described well the order of partitioning strength of hydrophobic organic pollutants between MPs and environmental water. Accordingly, the highest partitioning strength was observed in LDPE and HDPE, while the highest attraction forces were observed in PET. And the partitioning forces of PVC and PS layers were measured between the above extremes. Llorca et al. (2018) concluded that PP, PE, PS, and PET are the topmost vectors of POPs or hydrophobic organic chemicals, PCBs, PAFSs, PAHs, HCHs, pesticides (e.g., DDT and OCP), and BPAs. Other common chemical pollutants that have good associations with MP surfaces are pesticides and phthalates. The risk level of all chemical contaminants to human and aquatic life can be estimated from their concentration magnitude on MP surfaces. For example, the report of Mato et al. (2001) showed that the magnitude of the concentration of chemical contaminants was several folds higher in their respective vectors than surrounding waters.

3.4 Role of Climate Change on the Destination of MPs

3.4.1 Overview of Climate Change Crisis

Nowadays, the definition of climate crisis for most people is becoming as clear as water. Climate change is a sort of alteration of the world's climate through the addition of greenhouse gases (GHGs) to the air around the earth. UNFCCC and IPCC described climate change differently but with similar concepts. UNFCC defines climate change as an alteration in the chemical composition of the global atmosphere and observation of natural climate variability as the result of "directly or indirectly human activity." And IPCC Fourth Assessment Report, Working Group I (2007) defines climate change as a long-term (may be 10 years or longer) change in the state and characteristics of climate by noticing changes in the mean or variability. Again, many people get confused to understand the difference between climate and weather. But their difference can be stated by the popular statement "*climate is what you expect, weather is what you get.*" Weather is simply short-term changes (meteorological conditions) in the air around the globe, while climate is average of climatic variables over an extended period of time, may be decades or 100 years or millions of years, in a specific area. In short, climate is a long-term summary of average weather conditions as well as variability of these conditions. A popular climatological period or climatic normal is a 30-year period, as defined by the world meteorological organization (WMO).

The atmosphere, terrestrial bodies, cryosphere, and marine bodies, along with organisms living in them, constitute a complex and interactive climate system. The climate system is seen to evolve over time by internal and external factors or forcing. External factors include natural phenomena like solar conditions (including solar wind and flares, sun spots, coronal mass discharges) and anthropogenic practices (emission of GHGs from land use and agriculture, deforestation, industrial wastes, etc. Le Treut et al. (2007) list the main ways that change the attribute to change of radiation balance of the earth. Any change in the Earth's orbit may affect incoming sunlight. The second one is that changes in cloud cover, atmospheric particles, or vegetation alter the fraction of reflected solar radiation (or albedo). The third most important is that any change in greenhouse gas concentrations results in a change in the longwave radiation from the Earth back toward space. Greenhouse gases absorb the longwave radiation like a blanket (but partially). Water vapor and carbon dioxide are the most important greenhouse gases, which have larger blanketing effects and keep the earth warm so that it is comfortable to its creatures.

Nowadays, human activities such as combustion of fossil fuels and biomass, clearing forests, transportation, and other energy production processes add more GHGs to the atmosphere and ultimately increase their blanking effects. Based on the geological evidence of ice ages, tangible sea-level changes, and meteorological records, past climate change is thought to be attributed to volcanic eruptions, ocean currents, solar activity, and others. Currently, the IPCC fourth assessment report delivered strong evidence on the increase of average global and ocean temperatures, widespread melting of snow and ice, extreme weather events occurrences, and rising average global sea levels. Though detecting, understanding, and accurately quantifying climate feedbacks are becoming continuous tasks of climate scientists, still more integrated approach is very crucial to unravelling the complexities of Earth's climate.

IPCC, one of the United Nations bodies, was founded in 1988 to take care of climate change based on current and future knowledge of climate change. The latest IPCC (2023) report indicates that the global average surface temperature in 2011–2020 is 1.1 °C higher than the industrial period (1850–1900). Under different emission scenarios, the global average surface temperature is expected to increase by 1.5 °C–4.4 °C by the end of the twenty-first century as compared to the pre-industrial period due to a host of anthropogenic activities (IPCC 2023). The concentration of major greenhouse gases (GHGs) such as atmospheric CO_2, CH_4, and nitrous oxide (N_2O) reached 415 parts per million (ppm), 1896 parts per billion (ppb), and 385 ppb, respectively, in 2021. The report also predicted the rise of global mean sea level under different emission scenarios. The rise will reach 0.28–0.55 m by 2100 for a low emission scenario (SSP1–1.9) and by 0.63–1.01 m for a high emission scenario (SSP5–8.5). IPCC (2023) predicted the decrease of average surface ocean pH to be 0.14–0.32 units by 2081–2100, relative to 2006–2015, and this will enhance ocean acidity. The magnitude of ocean acidification differs from region to region depending up on temperature status and uptake of CO_2. Climate change has shown significant impacts and resulted in extreme droughts, rainfall events, and flooding. These changes will be more intense and frequent if global surface temperature goes up to 2 °C or more above the level of global warming in 1850–1900.

3.4.2 Aggravation of the Climate Change Crisis by MPs

Not a few studies show the links that exist between global warming and plastic contamination. Unfortunately, policymakers, decision-makers, and some scientists around the globe did not grasp the relationship clearly. Climate change heavily affects the dynamics of plastic pollution, and the same is true with the influences of microplastics on climate change aggravations.

Microplastics are known to be continuous emitters of greenhouse gases throughout their lives (Zhang et al. 2020; Tang 2023) as they are made up of mainly fossil fuels, coal, petroleum, and natural gas (Li et al. 2024). Billions of tons of greenhouse gases have been deposited in the atmosphere as a result of worldwide MP pollution and plastic products. MPs in the air aggravate the climate change crisis by interfering with cloud formation and atmospheric cooling. Their accumulation in the atmosphere also aggravates the climate crisis by disturbing the solar energy balance and radiation. Carbon dioxide and methane may also enter into atmosphere through the incineration of plastic materials. Oceans are called the most reliable friends of the globe against climate change. However, MPs hinder the role of oceans to mitigate the strong effects of climate variability through their positive contributions to the marine ecosystem. They negatively affect photosynthesis and carbon sequestration roles, played by phytoplankton and the blue carbon ecosystem (Tang 2023). MPs in soil also fuel the climate change crisis. They encourage GHG emissions from the soil into the atmosphere. Both MPs and climate change together heavily stress soil health.

3.4.3 Microplastic Pollution as Affected by Climate Change

Scientific evidence ensured the intricate connections between MP contamination and global heating (Sharma et al. 2023; Tang 2023). And these connections generate an unhealthy cycle and endanger the ecosystem and environmental health greatly. Global warming and climate change worsen the microplastic crisis in many ways. Global warming, along with other degrading factors, facilitates the infinite breakdown of plastics. The significant effects of climate change on plastic contamination have been intensified via increasing surface temperature, sea level movement, frequent variations of climate variables, and decrease in pH of marine waters (Tang 2023).

As global surface temperature increases, ocean surface temperature will also increase the extent of plastic degradation. This plastic degradation magnifies MPs pollution by enhancing the formation of micro-sized plastic materials in all ecosystems to a level of microplastic contamination difficult to mitigate its effects. Climate change imposes microplastic degradation and migration, and it is the main reason for the spread of microplastic from affected to unaffected areas, by way of its effects on ocean tides, weather events, and rising sea levels. Climate change plays a great role in the mechanical degradation and dispersion of plastics through extreme weather events mainly hurricanes and storms. The extreme weather events, which are induced by global warming and climate change, exacerbate plastic pollution—horizontal and vertical dimensions and form long staying feedback loop situations.

Higher temperature has the possibility to hasten the chemical degradation of microplastics through enhancing oxidation, hydrolysis, and photodegradation reactions. Runoff and melting glaciers carry huge volumes of plastics to lakes and oceans. Ultimately, with time, the plastics get fragmented and chemically degraded and converted into microplastics and harmful chemicals. Abiotic degradation process is one of the leading mechanisms that facilitate microplastic pollution and its related consequences. Climate change disrupts marine ecosystem balance through its effects on microplastics. That is, time-to-time increment of temperature of marine ecosystem results in not only acidification of oceans but intensification of negative interferences of MPs on reproduction and behaviors of marine species. These interferences encourage the increase of the concentration of MPs and their derivatives in the food chain, which is a very critical threat to both marine organisms and human beings.

3.5 Conclusion

No sector and ecosystem (coastal, terrestrial, and aquatic) are free from plastic waste. Thus, plastics pose significant threats to the ecosystem and animal health. MPs are vectors for heavy metals and persistent organic pollutants (POPs), such as hexachlorocyclohexane (HCHs) and polycyclic aromatic hydrocarbons (PAHs). Nowadays, it has become evident that microplastic pollution and climate change are interlinked and serious global problems. Plastic and microplastics are exacerbating the climate change crisis by nonstop GHG emissions, causing considerable damage to all global ecosystems. Climate change also aggravates the MP pollution crisis and worldwide dispersion of harmful chemicals carried by MPs. Climate change plays a significant role in degrading plastics, so that toxic derivatives of MPs and dangerous pollutants will leach into the environment and the food web.

References

Brennecke D, Duarte B, Paiva F, Caçador I, Canning-Clode J (2016) Microplastics as vector for heavy metal contamination from the marine environment. Estuar Coast Shelf Sci 178:189–195. https://doi.org/10.1016/j.ecss.2015.12.003

Godoy V, Blazquez G, Calero M, Quesada L, Martín-Lara MA (2019) The potential of microplastics as carriers of metals. Environ Pollut 255:113363. https://doi.org/10.1016/J.ENVPOL.2019.113363

Gorman D, Moreira FT, Turra A, Fontenelle FR, Combi T, Bícego MC, de Castro Martins C (2019) Organic contamination of beached plastic pellets in the South Atlantic: risk assessments can benefit by considering spatial gradients. Chemosphere 223:608–615. https://doi.org/10.1016/j.chemosphere.2019.02.094

Guo X, Wang J (2019) The phenomenological mass transfer kinetics model for Sr2+ sorption onto spheroids primary microplastics. Environ Pollut 250:737–745. https://doi.org/10.1016/J.ENVPOL.2019.04.091

IPCC (2023) Sections. In: Lee H, Romero J (eds) Climate change 2023-synthesis report. IPCC, Geneva, pp 35–115. https://doi.org/10.59327/IPCC/AR6-9789291691647

IPCC Fourth Assessment Report, Working Group I (2007) Glossary of terms. http://ipccwg1.ucar.edu/wg1/Report/AR4WG1_Print_Annexes

Kanungo J, Sahoo T, Bal M, Behera ID (2023) Performance of bioremediation strategy in waste lubricating oil pollutants: a review. Geomicrobiol J. https://doi.org/10.1080/01490451.2023.2245395

Kutralam-Muniasamy G, Perez-Guevara F, Martínez IE, Shruti VC (2021) Overview of microplastics pollution with heavy metals: analytical methods, occurrence, transfer risks and call for standardization. J Hazard Mater 415:125755. https://doi.org/10.1016/j.jhazmat.2021.125755

Le Treut H, Somerville R, Cubasch U, Ding Y, Mauritzen C, Mokssit A, Peterson T, Prather M (2007) Historical overview of climate change. In: Solomon S, Qin D, Manning M, Chen Z, Marquis M, Averyt KB, Tignor M, Miller HL (eds) Climate change 2007: the physical science basis. Cambridge University Press, Cambridge

Lee H, Shim WJ, Kwon JH (2014) Sorption capacity of plastic debris for hydrophobic organic chemicals. Sci Total Environ 470–471:1545–1552. https://doi.org/10.1016/j.scitotenv.2013.08.023

Li J, Zhang K, Zhang H (2018) Adsorption of antibiotics on microplastics. Environ Pollut 237:460–467. https://doi.org/10.1016/j.envpol.2018.02.050

Li K, Du L, Qin C, Bolan N, Wang H, Wang H (2024) Microplastic pollution as an environmental risk exacerbating the greenhouse effect and climate change: a review. Carbon Res 3:9. https://doi.org/10.1007/s44246-023-00097-7

Llorca M, Schirinzi G, Martínez M, Barcelo D, Farre M (2018) Adsorption of perfluoroalkyl substances on microplastics under environmental conditions. Environ Pollut 235:680–691. https://doi.org/10.1016/j.envpol.2017.12.075

Luo H, Liu C, He D, Xu J, Sun J, Li J, Pan X (2022) Environmental behaviors of microplastics in aquatic systems: a systematic review on degradation, adsorption, toxicity and biofilm under aging conditions. J Hazard Mater 423:126915. https://doi.org/10.1016/j.jhazmat.2021.126915

Magadini DL, Goes JI, Ortiz S, Lipscomb J, Pitiranggon M, Yan B (2020) Assessing the sorption of pharmaceuticals to microplastics through in-situ experiments in New York city waterways. Sci Total Environ 729:138766. https://doi.org/10.1016/j.scitotenv.2020.138766

Massos A, Turner A (2017) Cadmium, lead and bromine in beached microplastics. Environ Pollut 227:139–145. https://doi.org/10.1016/J.ENVPOL.2017.04.034

Mato Y, Isobe T, Takada H, Kanehiro H, Ohtake C, Kaminuma T (2001) Plastic resin pellets as a transport medium for toxic chemicals in the marine environment. Environ Sci Technol 35(2):318–324. https://doi.org/10.1021/es0010498

Melo-Agustín P, Kozak ER, de Jesús Perea-Flores M, Mendoza-Perez JA (2022) Identification of microplastics and associated contaminants using ultra high resolution microscopic and spectroscopic techniques. Sci Total Environ 828:154434. https://doi.org/10.1016/j.scitotenv.2022.154434

Miao M, Yu B, Cheng X, Hao T, Dou Y, Zhang M, Li Y (2023) Effects of chlorination on microplastics pollution: physicochemical transformation and chromium adsorption. Environ Pollut 2023(323):121254. https://doi.org/10.1016/j.envpol.2023.121254

Mai L, Bao LJ, Shi L, Liu LY, Zeng EY, (2018) Polycyclic aromatic hydrocarbons affiliated with microplastics in surface waters of Bohai and Huanghai Seas, China. Environmental Pollution 241, 834–840.

Murphy J (2001) Additives for plastics handbook. Elsevier Science, Oxford

O'Brien AM, Lins TF, Yang Y, Frederickson ME, Sinton D, Rochman CM (2022) Microplastics shift impacts of climate change on a plant-microbe mutualism: temperature, CO2, and tire wear particles. Environ Res 203:111727

Pozo K, Urbina W, Gomez V, Torres M, Nunez D, Pribylova P, Audy O, Clarke B, Arias A, Tombesi N, Guida Y, Klanova J (2019) Persistent organic pollutants sorbed in plastic resin pellet—"Nurdles" from coastal areas of central. Environ Pollut 330:121829. https://doi.org/10.1016/j.marpolbul.2019.110786

Ribeiro F, O'Brien JW, Galloway T, Thomas KV (2019) Accumulation and fate of nano- and micro-plastics and associated contaminants in organisms. Trends Anal Chem 111:139–147

Rochman CM, Hentschel BT, The SJ (2014) Long-term sorption of metals is similar among plastic types: implications for plastic debris in aquatic environments. PLoS One 9(1):e85433. https://doi.org/10.1371/journal.pone.0085433

Santos-Echeandía J, Zeler A, Gago J, Lacroix C (2021) The role of cigarette butts as vectors of metals in the marine environment: could it cause bioaccumulation in oysters? J Hazard Mater 416:125816

Sharma S, Sharma V, Chatterjee S (2023) Contribution of plastic and microplastic to global climate change and their conjoining impacts on the environment - a review. Sci Total Environ 875:162627. https://doi.org/10.1016/j.scitotenv.2023.162627

Tang KHD (2023) Climate change and plastic pollution: a review of their connections. Trop Environ Biol Technol 1(2):110–120. https://doi.org/10.53623/tebt.v1i2.341

Thompson RC, Olsen Y, Mitchell RP, Davis A, Rowland SJ, John AWG, McGonigle D, Russell AE (2004) Lost at sea: where is all the plastic? Science 304(5672):838–838. https://doi.org/10.1126/science.1094559

Tivefalth M, Carney Almroth B (2018) Microplastics as a vector for exposure to hydrophobic organic chemicals in fish: a comparison of two polymers and silica particles spiked with three model compounds at sven loven Centre kristineberg, Sweden view project trophic transfer of nanoparticles in a microcosm system view project giedre asmonaite. Ecol Ecotoxicol 35:11–13. https://doi.org/10.3389/fenvs.2020.00087

Tumwasigye E, Nnadozie C, Akamagwuna FC, Noundou XS, Nyakairu GW, Odume ON (2023) Review microplastics as vectors of chemical contaminants and biological agents in freshwater ecosystems: current knowledge status and future perspectives. Environ Pollut 330:121829. https://doi.org/10.1016/j.envpol.2023.121829

Turner A, Wallerstein C, Arnold R (2019) Identification, origin and characteristics of bio-bead microplastics from beaches in western Europe. Sci Total Environ 664:938–947. https://doi.org/10.1016/j.scitotenv.2019.01.281

Vedolin MC, Teophilo CYS, Turra A, Figueira RCL (2018) Spatial variability in the concentrations of metals in beached microplastics. Mar Pollut Bull 129(2):487–493. https://doi.org/10.1016/J.MARPOLBUL.2017.10.019

Wiesinger H, Wang Z, Hellweg S (2021) Deep dive into plastic monomers, additives, and processing aids. Environ Sci Technol 55:9339–9351

Yeo BG, Takada H, Yamashita R, Okazaki Y, Uchida K, Tokai T, Tanaka K, Trenholm N (2020) PCBs and PBDEs in microplastic particles and zooplankton in open water in the Pacific Ocean and around the coast of Japan. Mar Pollut Bull 151:110806. https://doi.org/10.1016/j.marpolbul.2019.110806

Zhang Y, Liang J, Zeng G, Tang W, Lu Y, Luo Y, Xing W, Tang N, Ye S, Li X, Huang W (2020) How climate change and eutrophication interact with microplastic pollution and sediment resuspension in shallow lakes: a review. Sci Total Environ 705:135979. https://doi.org/10.1016/j.scitotenv.2019.135979

Zhou Q, Tu C, Liu Y, Li Y, Zhang H, Vogts A, Plewe S, Pan X, Luo Y, Waniek JJ (2022) Biofilm enhances the copper (II) adsorption on microplastic surfaces in coastal seawater: simultaneous evidence from visualization and quantification. Sci Total Environ 853:158217. https://doi.org/10.1016/j.scitotenv.2022.158217

From the Depth to the Shore Microplastics in the Food Chain

4

Md. Mostafizur Rahman, Mohammad Toha, Tareq Rahman, and Sadia Sikder

Abstract

Microplastics (MPs) are present in terrestrial and marine environments across the globe. Moreover, MPs in benthic sediments and surface water concern the food chain. Coastal settings and commercial seafood have been found to contain MPs. The massive presence of MPs in aquatic bodies has already created the issue of global food security and safety. MPs are passed between species at different food chain trophic levels by food consumption, which has a tremendous ecotoxicological effect. As a result, this chapter explores the widespread problem of MP pollution and how it affects the food chain. Moreover, this chapter clarifies how these tiny particles go from the ocean's depths to the shore. This chapter will

M. M. Rahman (✉)
Laboratory of Environmental Health and Ecotoxicology, Department of Environmental Sciences, Jahangirnagar University, Dhaka, Bangladesh

Department of Environmental Science, Bangladesh University of Professionals (BUP), Dhaka, Bangladesh
e-mail: rahmanmm@juniv.edu

M. Toha · S. Sikder
Laboratory of Environmental Health and Ecotoxicology, Department of Environmental Sciences, Jahangirnagar University, Dhaka, Bangladesh

Department of Environmental Science, Bangladesh University of Professionals (BUP), Dhaka, Bangladesh

Department of Environmental Science and Disaster Management, Daffodil International University, Dhaka, Bangladesh

T. Rahman
Department of Environmental Science, Bangladesh University of Professionals (BUP), Dhaka, Bangladesh

Department of Project Management, Charles Darwin University, Sydney, NSW, Australia

M. Oves et al. (eds.), *Marine Microplastics and the Quest for Remediation*,
https://doi.org/10.1007/978-981-96-8979-8_4

thoroughly analyze case studies and scientific literature to investigate MPs' origins, distribution, and food web pathways from the marine environment to the coast. This chapter aims to improve our understanding of the complex interaction between MPs and the food web while highlighting the urgency of taking proactive measures to address this pressing environmental issue.

Keywords

Microplastics · Marine Pollution · Food chain · Food web · Ingestion

4.1 Introduction

With time, human civilization went through the stone age, bronze age, and iron age (Praet et al. 2024). Currently, they are living in the Age of Plastics (Zhang et al. 2025). According to Praet et al. (2024), these plastics comprise a series of molecular monomers that are joined to form macromolecules, which have numerous applications in contemporary society. Over the decades, substantial quantities of plastic have been used worldwide due to efficient manufacturing processes, socioeconomic development, urbanization, population growth, and modernization (Toha et al. 2024a). In recent years, plastic manufacturing has been determined to increase by 3% annually, reaching an expected 400 million tons annually by 2025 (Toha et al. 2024a). Furthermore, as only 9% of plastics are recycled, if this rate of plastic manufacturing continues, billions of tons of plastic will be accumulated in our environment by 2050, posing a serious concern and a topic of current research (Sikder et al. 2024; Toha et al. 2024b). China accounted for 30% of global plastic production in 2018, followed by Europe and North America (Rahman et al. 2024). According to Oni et al. (2025), this increasing trend indicates that the microplastics (MPs) level in the Pacific Ocean's surface water could surpass the standard level if the rate of plastic discharge into oceans remains unchanged. This would have detrimental effects on both aquatic and terrestrial organisms. After entering environmental compartments, plastic waste degrades continuously through physical and chemical processes, including ultraviolet (UV) light exposure, resulting in microplastics (MPs) formation. This pollution is one of the ten rising concerns listed in the UNEP Year Book 2014 (Mandal and Ghosh 2025; Hossain et al. 2025; Zhai et al. 2024; Horton et al. 2024; Hasnine et al. 2024). Additionally, it has been recognized as a significant factor contributing to biodiversity loss (Toha et al. 2024a). However, these MPs often come from two different sources. One is classified as primary MPs, while the other is classified as secondary (Anik et al. 2021; Huang et al. 2021; Kumar et al. 2022). Primary MPs are mainly presented as plastic beads (Anik et al. 2021). Furthermore, it can be released indirectly from secondary sources when more significant pieces of plastic break down due to photocatalytic breakdown, oxidation, and mechanical aging (Liu et al. 2019). It can be identified to a variety of sources, such as household dust, road littering, urban runoff, sewage treatment plants, atmospheric deposition, washing and drying of clothes, synthetic fibres used in clothing, plastic carry bags, exfoliants in cosmetics and as abrasives in cleaners (Rai et al. 2024; Luqman et al. 2023).

Consequently, a significant amount of MPs have been detected in various environmental compartments (Rahman et al. 2024). Moreover, MPs can interact with other toxic pollutants in environmental compartments, altering the behavior of pollutants (Zhu and Parker 2025). With the help of stormwater, wind, and snow, these MPs can be deposited into aquatic and terrestrial media (Toha et al. 2024b). The abundance of microplastic in varying depths and points of confluence was found in the water and sediments of the Yangtze River, one of the largest carriers of microplastic debris to the sea (Xiong et al. 2019). A thorough literature review found a global set of rivers and oceans containing MPs (Toha et al. 2024b). These include the Rhine River sediment in Germany (Klein et al. 2015), the St. Lawrence River in North America (Castañeda et al. 2014), Ciwalengke River water and sediment Indonesia, sediment and surface water of West River south of China, in five different rivers in the Tibet Plateau, Beijiang river surface sediment in China, sediments of Sabarmati River of India, in freshwater system of Antua River Portugal, Qin River China, Vistula river Poland, and Brisbane river sediments Australia (Alam et al. 2019; Huang et al. 2021; Jiang et al. 2019; Wang et al. 2017). The results of all these studies showed different characteristics of the MPs (Islam et al. 2022; Sekudewicz et al. 2021; Zhang et al. 2020a, b; Rodrigues et al. 2018). The polymer types are usually low-density polyethylene (LDPE), polypropylene (PP), PEVA, polyamide (PA), high-density polyethylene (HDPE), PBA, Polymethyl Methacrylate (PMMA), polyester (PET), Polycarbonate (PC), polystyrene (PS), polyvinyl alcohol (PVA), polyethylene (PE), polyethylene terephthalate (PETE), polyvinyl chloride (PVC), Cellulose Acetate (CA), and acrylonitrile butadiene styrene (ABS) (Toha et al. 2024a, b). A study in Germany indicated the influence of rivers as channels for MP transmission into the ocean (Klein et al. 2015). Moreover, MP particles were discovered in the water, sediment, and sponges of protected and nonurban mangroves in Mexico (Rodrigues et al. 2018). Hence, freshwater and estuary systems are recognized as sources and conduits for plastics entering marine environments (Toha et al. 2024b). The highest concentration of MP contamination is around the city center and the estuary, suggesting a trend of escalating MP quantities downstream (Ali et al. 2024). River networks are anticipated to transport MPs from pollution sources to deposition sites (Xiang et al. 2025). This study recommended prioritizing MP monitoring in river networks rather than specific water body assessments (Ali et al. 2024). These investigations prompted the hypothesis that coastal rivers, despite being surrounded predominantly by rural human communities with no industrial presence, may still be contaminated by MPs originating from upstream sources (Patel et al. 2020; Wang et al. 2017). Furthermore, a network of rivers converging at the beach may offer insight into the potential transit of MP particles (Luo et al. 2019). However, exposure to MPs results in aberrant behavior and neurotoxic effects affecting human health (Toha et al. 2024c). In humans, exposure to MPs may lead to cellular damage, neurotoxicity, cytotoxicity, disruption of the immunological response, oxidative stress, and translocation of MPs to other organs (Islam et al. 2022). Multiple research have examined cytotoxicity, oxidative stress, and cell viability to ascertain the potentially harmful effects of MPs and their components at the cellular level (Toha et al. 2024b; Ali et al. 2024). A further investigation revealed that the positive control inflicted considerable damage in all in vitro assays involving direct contact (Ali et al. 2024). According to a study (Hwang et al. 2019), direct interaction between polypropylene MPs and human cells may induce the synthesis of cytokines and histamines. Moreover, prolonged exposure to MPs may result in discomfort, edema, cellular proliferation, cellular apoptosis, and a diminished capacity of the animal's immune system to combat infections (Ali et al. 2024), disrupting the mitochondrial membrane potential. According to a recent study, various types of bacteria can be spread through MPs (Tehrani et al. 2025). These organisms can assimilate matrix-ejected chemicals and ambient molecules (Megha et al. 2025). Nonetheless, environmental MPs were adequate to provoke immunosuppression or autoimmune disorders in physiologically susceptible persons (Prata et al. 2024). Consequently, MP accumulations vary according to location, polymer type, and proximity to the source.

Hence, it is crucial to understand the characteristics and spatial distribution dynamics of MPs in global environmental reservoirs to assess the ecotoxicological dangers and develop appropriate management strategies. Furthermore, it is essential to perform comprehensive investigations of MPs to reveal their properties, identification, and quantification in oceanic pollution and its primary causes in developing countries. This chapter provides an overview of the current status of MP pollution in terrestrial and coastal systems. This chapter also delineates the various paths of MPs in ocean systems and their impacts and highlights future research requirements. Lastly, this chapter will discuss MP management in environmental compartments.

4.2 Source, Fate, and Transport of Microplastics

4.2.1 Types of Microplastic Polymer and their Application

Plastics, classified as polymers, can be molded by applying heat and pressure. Polymer and synthetic materials significantly contribute to microplastics (MPs) (Rochman et al. 2019). HDPE, LDPE, PP, PE, and PET are synthetic polymers utilized in domestic and commercial settings (Hartmann et al. 2019). Moreover, PVC, PP, and PE polymers have been found in stormwater runoff, which can potentially be released into the aquatic environment (Toha et al. 2024a). Recent studies have identified the predominant types of polymers found in stormwater discharges in Mexico. According to Monira et al. (2022), five main polymers were detected, with polyethylene (PE) and polystyrene (PS) being the most abundant. PE accounted for 45.2% and PS for 27.0% of the total polymer content. These findings align with those reported by Pramanik et al. (2021), which also emphasize the significant presence of PE and PS in stormwater runoff. Moreover, SR polymers were found in Australia's wetlands (Ziajahromi et al. 2020). Depending on the density of the substance, MPs can float or sink in water (Toha et al. 2024a). Generally, PET and PVC exhibit greater density than lighter plastics like LDPE, HDPE, and PP, resulting in a higher likelihood of sinking in water. These lighter plastics are commonly present in waste (Rochman et al. 2019). However, the type of polymer and the manufacturing process impact the density of plastics.

4.2.2 Morphological Characteristics of MPs

MPs exhibit diverse shapes, origins, and sizes, reflecting the extensive variety of their sources (Hartmann et al. 2019; Zhang et al. 2020a, b). Research indicates that most MP particles are observed within the size range of (50–350 µm), accounting for 85% of the total MPs. Moreover, fragments of synthetic rubber are present in sizes exceeding 500 µm. It was found that beads are created through primary sources, while foam, fibers, fragments, and films are generated by the fragmentation of large plastic litters and are considered secondary sources (Ali et al. 2024). Different shapes of these tiny particles have different properties that depend on the types of primary MPs used, how quickly and easily plastic particles break down and wear down, and how long they stay in the environment (Zhang et al. 2020a, b). In Chennai, India, most of the MPs were fragments, accounting for 92.46%, while in Melbourne, Australia, the percentage of fragments was 68.5% (Patchaiyappan et al. 2021; Townsend et al. 2019). This occurred due to the fragmentation of larger plastics. On the other hand, fibers were identified as the most prevalent shape in Iran, accounting for 75.87%, while in Mexico, their prevalence ranged from 68% to 87% (Teresita et al. 2020). Furthermore, fibers constituted the predominant category of MPs, at 80.9%, while fragments and films accounted for only 18.9% and 2.2%, respectively, in China (Liu et al. 2019). A substantial quantity of fiber resulted from the extensive manufacturing of textile products. It has been noted that colors significantly influence the identification of MPs (Hartmann et al. 2019). Various colors of MPs have already been identified in aquatic bodies (Rochman et al. 2019). The fiber type MPs predominantly exhibit red and blue characteristics (Rahman et al. 2024). MPs were discovered in aquatic samples in a prior study in the following colors: red and pink (56.20%), white (11.90%), black and grey (20.20%), blue and green (7.80%), and yellow and orange (3.90%). Furthermore, significant quantities of black particles have been identified in Australia and Sweden (Roychand and Pramanik 2020) (Table 4.1).

4.2.3 Pathways of Terrestrial Microplastics to the Ocean Environment

Microplastics (MPs) play a crucial role in the transformation and transportation of pollutants in the terrestrial ecosystem (Ali et al. 2024). Recent studies have identified the presence of antibiotics, persistent organic pollutants (POPs), and heavy metals on the surfaces of MPs (Toha et al. 2024c). The collective influence of MPs, acting as efficient carriers for these substances, on the soil ecosystem and its

Table 4.1 Characteristics of MPs in ocean systems

Shapes	Clarity	Types	Color	Size	References
Fibers	Thin and soft (filaments, microfibers, threads, strings)	Polyester (PET), acrylic, PA, PVC, PAM, PE, LDPE, nylon, PS	Black, red, blue, purple, clear, green, white	20 μm–1 mm	Toha et al. (2024a), Ali et al. (2024), Rahman et al. (2024), Toha et al. (2024b)
Fragments	Irregular-shaped particles (round, angular), crystals, fluff, powder, granules	Synthetic rubber, SBR, NBR, PE, LDPE, HDPE, PP, PS	Black, red, blue, clear, green, white, yellow, orange	Up to 5 mm	
Foams	Soft, compressible, lightweight, and sponge/cloud-like	PS, EPS	Any color	3 mm in size	
Films	Films are solid and flat	PE, LDPE, HDPE	Partially or fully transparent and can be any color	1 μm–1.7 mm	
Spheres/microbeads	Round shaped with smooth surface	PE (primary), PP, PS, PET	Any color	100 μm–2 mm	

organisms has sparked significant interest within the contemporary scientific community (Rabin et al. 2023). A variety of elements influence the dynamics of MPs in terrestrial ecosystems. The migration of MPs is influenced by various factors, including their morphological characteristics (size, density, and shape), external climatic conditions such as wind and rain, and the presence of soil organisms (Abbasi et al. 2017). Microorganisms, clay minerals, metal oxides, hydroxides, and humus can interact with environmental pollutants. The observation of polyethylene (PE) particles in the middle and lower soil layers after 21 days indicates the involvement of earthworms in facilitating the migration of MPs through the soil profile, with smaller particle sizes demonstrating a higher propensity for migration. The correlation between particle size and low-density polyethylene (LDPE) migration indicates that smaller particles increase biological issues (Huang et al. 2021). Nonetheless, MPs can find their way into aquatic environments through several pathways (Liu et al. 2019). The fate of MPs in freshwater is influenced by their dimensions, morphology, density, chemical makeup, surrounding conditions, and interactions with organisms. MPs can either settle into sediment or remain suspended in the water column for extended periods. Figure 4.1 illustrates the transition of MPs from terrestrial ecosystems into the aquatic food chain.

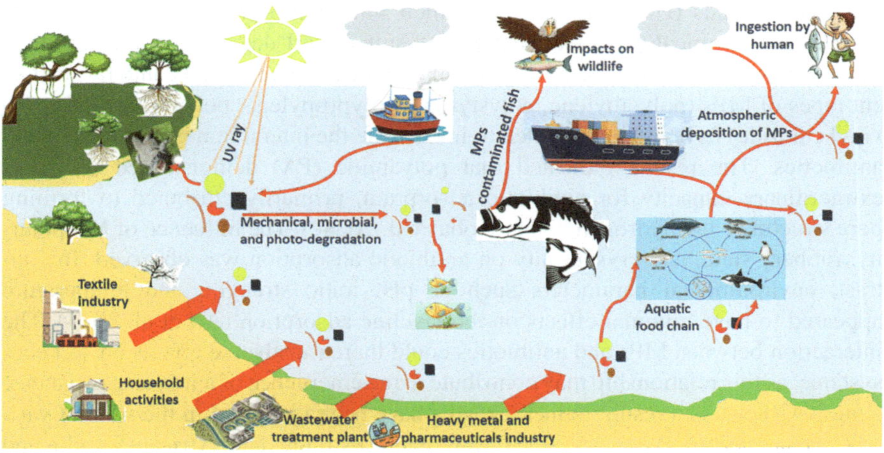

Fig. 4.1 Microplastics in food chain

MPs may remain in sediment or be resuspended. Aquatic organisms can consume MPs, which can disrupt food webs and harm the environment. MPs may be broken apart by mechanical and chemical processes, increasing their frequency and possible adverse effects in freshwater environments (Toha et al. 2024b). Urban runoff and wastewater treatment can introduce MPs into freshwater ecosystems (Monira et al. 2022). In freshwater environments, MPs can build up in the sediment, be eaten by aquatic organisms, or be carried downstream to other areas of water. The movement and fate of MPs in freshwater environments are influenced by flow, sedimentation, and biological interactions (Ali et al. 2024). Notably, MPs can persist in aquatic environments for up to 100 years owing to their gradual degradation and hydrophobic properties (Rochman et al. 2019). Large MP particles are often deposited on road surfaces, while small particles are typically released into the atmosphere. However, certain particles may be retained, while others can travel into soils, sewers, and surface waters by stormwater runoff (Abbasi et al. 2017). Coarser MP particles from road runoff generally accumulate along the roadside or in gully pots, whereas only the finer fractions (50 μm) are deposited.

4.3 Microplastics in Terrestrial Ecosystems

In terrestrial ecosystems, microplastics (MPs) are effective carriers for harmful pollutants (Monira et al. 2022). Research indicates that higher hydrophobicity plastics attract more hydrophobic organic pollutants (Toha et al. 2024c). Furthermore, it is contingent upon the polarity of the organic pollutants. Subsequently, they often form stronger connections, leading to the development of harmful environments. Furthermore, the absorbance of MPs is significantly influenced by factors such as pH, temperature, moisture, and water flow (Myszka et al. 2023). For instance, environments with lower pH levels promote enhanced adsorption. The release of MPs and heavy metals (HMs) indicates that their introduction into soil environments initiates geochemical processes that may result in soil degradation (Rabin et al. 2023). Studies examining the adsorption of five different antibiotics onto five different types of MPs (polyethylene, polystyrene, polypropylene, polyamide, and polyvinyl chloride) have been conducted to investigate the interaction between MPs and antibiotics. The results indicated that polyamide (PA) demonstrated the most extraordinary capacity for antibiotic adsorption, primarily attributed to forming pore structures and hydrogen bonds (Toha et al. 2024c). The influence of MP polarity, rubbery state, and crystallinity on antibiotic adsorption was observed. In contrast, environmental parameters such as pH, ionic strength, and temperature appeared to have minimal effects on tetracycline adsorption (Ali et al. 2024). The interaction between MPs and antibiotics could increase adverse effects on soil ecosystems, as this relationship may contribute to the emergence of antibiotic-resistance genes (ARGs). As a result, numerous MPs have been identified in the soils of various Asian nations, Europe, North America, Australia, and Africa (Sa'adu and Farsang 2023). Table 4.2 presents the concentrations of microplastics found in diverse terrestrial ecosystems across the globe.

As MPs are infiltrating agricultural soil, posing a growing threat to food safety, ecosystem health, and soil sustainability, understanding their distribution and accumulation patterns is critical for combating this pervasive issue.

Table 4.2 Recent status of microplastics in terrestrial ecosystem

Location	Concentration (MPs/kg)	Year	References
Agricultural land, Germany	8000–2500	2024	Zeb et al. (2024)
Agricultural land, USA	4500–1200	2024	Mai et al. (2024)
Urban land, China	45,000–15,000	2024	Liu et al. (2019)
Grassland, Australia	2500–700	2024	Horton et al. (2024)
Agricultural land, India	7000–1500	2023	Kumar et al. (2022)
Forest land, Brazil	2300–500	2024	Zhang et al. (2025)
Urban area, UK	38,000–11,000	2025	Horton et al. (2024)
Forest area, Canada	1900–300	2023	Rai et al. (2024)
Russia	2400–600	2024	Huang et al. (2021)
Agricultural land, France	7500–3000	2025	Rillig et al. (2025)
Urban area, Japan	42,000–15,000	2024	Liu et al. (2019)
Agricultural land, Italy	6500–2500	2025	Horton et al. (2024)
Urban area, Mexico	9500–27,000	2024	Horton et al. (2024)
Grassland, South Korea	3500–1000	2025	Huang et al. (2021)
Forest area, Argentina	2200–600	2023	Zhang et al. (2025)
Desert area, Saudi Arabia	550–100	2024	Huang et al. (2021)
Urban area, Egypt	24,000–9000	2025	Rai et al. (2024)
Agricultural land, Spain	7500–3000	2024	Rillig et al. (2024)
Urban land, the Netherlands	42,000–18,000	2023	Liu et al. (2019)
Grassland, Kenya	2800–800	2024	Huang et al. (2021)
Soil of Dhaka, Bangladesh	2.13×10^4	2024	Toha et al. (2024c)
Soil of Narayanganj, Bangladesh	151	2024	Toha et al. (2024c)
Sagian, Pakistan	876 ± 194	2023	Luqman et al. (2023)

4.4 Microplastics in Shorelines and Ocean Systems

Microplastics (MPs) generally come from river runoff, atmospheric deposition, and coastal and marine activities in shorelines and oceans (Huang et al. 2021). Generally, river runoff is a significant transport medium for MPs from inland areas to marine environments. This river's runoff has been found to contain various types of plastic litter and untreated wastewater that can finally accumulate along shorelines and in ocean systems (Luqman et al. 2023). Notably, low-density types of MPs can be floated through wind and atmospheric currents, depositing them onto shorelines and beach areas (Toha et al. 2024b). According to recent studies, fishing, shipping, and aquaculture activities are also responsible for MP deposition in shoreline areas (Toha et al. 2024b). However, various environmental and anthropogenic activities distribute MPs in shoreline areas. For example, shorelines in urbanized and industrialized areas contain comparatively more MPs than in other areas. Moreover, sandy beaches are considered to be more vulnerable to MP pollution due to their porous nature (Ali et al. 2024; Smith et al. 2023). On the other hand, water columns, ocean currents, and gyres are responsible for distributing MPs in ocean systems. It has been found that the North Pacific Gyre shows rapid hotspots for MP accumulation compared to the Mediterranean Sea (Tanaka et al. 2024). The MP concentration levels in various shorelines and oceans are given in Tables 4.3 and 4.4.

The fast increase in MPs presents significant hazards to coastal and marine ecosystems. MPs can modify sandy sediments' physical and chemical characteristics,

Table 4.3 MPs concentration level in various shorelines in the world

Location	Concentration (Items/m^2)	Source	References
Sandy Beach, USA	1200	Discharge of industry	Smith et al. (2023)
Rocky Shore, Australia	950	Fishing gear	Johnson and Lee (2022)
Urban Coastal Beach, Japan	2500	Urban runoff	Tanaka et al. (2024)
Tropical Sandy shore, Brazil	1100	Tourism	Pereira et al. (2023)
Estuarine shoreline, South Africa	1800	Fishing industry	Dlamini (2020)
Pebble Beach, UK	1350	Industrial waste	Ribó et al. (2023)
Mangrove shoreline, India	3000	Riverine transport and urban waste	Kumar et al. (2022)
Coral Reef Beach, Indonesia	2800	Tourism and lack of recycling	Daud et al. (2024)
Sandy beach, Kuakata, Bangladesh	232 ± 52 items/kg	Tourism and urban waste	Banik et al. (2022)
Sandy beach, southeastern coastal beaches, Bangladesh	242.86 items/kg	Tourism and urban waste	Hossain and Islam (2023)

Table 4.4 MPs concentration level in various countries ocean systems

Location	Concentration	Description	References
North Pacific Ocean, USA	0–349 items/m^3	Rapid fluctuation of MPs has been observed surrounding the area. Concentrations in certain areas	Egger et al. (2020)
Mediterranean Sea, France	10^6 items/km^2	Increase of MPs with alarming rate	Le Monde (2024)
South Pacific Ocean, Chile	Not specified	Easter Island, near South Pacific Gyre, experiences severe plastic pollution due to ocean currents	Reuters (2024)
Bay of Bengal, Bangladesh	232 ± 52 items/kg	Kuakata area, Bay of Bengal, showed rapid MPs pollution of fibers and fragments	Banik et al. (2022)

drastically diminishing permeability, water drainage, and gas exchange (Ciocan et al. 2024). Consequently, the livelihoods of burrowing animals, such as crabs and mollusks, have grown increasingly precarious (Bos et al. 2023). Furthermore, due to the diminutive size of MPs, they can readily adsorb organic and metallic pollutants, potentially leading to bioaccumulation within shoreline ecosystems (Toha et al. 2024b). Conversely, several marine creatures, including fish and plankton, may inadvertently ingest these MPs, resulting in obstruction and exposure to hazardous substances (Kim et al. 2022). Furthermore, floating macrophytes can serve as a conducive habitat for invasive species, which may further disturb ecosystems. Nonetheless, the ecotoxicological impacts of MPs can differ across trophic levels (Toha et al. 2024b). Research indicates that MPs can swiftly diminish aquatic populations by modifying species' reproductive behaviors and survival rates. Furthermore, it significantly influences food webs and ecosystem services (Ali et al. 2024). Particular creatures, including coral reefs, mangroves, and seagrass beds, are imperiled by MP contamination, which impacts ecosystem functions (Smith et al. 2023). Numerous studies indicate a correlation between coral bleaching and reduced development rates in seagrass and MP pollution (Smith et al. 2023; Huang et al. 2021). Consequently, tourism and esthetic and cultural values have eroded worldwide (Banik et al. 2022). Recent investigations have identified a significant concentration of MPs in mussels and oysters, adversely affecting human health.

Furthermore, MPs are recognized as endocrine disruptors with probable carcinogenic effects (Smith et al. 2023). Therefore, it is imperative to regulate this pollution. In addition to coastal areas and surface water, microplastics have also been detected in deep-sea water samples and polar regions (Gola et al. 2021). Nearly 700 aquatic species globally, such as sea turtles, penguins, and diverse crustaceans, were adversely affected by the proliferation of microplastics (Marn et al. 2020). Tables 4.5, 4.6, and 4.7 show microplastics' effects on flora, wildlife, and vegetation. However, MP toxicity impacts aquatic plants through direct and indirect mechanisms.

Direct effects can involve physical entanglement or covering plant surfaces, which prevents light and gas exchange. Chemical additives and contaminants associated with MPs might further hinder the growth and health of aquatic plants. MPs indirectly affect aquatic plants by altering the water's pH, dissolved oxygen level,

Table 4.5 Microplastics in oceanic flora and fauna

Organism	Country	Concentration	References
Oysters	UK	11,220 items/kg	Ciocan et al. (2024)
Mussels	UK	2740 items/kg	Ciocan et al. (2024)
Fish and crustaceans	Gulf of Mexico	26% of fish and 29% of crustaceans contained microplastics	Bos et al. (2023)
Plankton	Global oceans	Variable concentrations	Mongabay News (2023)
Seagrass	Turneffe atoll, Belize	75% contamination rate	McAlpine (2017)

and nutritional availability. These changes may impact the delicate balance plants need to develop and survive as best they can. Studies have shown that MPs can accumulate in seagrass beds and harm the plants' functioning and general health (Thacharodi et al. 2024).

Overall, the impact of MPs on aquatic flora and fauna varies depending on the species and the specific stress caused, highlighting the complexity of this issue in the plant kingdom.

4.5 Microplastic Pollution Management Strategies and Future Directions

As microplastics (MPs) substantially impact ocean and shoreline systems, it is imperative to implement effective management strategies to regulate this contamination. Source reduction, mitigation strategies, and remediation methodologies comprise the standard management system of MPs (Praveena 2024). The following is a description of these sections.

4.5.1 Source Reduction of Microplastics

The proactive approach of minimizing the production and release of microplastics (MPs) is commonly referred to as source reduction. The influence of MPs can be promptly tackled at its origin (Wu et al. 2017). Multiple approaches are implemented to decrease MPs at their source (Tariq et al. 2024). Opting for sustainable alternatives in industrial processes, advocating for eco-friendly materials, and implementing policy interventions are recognized strategies for reducing MP at the source (Zeb et al. 2024). Furthermore, advancements in material science, educational initiatives, and understanding consumer behavior may be crucial in mitigating MPs (Rai et al. 2024). Therefore, the development of biodegradable plastic polymers and the cooperation among stakeholders, industries, and policymakers are essential to improve the effectiveness of this approach.

Table 4.6 Impacts of microplastics on aquatic fauna

Organisms	Types of MPs	MPs amount/ concentration	Effects	References
Zebrafish	High-density polyethylene (HDPE)	20 mg/L	– Large MPs affected fish's lateral line system's mechanosensory receptors and hair cells – Gastrointestinal system exhibited morphological alterations because of larvae consuming MPs smaller than 100 μm – All larvae exposed to MPs displayed altered behavioral patterns	Kim et al. (2022)
Grass carp	80 nm and 8 μm microplastics	5, 15, and 45 mg/L	– Fish may re-consume excreted microplastics and store them in the mouth cavity, whereas the nasal cavity may retain 1 μm of MPs	Zhang et al. (2022)
Shrimp (*Litopenaeus vannamei*)	Polystyrene, 1.1 μm	40 and 400 μg MP/kg of ration	– Damage to DNA and lipids was visible in the hepatopancreas – A histological examination showed edema in the muscle, hepatopancreas, and gut	Seta et al. (2023)
Clam (*Ruditapes decussatus)*	Polyethylene (PE) (40–48 μm)	10 μg/L (low), 100 μg/L (medium), and 1000 μg/L (high)	– Increased quantity of hemocyte cells – Clams' overall weight decline – Substantial decline in hemoglobin viability.	Abidli et al. (2023)
Japanese quail (*Coturnix japonica*)	Polypropylene and polyethylene	600	– Ingesting microplastic did not directly change body mass	Monclús et al. (2022)
Coral (*Goniopora columna*)	Polyethylene (40–48 μm)		– Affect the antioxidant system – Impair the activity of the enzymes that detoxify G. Column – Raise the concentration of lipid peroxide – Harm tissue by building up on the coral mesenteric	Chen et al. (2022)

Table 4.7 Effects of microplastics on different kinds of aquatic plants

Plant types	MPs types	MPs amount/ concentration	Incubation time	Effects	References
Lemna minor	Polyethylene microbeads (4–12 μm)	0, 10, 50 and 100 mg/L	7 days	The root development was notably impacted by mechanical obstruction	Kalčíková et al. (2017)
Lemna minuta	Polystyrene	50, 100 mg/L	28 days	Chlorophyll levels and photosynthetic capacity were reduced	Ceschin et al. (2023)
Thalassiosira pseudonana	0.055, 1 and 6 μm polystyrene	10^{-4} to 250 mg/L	2 days	Actively changes the chemical composition of EPS to withstand the strain of pollution	Shiu et al. (2020)
Utricularia vulgaris	1, 2, and 5 μm Polystyrene	15, 70, and 140 m/L	7 days	Substantial ecotoxicity observed	Yu et al. (2020)
Spirodela polyrhiza	PVC	0, 10, 100, and 1000 mg/L		Impact on the nitrogen fate of aquatic plant habitats	Wang et al. (2023)

4.5.1.1 Reduction of Single Use Plastics (SUPs)

Single-use plastics (SUP) are defined as plastics that are intended for one-time use and are often discarded or recycled after a single usage (Praveena 2024). It includes drinking straws, plastic cotton buds, sachets, food packaging, and polyethylene bags. During the gradual disintegration of plastics, they emit harmful compounds that are now being identified in human bloodstreams and can potentially induce cancer, infertility, birth abnormalities, and several other illnesses (Ali et al. 2024). Their composition mainly consists of PET, which is potentially carcinogenic when exposed to heat (Toha et al. 2024c). They fall under recycling code 1, indicating that they may be recycled but not reused. However, it is worth noting that less than 40% of SUP waste undergoes recycling, while the remaining portion contributes to environmental pollution by polluting both land and ocean areas (Nafea et al. 2024; Acharjee et al. 2023). Hence, it is crucial to implement effective recycling techniques to reduce SUPs.

4.5.1.2 Industrial Innovations and Sustainable Materials

Sustainable industrial practices are highly effective in reducing microplastic (MP) pollution. Major industries reliant on plastic can be crucial in addressing this challenge. The shift of industries from conventional plastic materials to biodegradable options has the potential to significantly reduce the likelihood of MP formation

(Tariq et al. 2024). Industries are increasingly adopting biopolymers like polylactic acid (PLA) and novel materials like mycelium-based packaging to reduce dependence on traditional plastics (Rani et al. 2024). The contribution of synthetic fibers to MP contamination through fiber shedding is substantial. Hence, developing textiles with reduced shedding properties could help mitigate this issue (Shruti and Kutralam-Muniasamy 2019). At the same time, innovations in technology, such as washing machine filters designed to capture microfibers, could improve efforts to reduce pollution sources.

4.5.1.3 Application of Policy and Legislative Measures

Policies emphasize the importance of stringent regulatory frameworks and robust enforcement mechanisms (Praveena 2024). For instance, the Basel Convention's amendment on plastic waste ensures that countries adhere to strict controls on transboundary movements of plastic waste. Similarly, the EU SUP Directive mandates restrictions and extended producer responsibility (EPR) schemes, providing a comprehensive regulatory approach. National policies like Germany's Packaging Act and Japan's Containers and Packaging Recycling Law reinforce these principles by setting clear guidelines and recycling targets (Rai et al. 2024). Hence, governments and regulatory bodies are pivotal in driving source reduction through policy interventions.

Extended producer responsibility (EPR): Manufacturers are responsible for collecting and recycling plastic waste generated from their products. This can be enforced through government regulations and periodic compliance checks (Rani et al. 2024).

Bans and restrictions: A phased ban should be implemented on single-use plastics, such as plastic bags and straws, and incentives for adopting biodegradable alternatives.

Regulatory framework: A comprehensive legal framework should be developed that includes penalties for noncompliance and rewards for adherence to sustainable practices (Praveena 2024).

4.5.1.4 Stakeholder Engagement and Public–Private Partnerships

Effective plastic waste management requires the involvement of all stakeholders. The African Union's action plan and Japan's waste management policies highlight the significance of public–private partnerships (Helm 2017; An et al. 2020; Calero et al. 2021). These collaborations foster innovation and ensure that policies are effectively implemented. The UNEP Clean Seas Campaign exemplifies global efforts to engage multiple stakeholders, including governments, businesses, and civil society, in addressing plastic pollution (Praveena 2024; Nafea et al. 2024; Acharjee et al. 2023; Tariq et al. 2024). The role and importance of stakeholder are given in Table 4.8.

Table 4.8 The role of stakeholder

Stakeholder	Role	Importance
Government institutions	– Responsible for creating and enforcing waste management policies and regulations	– The success of circular economy initiatives depends on the government's ability to implement effective policies and regulations
NGOs	– Raise awareness, promote source segregation, and advocate for sustainable waste management practices	– NGOs bridge the gap between government initiatives and community needs, fostering public engagement and responsible waste management practices
Industries	– Adopting eco-friendly product	– Industries play a critical role in reducing plastic waste by collaborating with other stakeholders and taking responsibility for the waste they generate
Communities	– Engage in source segregation, participate in awareness programs, and promote responsible waste disposal	– Community participation is essential for effective waste segregation at the household level, which improves the quality of recyclable materials
Informal waste pickers	– Collect and sort plastic waste from various sources, significantly contributing to waste management efforts	– Formalizing the informal waste sector by providing training and integrating them into the formal system enhances waste management efficiency

4.6 Educational Programs and Public Awareness

Public awareness and education are fundamental to changing consumer behavior and promoting sustainable practices (Praveena 2024). Global initiatives like the UNEP Clean Seas Campaign and national programs in Japan and Sweden emphasize the importance of educational programs (Nafea et al. 2024). These initiatives aim to inform and engage the public, fostering a culture of sustainability and responsible waste management.

4.6.1 Interministerial Collaboration

Interministerial collaboration ensures that plastic waste management policies are comprehensive and cohesive (Helm 2017; Rai et al. 2024). National policies in Germany and Japan demonstrate the effectiveness of coordinated efforts among various ministries (Zeb et al. 2024). These collaborations facilitate the development of integrated strategies that address all aspects of plastic waste management, from production to disposal.

4.6.2 Sustainable Design and Circular Economy Approaches

Carbon nanoparticles derived from waste play a crucial role in the circular economy by transforming waste into valuable resources, enhancing sustainability, and mitigating environmental impacts for various reasons (Nafea et al. 2024; Rai et al. 2024). Carbon content can be extracted from waste streams by producing carbon nanoparticles derived from various waste materials, including plastics, organic waste, and agricultural residues (Rani et al. 2024). The conversion of waste materials into carbon nanoparticles significantly reduces greenhouse gas emissions associated with waste disposal, the production of traditional carbon products, and landfill accumulation (Toha et al. 2024c). Carbon nanoparticles hold considerable economic value owing to their applications in innovative materials, energy storage, and environmental cleanup (Sikder et al. 2024). Creating value from waste provides economic motivations for recycling and reusing materials. Industries can enhance resource efficiency, close the material use loop, and promote a more sustainable production-consumption cycle by integrating carbon nanoparticles into their processes (Praveena 2024). The synthesis of carbon nanoparticles from waste enhances waste management and drives innovation in material science, fostering sustainable practices and developing new technologies.

4.6.3 Focus on Successful Strategies

Various types of strategies have been applied to reduce microplastic pollution around the world (Table 4.9). Hence, it is necessary to focus on successful strategies.

4.6.4 Mitigation and Remediation Approaches

Addressing pollution in aquatic environments necessitates the elimination of microplastics (MPs). It is essential to develop suitable treatment technologies to prevent the introduction and spread of MPs in the environment, informed by a deeper understanding of the characteristics of MPs (Moffett 2024). Various technologies have already been developed to remove MPs (Table 4.10). The available technologies for removal of MPs from water are given in Table 4.10.

Biochar, sol-gel, advanced oxidation process, and metal-organic framework represent various upgrading removal techniques and mitigation approaches that may serve as potential strategies for removing MPs from environmental compartments (Toha et al. 2024c). In recent years, numerous bacterial and fungal species have been effectively employed in removing MPs (Thacharodi et al. 2024). Research indicates that specific bacteria and fungi can utilize hydrocarbons present in plastic, leading to the degradation of plastic fragments (Toha et al. 2024c).

Microorganisms engage in the plastic degradation process, decomposing and consuming various plastic molecules through the action of active enzymes (Table 4.11). Microorganisms can utilize plastic products as substrates in aerobic

Table 4.9 Overview of global best practices and recommendations for enhancing plastic waste management in developing nations

Aspect	Global best practices	Recommendations for developing nations
Regulatory framework	Basel Convention regulates transboundary plastic waste (Article 4, 2019 amendment); EU SUP Directive restricts single-use plastics (Articles 5, 8); Germany's Packaging Act mandates EPR (Section 15); Japan's Recycling Law sets clear recycling obligations (Article 5) (Basel Convention, 1989; EU Directive, 2019; Germany Packaging Act, 2019; Japan Recycling Law, 1997) (Yang et al. 2024; Alam and Rahman 2025)	Develop and enforce a national circular economy framework, integrating EPR schemes and clear recycling targets
Stakeholder engagement	Public–private partnerships fostered by African Union Action Plan and Japan's collaboration with industry and NGOs (African Union, 2020; Ministry of the Environment, Japan, 2020) (Praveena 2024)	Foster multi-stakeholder platforms for collaboration among government, private sector, and civil society to enhance policy formulation and implementation
Innovation and technology	Germany and Japan invest heavily in advanced recycling technologies and R&D for sustainable materials (OECD 2018; BMU 2020)	Invest in affordable sorting and recycling technologies, leveraging international collaboration for knowledge transfer and funding
Financial incentives	EU SUP Directive includes EPR schemes (Article 8); Germany's Packaging Act provides grants and subsidies for compliance (EU Directive 2019; BMU 2020)	Introduce financial mechanisms like tax breaks, subsidies, and grants to encourage private sector investment in sustainable solutions
Educational programs	UNEP Clean Seas Campaign promotes global awareness; Japan integrates waste management into school curricula (Ministry of the Environment, Japan 2020; UNEP 2017)	Launch national education and awareness campaigns, integrating waste management topics into school curricula and leveraging media platforms
Interministerial collaboration	Germany and Japan demonstrate coordinated efforts among environment, industry, and education ministries for cohesive strategies (BMU 2020; Ministry of the Environment, Japan 2020)	Establish interministerial committees to align policies across environment, industry, and education sectors for integrated plastic waste management strategies

(continued)

Table 4.9 (continued)

Aspect	Global best practices	Recommendations for developing nations
Waste management infrastructure	Sweden and South Korea invest in robust collection systems and advanced sorting facilities (Swedish Environmental Protection Agency 2019; Ministry of Environment, South Korea 2020)	Develop infrastructure for waste collection and sorting, prioritizing plastic waste segregation at the source for higher recycling efficiency
Innovation in policies	EU's restriction on SUP, EPR, and waste-to-energy integration in developed countries (EU Directive 2019; OECD 2020)	Adopt policies to phase out SUP and promote sustainable alternatives, aligning with international conventions and directives

Table 4.10 Available technologies for removal of MPs from water

Approaches	Technologies	Efficiency (%)	References
Physical	Filtration	95	Toha et al. (2024c)
	Adsorption/magnetic separation		
	Density separation		
Chemical	Coagulation	90	Cui et al. (2024)
	Photocatalysis		
	Oxidation treatment		
Biological	Microbial degradation	96	Thacharodi et al. (2024)
	Bioreactors		

Table 4.11 The efficiency of various bioremediation techniques

Strain	Degrading polymers	Efficiency (%)	Time span (days)	References
Pseudomonas aeruginosa	PS-PLA	10	21	Sharma and Neelam (2023)
Enterobacter asburiae (strain YT1)	PE	6.10	28	Nabi et al. (2023)
Comamonas acidovorans (TB-35)	PUR	100	8	Rajan and Vijayalakshmi (2023)
Aspergillus sp. (VRKPT1)	PE	6.02	30	Viel et al. (2023)
Gloeophyllum trabeum	PS	50	20	Nabi et al. (2023)
Pestalotiopsis microspora	PUR	–	14	Xiang et al. (2023)
Penicillium pinophilum	LDPE	0.64	930	Nabi et al. (2023)
Penicillium simplicissium (YK)	PE	0.50	90	Nabi et al. (2023)

environments, producing carbon dioxide, water, and biomass (Nabi et al. 2023). Conversely, microorganisms can transform the carbon chain of plastics into acetic acid. Subsequently, methanogens can further decompose, leading to methane generation (Shen et al. 2022).

4.6.5 Future Directions

There is still a limited amount of research concentrating on the techniques for the remediation of microplastics (MPs). Exploring the role of microorganisms in the removal of MPs from environmental compartments is crucial for advancing upgrading technologies. Furthermore, it is essential to apply MP remediation technologies at suitable sources to effectively prevent contamination across various ecosystems and atmospheres.

Although numerous nations have implemented different legislative strategies, there remains a significant limitation in the standard legislative frameworks addressing MP pollution. Therefore, the development of effective policy and management tools may mitigate the effects of MPs on environmental compartments.

The occurrence and spread of microplastics (MPs) in various environmental compartments have been extensively studied. However, there remains a significant gap in our understanding of the seasonal variations in MP contamination within oceanic systems. Therefore, additional investigation is required regarding the seasonal variation of MPs in these regions, as the distribution of MPs may vary based on current flow rates, runoff, and changes in human activity on land influenced by seasonal shifts.

While many studies have focused on the quantity and morphological characteristics of MP contamination in aquatic environments, there remains a need for a more thorough investigation into the geographical and temporal distribution of these particles. Moreover, sediments act as a repository for MPs in aquatic environments and retain crucial information regarding human activities. Consequently, it is essential to carry out investigations into the vertical distribution of MPs to assess their life cycle and the historical prevalence of MP pollution in a particular aquatic setting. Future investigations should concentrate on developing a mathematical framework that enhances our comprehension of the locations and distribution of MPs across various environmental contexts.

4.7 Conclusion

People around the world are worried about the potential health effects and growing volume of microplastics in the shorelines. As landmass is one of the primary sources of MP, the large quantity of MPs that have been discovered in the major coastal rivers and beaches of our planet suggests that these MPs are being transported to the ocean. This large amount of MP in water and sediment could have an impact on the benthic and surface flora and fauna, as well as on the human habitat that relies on these ecosystems. In future studies, it is important to determine the potential origins of the MPs and these elements, as well as to investigate how they cumulatively affect health and biota. Furthermore, since the majority of the MPs were fibers, it can be assumed that they were primarily obtained from clothing. This is similar to a study that indicated that more fiber form may be formed from the immediate garments of riverfront people as well as the washing of materials. This could originate from the localities on both sides of the river and inland. Nevertheless, the sources of the other particles may have been from aquaculture, fishing nets, paint on vessels and other vehicles, particularly in the riverbody, the tourist zones, and the rivers upstream. Hence, discovering more advanced and efficient technologies to remove these pollutants from the environment is essential. Moreover, priority should be given to environmentally sustainable technology and strategies to reduce this alarming concern of MP.

References

Abbasi S, Keshavarzi B, Moore F, Delshab H, Soltani N, Sorooshian A (2017) Investigation of microrubbers, microplastics and heavy metals in street dust: a study in Bushehr city. Iran Environ Earth Sci 76:1–19

Abidli S, Zaidi S, Ben Younes R, Lahbib Y, Trigui El Menif N (2023) Impact of polyethylene microplastics on the clam Ruditapes decussatus (Mollusca: Bivalvia): examination of filtration rate, growth, and immunomodulation. Ecotoxicology 32(6):746–755. https://doi.org/10.1007/s10646-023-02683-2

Acharjee SA, Bharali P, Gogoi B, Sorhie V, Walling B, Alemtoshi. (2023) PHA-based bioplastic: a potential alternative to address microplastic pollution. Water Air Soil Pollut 234(1):21

Alam MJ, Rahman MM (2025) Various conventional and advanced management techniques and policies adopted at the global level for microplastics. In: Microplastics. Elsevier, pp 429–440

Alam FC, Sembiring E, Muntalif BS, Suendo V (2019) Microplastic distribution in surface water and sediment river around slum and industrial area (case study: Ciwalengke River, Majalaya district, Indonesia). Chemosphere 224:637–645. https://doi.org/10.1016/j.chemosphere.2019.02.188

Ali N, Liu W, Zeb A, Shi R, Lian Y, Wang Q, Liu J (2024) Environmental fate, aging, toxicity and potential remediation strategies of microplastics in soil environment: current progress and future perspectives. Sci Total Environ 906:167785

An L, Liu Q, Deng Y, Wu W, Gao Y, Ling W (2020) Sources of microplastic in the environment. In: Microplastics in terrestrial environments: emerging contaminants and major challenges. Springer, Cham, pp 143–159

Anik AH, Hossain S, Alam M, Sultan MB, Hasnine MT, Rahman MM (2021) Microplastics pollution: a comprehensive review on the sources, fates, effects, and potential remediation. Environ Nanotechnol Monitor Manage 16:100530

Banik P, Hossain MB, Nur A-AU, Choudhury TR, Liba SI, Yu J, Noman MA, Sun J (2022) Microplastics in sediment of Kuakata Beach, Bangladesh: occurrence, spatial distribution, and risk assessment. Front Mar Sci 9:860989. https://doi.org/10.3389/fmars.2022.860989

Bos J et al (2023) Microplastics in deep-sea stomachs: study finds plastic eaten by fish and crustaceans increases with depth in the Gulf of Mexico. NOAA RESTORE Science Program

Calero M, Godoy V, Quesada L, Martín-Lara MÁ (2021) Green strategies for microplastics reduction. Curr Opin Green Sustain Chem 28:100442

Castañeda RA, Avlijas S, Simard MA, Ricciardi A (2014) Microplastic pollution in St. Lawrence River sediments. Can J Fish Aquat Sci 71:1767–1771. https://doi.org/10.1139/cjfas-2014-0281

Ceschin S, Mariani F, Di Lernia D, Venditti I, Pelella E, Iannelli MA (2023) Effects of microplastic contamination on the aquatic plant Lemna minuta (least duckweed). Plan Theory 12(1):207. https://doi.org/10.3390/plants12010207

Chen YT, Ding DS, Lim YC, Singhania RR, Hsieh S, Chen CW, Hsieh SL, Dong CD (2022) Impact of polyethylene microplastics on coral Goniopora columna causing oxidative stress and histopathology damages. Sci Total Environ 828:154234. https://doi.org/10.1016/j.scitotenv.2022.154234

Ciocan C et al (2024) High levels of fiberglass discovered in oysters and mussels. J Hazard Mater

Cui B, Rong H, Tian T, Guo D, Duan L, Nkinahamira F, Naidu R (2024) Chemical methods to remove microplastics from wastewater: a review. Environ Res 249:118416

Daud A, Birawida AB, Maming M, Mukono HJ, Arsyad DS (2024) Microplastic polymers in shellfish and fish in the coastal area. Glob J Environ Sci Manage 10(4):1477–1500

Dlamini TU (2020) The feasibility of wood plastic composites as building material for RDP houses in South Africa (Doctoral dissertation, Stellenbosch: Stellenbosch University)

Egger M, Sulu-Gambari F, Lebreton L (2020) A near-synoptic survey of ocean microplastic concentration along an around-the-world sailing race. PLoS One 15(8):e0237264

Gola D, Tyagi PK, Arya A, Chauhan N, Agarwal M, Singh SK, Gola S (2021) The impact of microplastics on marine environment: a review. Environ Nanotechnol Monit Manage 16:100552. https://doi.org/10.1016/j.enmm.2021.100552

Hartmann NB, Hüffer T, Thompson RC, Hassellöv M, Verschoor A, Daugaard AE, Rist S, Karlsson T, Brennholt N, Cole M, Herrling MP, Hess MC, Ivleva NP, Lusher AL, Wagner M (2019) Are we speaking the same language? Recommendations for a definition and categorization framework for plasticdebris. Environ Sci Technol 53:1039–1047

Hasnine MT, Anik AH, Alam M, Yuan Q (2024) Navigating microplastic challenges: separation and detection strategies in wastewater treatment. In: Microplastics pollution and its remediation. Springer, Singapore, pp 1–23

Helm PA (2017) Improving microplastics source apportionment: a role for microplastic morphology and taxonomy? Anal Methods 9(9):1328–1331

Horton AA et al (2024) Microplastics in terrestrial ecosystems: a review of their occurrence and impacts. Environ Sci Technol 58(3):1456–1465

Hossain KB, Islam MS (2023) Spatiotemporal distribution of microplastic debris in the surface beach sediment of the southeastern coastal region of Bangladesh. Heliyon 9(11):e12820. https://doi.org/10.1016/j.heliyon.2023.e12820

Hossain R, Ghose A, Sahajwalla V (2025) Microplastics pollution in Australia: sources, impacts, and mitigation strategies. Case Stud Chem Environ Eng 11:101036

Huang D, Li X, Ouyang Z, Zhao X, Wu R, Zhang C, Lin C, Li Y, Guo X (2021) The occurrence and abundance of microplastics in surface water and sediment of the West River downstream, in the south of China. Sci Total Environ 756:143857. https://doi.org/10.1016/j.scitotenv.2020.143857

Hwang J, Choi D, Han S, Choi J, Hong J (2019) An assessment of the toxicity of polypropylene microplastics in human derived cells. Sci Total Environ 684:657–669

Islam MS, Islam Z, Hasan MR (2022) Pervasiveness and characteristics of microplastics in surface water and sediment of the Buriganga River, Bangladesh. Chemosphere 307:135945. https://doi.org/10.1016/j.chemosphere.2022.135945

Jiang C, Yin L, Li Z, Wen X, Luo X, Hu S, Yang H, Long Y, Deng B, Huang L, Liu Y (2019) Microplastic pollution in the rivers of the Tibet plateau. Environ Pollut 249:91–98. https://doi.org/10.1016/j.envpol.2019.03.022

Johnson M, Lee S (2022) Rocky shoreline microplastic concentrations in Australia. Environ Sci J 45:98–105

Kalčíková G, Gotvajn AŽ, Kladnik A, Jemec A (2017) Impact of polyethylene microbeads on the floating freshwater plant duckweed Lemna minor. Environ Pollut 230:1108–1115. https://doi.org/10.1016/j.envpol.2017.07.050

Kim SA, Kim L, Kim TH, An YJ (2022) Assessing the size-dependent effects of microplastics on zebrafish larvae through fish lateral line system and gut damage. Mar Pollut Bull 185:114279. https://doi.org/10.1016/j.marpolbul.2022.114279

Klein S, Worch E, Knepper TP (2015) Occurrence and spatial distribution of microplastics in river shore sediments of the Rhine-main area in Germany. Environ Sci Technol 49:6070–6076. https://doi.org/10.1021/acs.est.5b00492

Kumar R, Sinha R, Rakib MRJ, Padha S, Ivy N, Bhattacharya S, Sharma P (2022) Microplastics pollution load in Sundarban delta of Bay of Bengal. J Hazard Mater Adv 7:100099

Le Monde A (2024) Le "continent de plastique" du Pacifique Nord accumule de plus en plus de déchets

Liu F, Olesen KB, Borregaard AR, Vollertsen J (2019) Microplastics in urban and highway stormwater retention ponds. Sci Total Environ 671:992–1000

Luo W, Su L, Craig NJ, Du F, Wu C, Shi H (2019) Comparison of microplastic pollution in different water bodies from urban creeks to coastal waters. Environ Pollut 246:174–182. https://doi.org/10.1016/j.envpol.2018.11.081

Luqman M, Shahid T, Awan MUF, Kashif SUR, Arooj F, Awan AR (2023) Quantification and characterization of microplastics (MPs) pollution in peri-uburban agricultural lands of Lahore, Pakistan. PLoS One 18(10):e0291760

Mai L, Bao LJ, Wong CS, Zeng EY (2024) Microplastics in the terrestrial environment. In: Microplastic contamination in aquatic environments. Elsevier, Amsterdam, pp 229–247

Mandal A, Ghosh AR (2025) Impact of microplastics in terrestrial food production. In: Microplastics in the terrestrial environment. CRC Press, Boca Raton, pp 121–135

Marn N, Jusup M, Kooijman SA, Klanjscek T (2020) Quantifying impacts of plastic debris on marine wildlife identifies ecological breakpoints. Ecol Lett 23(10):1479–1487. https://doi.org/10.1111/ele.13574

McAlpine KJ (2017) Microplastics found on seagrass in Belize. Marine Pollut Bull

Megha KB, Anvitha D, Parvathi S, Neeraj A, Sonia J, Mohanan PV (2025) Environmental impact of microplastics and potential health hazards. Crit Rev Biotechnol 45(1):97–127

Moffett J (2024) The disproportionate burden on vulnerable communities in the trade of plastic waste: how environmental justice should be integrated into the United Nations treaty on plastic pollution. UC Law Environ J 30(2):177

Monclús L, McCann Smith E, Ciesielski TM, Wagner M, Jaspers VL (2022) Microplastic ingestion induces size-specific effects in Japanese quail. Environ Sci Technol 56(22):15902–15911. https://doi.org/10.1021/acs.est.2c03878

Mongabay News (2023) Microplastics pose risk to ocean plankton, climate, other key Earth systems

Monira S, Roychand R, Bhuiyan MA, Hai FI, Pramanik BK (2022) Identification, classification and quantification of microplastics in road dust and stormwater. Chemosphere 299:134389

Myszka R, Enfrin M, Giustozzi F (2023) Microplastics in road dust: a practical guide for identification and characterisation. Chemosphere 315:137757

Nabi I, Zaheer M, Jin W, Yang L (2023) Biodegradation of macro-and micro-plastics in environment: a review on mechanism, toxicity, and future perspectives. Sci Total Environ 858:160108

Nafea TH, Chan FKS, Xu H, Wang C, Xiao H, He J (2024) Status of management and mitigation of microplastic pollution. Crit Rev Environ Sci Technol 54:1734–1756

Oni BA, Agu KC, Sanni SE (2025) Occurrence of microplastics in terrestrial habitats: hazards and pollution abatement. Pedosphere 1:10

Patchaiyappan A, Dowarah K, Zaki Ahmed S, Prabakaran M, Jayakumar S, Thirunavukkarasu C, Devipriya SP (2021) Prevalence and characteristics of microplastics present in the street dust collected from Chennai metropolitan city, India. Chemosphere 269:128757. https://doi.org/10.1016/j.chemosphere.2020.128757

Patel AK, Bhagat C, Taki K, Kumar M (2020) Microplastic vulnerability in the sediments of the Sabarmati River of India. In: Kumar M, Munoz-Arriola F, Furumai H, Chaminda T (eds) Resilience, response, and risk in water systems: shifting management and natural forcings paradigms. Springer, Singapore, pp 127–138. https://doi.org/10.1007/978-981-15-4668-6_7

Pereira LG, Ferreira GV, Justino AK, de Oliveira KMT, de Queiroz MT, Schmidt N, Lucena-Frédou F (2023) Exploring microplastic contamination in Guiana dolphins (Sotalia guianensis): insights into plastic pollution in the southwestern tropical Atlantic. Mar Pollut Bull 194:115407

Pramanik BK, Pramanik SK, Monira S (2021) Understanding the fragmentation of microplastics into nanoplastics and removal of nano/microplastics from wastewater using membrane, air flotation and nano-ferrofluid processes. Chemosphere 282:131053

Praet E, Schofield J, Tamoria RM (2024) Archaeological approaches to plastics and plastic pollution: a critical overview. Cambridge Prisms Plast 2:e32

Praveena SM (2024) Exploring public awareness, influencing factors and policy implications towards microplastic pollution: perspectives from Malaysia. Mar Policy 161:106042

Rabin MH, Wang Q, Enyoh CE, Xiao K, Sheuty TF (2023) Distribution, potential sources, and health risk of microplastics (MPs) in street dust during and after COVID-19 lockdown in Bangladesh. Environments 10:130. https://doi.org/10.3390/environments10070130

Rahman MM, Toha M, Sikder S (2024) Micro/nanoplastics pollution from landfill sites: a comprehensive review on the formation, distribution, effects and potential mitigation. In: Technical landfills and waste management: volume 1: landfill impacts, characterization and valorisation. Springer, Cham, pp 247–265

Rai R, Ranjan R, Kant C, Dhar P (2024) Microplastic and adhesive free, multifunctional, circular economy approach-based biomass-derived drinking straws. Iscience 27(5):109630

Rajan A, Vijayalakshmi S (2023) New insights on polyurethane biodegradation. AIP Conf Proc 2690:020060

Rani R, Malik S, Kumar D, Kumar R, Mukherjee S, Saharan BS, Duhan JS (2024) Advances in the role of microorganisms, waste management strategies and policies on microplastic abatement in the era of bio-circular economy. Sustain Chem Pharm 39:101595

Reuters (2024) Easter Island battles global vortex of plastic waste. https://energynews.oedigital.com/pollution/2024/10/16/easter-island-battles-global-vortex-of-plastic-waste

Ribó M, Watson SJ, Novikova NI, Deppeler S, Seabrook S, Hale R, Strachan LJ (2023) Microplastic pollution distribution: differences between marine reserves and urbanised areas. Cont Shelf Res 267:105115

Rochman CM, Brookson C, Bikker J, Djuric N, Earn A, Bucci K, Athey S, Huntington A, McIlwraith H, Munno KJ (2019) Rethinking microplastics as a diverse contaminant suite. Environ Toxicol Chem 38(4):703–711

Rodrigues MO, Abrantes N, Gonçalves FJM, Nogueira H, Marques JC, Gonçalves AMM (2018) Spatial and temporal distribution of microplastics in water and sediments of a freshwater system (Antuã River, Portugal). Sci Total Environ 633:1549–1559. https://doi.org/10.1016/j.scitotenv.2018.03.233

Sa'adu I, Farsang A (2023) Plastic contamination in agricultural soils: a review. Environ Sci Eur 35(1):13

Sekudewicz I, Dąbrowska AM, Syczewski MD (2021) Microplastic pollution in surface water and sediments in the urban section of the Vistula River (Poland). Sci Total Environ 762:143111. https://doi.org/10.1016/j.scitotenv.2020.143111

Seta AS, Müller L, Tavella R, da Silva Júnior FMR, Pedrosa V, Romano LA, Wasielesky W, Josende ME, Ventura-Lima J (2023) Oxidative effects of consuming microplastics in different tissues of white shrimp Litopenaeus vannamei. Mar Pollut Bull 193:115137. https://doi.org/10.1016/j.marpolbul.2023.115137

Sharma H, Neelam DK (2023) Understanding challenges associated with plastic and bacterial approach toward plastic degradation. J Basic Microbiol 63:292–307

Shen M, Xiong W, Song B, Zhou C, Almatrafi E, Zeng G, Zhang Y (2022) Microplastics in landfill and leachate: occurrence, environmental behavior and removal strategies. Chemosphere 305:135325

Shiu RF, Vazquez CI, Chiang CY, Chiu MH, Chen CS, Ni CW, Gong GC, Quigg A, Santschi PH, Chin WC (2020) Nano-and microplastics trigger secretion of protein-rich extracellular polymeric substances from phytoplankton. Sci Total Environ 748:141469. https://doi.org/10.1016/j.scitotenv.2020.141469

Shruti VC, Kutralam-Muniasamy G (2019) Bioplastics: missing link in the era of microplastics. Sci Total Environ 697:134139

Sikder S, Toha M, Mostafizur Rahman M (2024) An overview on municipal solid waste characteristics and its impacts on environment and human health. In: Technical landfills and waste management: volume 1: landfill impacts, characterization and valorisation. Springer, Cham, pp 135–155

Smith J, Doe A, Lee P (2023) Microplastic pollution on sandy beaches of the USA. Mar Pollut Bull 180:112–119

Tanaka S, Yukioka S, Li W, Imbulana S, Oluwoye I, Shim W, Takada H (2024) Towards a North Pacific Ocean long-term monitoring program for plastic pollution: a review of global occurrence of microplastics in the sea and deep-sea sediments. J Water Environ Technol 22(5):205–219

Tariq M, Iqbal B, Khan I, Khan AR, Jho EH, Salam A, Du D (2024) Microplastic contamination in the agricultural soil—mitigation strategies, heavy metals contamination, and impact on human health: a review. Plant Cell Rep 43(3):65

Tehrani RS, Yang X, van Dam J (2025) Rainfall-induced microplastic fate and transport in unsaturated Dutch soils. J Contam Hydrol 268:104456

Thacharodi A, Meenatchi R, Hassan S, Hussain N, Bhat MA, Arockiaraj J, Pugazhendhi A (2024) Microplastics in the environment: a critical overview on its fate, toxicity, implications, management, and bioremediation strategies. J Environ Manag 349:119433

Toha M, Rahman RR, Sikder S, Akbor MA, Rahman MM (2024a) Tracking of microplastics distribution patterns and their characterisation in deposited road dust from Dhaka city, Bangladesh. Emerg Contaminants 10:100381

Toha M, Sikder S, Rahman MM, Muhib MI (2024b) Microplastic occurrences in freshwater fish of Bangladesh

Toha M, Rayhan MRI, Rahman MM (2024c) Microplastics in terrestrial ecosystems. In: Microplastics pollution and its remediation. Springer, Singapore, pp 85–114

Townsend KR, Lu H-C, Sharley DJ, Pettigrove V (2019) Associations between microplastic pollution and land use in urban wetland sediments. Environ Sci Pollut Res Int 26(22):22551–22561

Viel T, Manfra L, Zupo V, Libralato G, Cocca M, Costantini M (2023) Biodegradation of plastics induced by marine organisms: future perspectives for bioremediation approaches. Polymers 15:2673

Wang J, Peng J, Tan Z, Gao Y, Zhan Z, Chen Q, Cai L (2017) Microplastics in the surface sediments from the Beijiang River littoral zone: composition, abundance, surface textures and interaction with heavy metals. Chemosphere 171:248–258. https://doi.org/10.1016/j.chemosphere.2016.12.074

Wang Y, Bai J, Wen L, Wang W, Zhang L, Liu Z, Liu H (2023) Phytotoxicity of microplastics to the floating plant Spirodela polyrhiza (L.): plant functional traits and metabolomics. Environ Pollut 322:121199. https://doi.org/10.1016/j.envpol.2023.121199

Wu WM, Yang J, Criddle CS (2017) Microplastics pollution and reduction strategies. Front Environ Sci Eng 11:1–4

Xiang P, Zhang T, Wu Q, Li Q (2023) Systematic review of degradation processes for microplastics: progress and prospects. Sustain For 15:12698

Xiang Y, Yao B, Peñuelas J, Sardans J, Nizzetto L, Li R, Li Y (2025) Microplastic effects on soil nitrogen cycling enzymes: A global meta-analysis of environmental and edaphic factors. J Hazard Mater 484:136677

Xiong X, Wu C, Elser JJ, Mei Z, Hao Y (2019) Occurrence and fate of microplastic debris in middle and lower reaches of the Yangtze River—from inland to the sea. Sci Total Environ 659:66–73. https://doi.org/10.1016/j.scitotenv.2018.12.313

Yang X, Zhou Y, Li Y, Wang S (2024) International law and regulation of marine microplastics: current situation, problems, and development. Sustain For 16(21):9337

Yu H, Zhang X, Hu J, Peng J, Qu J (2020) Ecotoxicity of polystyrene microplastics to submerged carnivorous Utricularia vulgaris plants in freshwater ecosystems. Environ Pollut 265:114830. https://doi.org/10.1016/j.envpol.2020.114830

Zeb A, Liu W, Ali N, Shi R, Wang Q, Wang J, Liu J (2024) Microplastic pollution in terrestrial ecosystems: global implications and sustainable solutions. J Hazard Mater 461:132636

Zhai Y, Bai J, Chang P, Liu Z, Wang Y, Liu G, Vijver MG (2024) Microplastics in terrestrial ecosystem: exploring the menace to the soil-plant-microbe interactions. TrAC Trends Anal Chem 174:117667

Zhang L, Liu J, Xie Y, Zhong S, Yang B, Lu D, Zhong Q (2020a) Distribution of microplastics in surface water and sediments of Qin river in Beibu Gulf, China. Sci Total Environ 708:135176. https://doi.org/10.1016/j.scitotenv.2019.135176

Zhang Y, Kang S, Allen S, Allen D, Gao T, Sillanpää M (2020b) Atmospheric microplastics: a review on current status and perspectives. Earth Rev 203:103118

Zhang C, Zuo Z, Wang Q, Wang S, Lv L, Zou J (2022) Size effects of microplastics on embryos and observation of toxicity kinetics in larvae of grass carp (Ctenopharyngodon idella). Toxics 10(2):76. https://doi.org/10.3390/toxics10020076

Zhang D, Li J, Sun C, Manullang CY, Yin J, Cao W, Jiang F (2025) Interface adsorption characteristics of microplastics on multiple morphological arsenic compounds. J Hazard Mater 488:137234

Zhu Z, Parker WJ (2025) Implementing differential recovery corrections enhances accuracy of mass balances on microplastics in wastewater treatment. Water Res 271:122912

Ziajahromi S, Drapper D, Hornbuckle A, Rintoul L, Leusch FDL (2020) Microplastic pollution in a stormwater floating treatment wetland: detection of Tyre particles in sediment. Sci Total Environ 713:136356. https://doi.org/10.1016/j.scitotenv.2019.136356

Unseen Predators: Marine Life and the Plastic Diet

5

Ana Isabel Hernandez-Soriano, Arely Areanely Cruz-Salas, Sheila Liliana Dafne Lobato-Rocha, Aylin Geraldine Rodríguez-Villa, Natalia Paulina Martínez-Toledo, and Juan Carlos Alvarez-Zeferino

Abstract

Plastic pollution in marine ecosystems has become an urgent environmental issue, affecting different marine organisms and their habitats. This chapter reviews the occurrence and effects reported of plastic and microplastic ingestion across marine species, highlighting significant findings from in situ and laboratory studies. Ingestion rates varied significantly among species and conditions; for marine birds, fish, and mammals, ingestion percentages varied widely, with some species recording the absence of microplastics and others reaching 100%. Cephalopods exhibited the highest ingestion rates, ranging from 93% to 100%. Notably, of the articles reviewed, only studies performed under controlled laboratory conditions reported adverse effects, and all reported effects were linked to cellular and internal damage. Mitigation strategies are essential for addressing the increasing problem of marine litter and include regulatory measures, technological innovations, public awareness campaigns, and international collaboration. Effective actions must be taken at local and global levels to protect marine biodiversity and ecosystems from the pervasive threat of plastic pollution. This comprehensive approach aims to reduce plastic waste and its impacts on marine

A. I. Hernandez-Soriano (✉) · A. A. Cruz-Salas · A. G. Rodríguez-Villa · N. P. Martínez-Toledo
Universidad Autónoma Metropolitana, Mexico City, Mexico
e-mail: al2173003535@azc.uam.mx

S. L. D. Lobato-Rocha
Universidad Autónoma de Baja California, Campus Mexicali, Mexicali, BC, Mexico

J. C. Alvarez-Zeferino
Universidad Autonoma Metropolitana, Alcaldia Azcapotzalco Ciudad de Mexico, Mexico City, Mexico

Secretaría de la Defensa Nacional, Universidad del Ejército y Fuerza Aérea. Escuela Militar de Ingeniería, Mexico City, Mexico

M. Oves et al. (eds.), *Marine Microplastics and the Quest for Remediation*,
https://doi.org/10.1007/978-981-96-8979-8_5

life, emphasizing the need for coordinated efforts in tackling this critical environmental challenge.

Keywords

Plastics · Marine pollution · Marine biodiversity · Plastic ingestion · Microplastics

5.1 Introduction

Recent years have revealed that plastic waste is a pervasive pollutant affecting the globe, attributed to the existing demand for plastics. Several studies have stated the negative impact on ecosystems, the economy, and people; the primary focus has been on the marine environment (Lebreton and Andrady 2019; Harris et al. 2021a). It is estimated that about 80% of plastics are of terrestrial origin, and 20% are associated with marine activities (Morales-Caselles et al. 2021), such as fishing vessels, recreational boats, and cruise ships (Galgani et al. 2013). In coastal areas, pollution of marine environments is influenced by river runoff, fishing activities, and tourism, with noticeable seasonal and temporal variations (Angiolillo et al. 2021; Pierdomenico et al. 2018), as well as high-energy events (Sheavly and Register 2007).

Plastic waste, once in the environment, can move between ecosystems; it has been observed in different marine environments such as surface water, beaches, estuaries, and deep parts of the oceans; for example, it is estimated that poorly managed land waste contributes with around 1.2 and 12.7 million tons of plastic that reaches the oceans, these are presented in plastic bottles, plastic sheets and fragments (Weiss et al. 2021; Yose et al. 2023) formed by fragmentation or degradation processes through UV light, pH and temperature conditions (Kalogerakis et al. 2017); 90% of plastic waste is composed of common polymers, including high-density polyethylene (HDPE), low-density polyethylene (LDPE), polystyrene (PS), polypropylene (PP), polyvinyl chloride (PVC), and polyethylene terephthalate (PET) (Andrady and Neal 2009).

Rivers are considered the most significant contributors of plastic waste to seas and oceans (Meijer et al. 2021). It is estimated that the 20 most polluted rivers account for two-thirds of all marine debris between May and October (Lebreton et al. 2017). The presence of this waste in rivers (river environment and sediment deposit) (Nyberg et al. 2023) is due to particle size, ecological processes, dams, hydrodynamics, and sedimentation (Waldschläger et al. 2022). Likewise, plastic pollution has effects on the mangrove forests because they are efficient particle traps (Martin et al. 2019); aerial roots for root oxygenation are vulnerable to asphyxiation by sediments and waste (van Bijsterveldt et al. 2021), generating an adverse impact on temperatures and, with it, reproductive failures and reduced growth rates (Mohamed Nor and Obbard 2014), it has been observed that the most common waste in mangroves are plastic bags (van Bijsterveldt et al. 2021; Kantharajan et al. 2018).

Similarly, plastic waste in lagoons makes it difficult for residents and migratory birds to feed. Likewise, some waste produce noise due to the action of the wind, disturbing birds and mammals in the area (Sahu et al. 2013). Lagoons are considered necessary due to their esthetic appeal and the lagoon–ocean interaction, where waste accumulate and are mobilized in high-energy events (Nukpezah et al. 2022). When plastics reach the oceans, they can be dispersed among ocean currents, with areas with greater concentration being the five oceanic gyres of each ocean basin (Cannon et al. 2016).

When plastic waste reaches the environment, the properties that allow persistence can have adverse consequences. This includes processes like the absorption of contaminants and the leaching of chemicals such as phthalates, colorants, and UV stabilizers, which are added during the manufacturing process to achieve specific properties. Due to their weight, they are capable of being transported over long distances, functioning as transports of plastic additives absorbed from the environment (Sühring et al. 2022; Wiesinger et al. 2021), and altering environmental matrices, including fauna (Bergmann et al. 2015). Plastic waste are considered a passive sampler due to its accumulation of hydrophobic pollutants; these undergo marine biofouling, where organisms such as fungi, algae, and eukaryotic taxa favor the development of the plastisphere (Nava and Leoni 2021; Pasqualini et al. 2023).

Several studies have determined that plastic waste adversely affect habitats; plastic waste cause physical damage to animals by becoming entangled or ingested, causing lacerations, suffocation, and starvation, all of which can cause blockages in the digestive tract, leading to false satiety (Schmaltz et al. 2020). The frequency in which the entanglement of birds in plastic waste occurs is greater than the ingestion of litter; most of the time, the birds suffer injuries, mobility issues, and drowning (Bergmann et al. 2015), and a portion of the birds that suffer entanglement are not detected because they die far from land (Laist 1997).

Plastics introduce endocrine-disrupting chemicals in birds, affecting the development of the neurological, cardiovascular, reproductive, metabolic, and immunological systems (Roman et al. 2019). Thinning of the eggshell, deformities in offspring, and death have also been determined in embryos (Giesy et al. 2003). The ingestion of plastics can occur in three ways: primary ingestion occurs when plastics are confused with something edible, usually in young animals. The ingestion of an animal that has consumed plastic is known as secondary ingestion, with a greater probability in predatory species, for example, birds, when consuming fish. Finally, accidental ingestion occurs when an animal looks for food, sucking or swallowing water and sediment (Bucci et al. 2020).

In fish, ingestion causes internal blockages, altered digestion, and contamination in tissues of persistent organic pollutants, changes in behavior, decreased activities, damage to liver functions and brain damage (Savoca et al. 2021; Jackson et al. 2000), and endocrine disruption (Cannon et al. 2016). In marine mammals such as dolphins, fur seals, otariids, and belugas, it can cause a decrease in energy, feeding capacity, and reproductive effects (Nabi et al. 2022). Likewise, it can alter kidney histomorphology, inflammation, and biomarkers. However, other effects observed are oxidative stress, cytotoxicity, and genes (Ugwu et al. 2021a; Meaza et al. 2021). It has been observed that marine or coastal turtles feed pelagically or benthically, so there is a difference in the probability of ingesting plastic waste. The study of the ingestion of plastic waste in live turtles is evaluated by analyzing feces or washing (Schuyler et al. 2014).

In this chapter, articles were reviewed to analyze the most current trends in research on marine organisms in the laboratory and in situ.

5.2 Sources and Types of Marine Plastic Pollution

At a global scale, estimates of plastic pollutants in bodies of water, such as rivers, lakes, and oceans, have ranged from 9 to 23 million tons annually, with projections that these figures will double by 2025 if current trends continue without making significant changes in lifestyles (MacLeod et al. 2021). Plastic pollution can be classified by its size; mesoplastics and macroplastics correspond to the bigger pieces, and their measurements depend on the author. Nevertheless, they can range from around 5 mm to 2.5 cm, and be larger than 2.5 cm in size, respectively (van Emmerik 2021). When plastics reach the environment, they can degrade through five mechanisms: photodegradation (sunlight), thermo oxidative degradation (high temperatures), hydrolytic degradation (contact with an aqueous environment), and, to a lesser extent, biodegradation (Webb et al. 2012), besides mechanical degradation (wind, waves, rough surfaces) (Castañeta et al. 2020). Microplastics (MP), plastics with a size smaller than 5 mm, can also pollute the environment. They can be classified as primary when they are intentionally made into that size for uses such as air-blasting media, facial cleansers and cosmetics, and medications, and as secondary when they originate from the fragmentation of larger plastics (Li et al. 2016). Breakage can be such that they can even become nanoplastics (1 nm to 1 μm) (Lai et al. 2022). In marine ecosystems, there are primarily two sources of plastic pollution: terrestrial sources, which refer to waste generated inland, and marine sources, which originate from human activities carried out offshore. Plastic pollution in aquatic systems comes mainly from terrestrial sources, responsible for 75–80% of this waste in marine ecosystems (Coyle et al. 2020; Andrady 2011). It originates from poor solid waste management, urban activities, domestic activities, industrial activities, tourism, agriculture, and other economic activities that are later transported through stormwater runoff, wind dispersion, littering, and waterways (Schmidt et al. 2017; Geyer et al. 2017). The remaining percentage of pollution is ocean-based, through recreational, commercial, military, and investigative activities (Pawar et al. 2016).

5.2.1 Land-Based Sources

Until 2015, 79% of plastic waste had been sent to landfill sites or freely disposed of in the environment, 9% was recycled, and only 12% was incinerated (Geyer et al. 2017). In landfills, polymers can take decades to degrade, and in doing so, they can generate MP, which can be transported to other sites through leachates and the wind, especially when the facilities are old or lack the necessary environmental protection measures (Wojnowska-Baryła et al. 2022). Landfills are essential sources of MP, considering that some studies have found concentrations in the range of

20,000–91,000 items/kg of landfill refuse, which are higher than those of sediments and soils (Golwala et al. 2021). While plastic incineration can reduce their mass and volume, after the process, unburned plastic may persist in the bottom ash or fly ash, which can be a risk of atmospheric pollution if dust collectors are absent (Shen et al. 2021) or can be at risk of being liberated into the environment when ashes are reused in activities like agriculture or construction (Yang et al. 2021) or poorly disposed of.

On the other hand, even though recycling may help reintegrate plastics into the production chain while avoiding incineration emissions and preventing their disposal in landfills, recycling processes involve plastic milling through friction, abrasion, or other similar methods, and their palletization, which implies that microplastics can be generated and potentially liberated through wash water that recycling facilities later discharge, particularly when they lack treatment or when the particles are relatively small (Brown et al. 2023). Composting is another recycling method for organic waste, but plastic waste often contaminates food waste. A review reported the presence of microplastics in compost within all studies, from 12 ± 8 to 46 ± 8 items/kg of dry-weight compost from garden waste to 300,000 items/kg of dry-weight compost from supermarket waste; thus, microplastics could be liberated into the environment through compost application (Porterfield et al. 2023). Wastewater is another critical source of microplastics due to its particle content originating from cosmetics and personal hygiene products, synthetic fibers like nylon and polyester (PEST) shed by clothing, and other bigger items, like single-use plastics, that reach the sewers and wastewater treatment plants (WWTPs) (Nafea et al. 2023), in which influents have presented concentrations from 0.28 ± 0.02 to 31,500 MP/L (Sheriff et al. 2023). WWTPs can remove these particles, with some facilities reporting 21.8% (Gao et al. 2022) and up to more than 99% efficiency (Okoffo et al. 2023). Nevertheless, treatment is deficient, with only 56% of the world's wastewater receiving treatment in 2020 (UN Habitat - United Nations Human Settlement Programme and WHO - World Health Organization 2021). Both treated and untreated water are discharged into the waterways (Raju et al. 2023) or used in crop irrigation. Furthermore, WWTP biosolids can be used in agriculture and constitute an additional source of microplastics (Radford et al. 2023). In the agriculture sector, plastics such as plastic covers, protection nets, pipelines, and irrigation tapes, polymer-covered fertilizers, growth plugs, and nursery pot trays are used, which are later generated or turned into waste. Transportation of microplastics to deeper grounds and groundwater is conditioned by their characteristics, soil properties, and plastic–particle interactions. Nonetheless, wind can also transport these particles, mainly in eroded soils that facilitate their propagation to water bodies (Hofmann et al. 2023). Meanwhile, the industrial sector can originate plastic pollutants when disposing of their waste inadequately or during the processes of charge, discharge, and transportation of goods, as has happened with accidentally released pellets during processing, transportation, and handling, becoming practically ubiquitous in the ocean, sediments, and beaches (Pawar et al. 2016). Road dust contains considerable amounts of MP, composed of tire particles, polymers used to modify bitumen and road marking paint, and fragments of road debris, like plastic containers, asphalt, and construction materials (Myszka et al. 2023). This dust can travel to the ocean through stormwater runoff (Monira et al. 2022). Coastal zones are usually densely populated, urbanized, and industrialized; thus, their plastic waste can be directly disposed of on the coast (Thushari and Senevirathna 2020). The tourism and hotel industries, particularly, could contribute to important pollutant concentrations because of their proximity to the coast and the wide use of disposable plastics, such as bags, food wraps, packages, cutlery, and beverage containers (Oceana 2023).

5.2.2 Ocean-Based Sources

About 20% of plastics in the ocean originate from activities within itself. It is considered that the majority, 18%, is due to fishing (Andrady 2011), generated when fishermen fail to recover their fishing gear, which can be comprised of ropes, lines, nets, holding straps, bate boxes, bags, and floats from gillnets or trawls, in addition to waste that can be thrown overboard (Pawar et al. 2016). The remaining plastic waste (2%) can have diverse origins (Andrady 2011). An estimated 6.5 million tons of plastic are disposed of in the sea yearly due to loss of goods or waste generation (Coyle et al. 2020). Recreational boats, cruise ships, and military and investigation vessels can generate domestic waste, such as bags, food containers, and fishing gear (Pawar et al. 2016; Kotrikla et al. 2021). Offshore infrastructure is also an important source of plastic waste in seas, as synthetic polymers and plastics are commonly used for protecting structures against corrosion, as is the case for oil and gas infrastructure and renewable energy platforms (Oluwoye et al. 2023). Moreover, activities within these platforms generate items that can be released accidentally or deliberately into the ocean, namely hard hats, gloves, storage drums, and other types of waste (Pawar et al. 2016).

5.3 Plastic Ingestion by Marine Organisms

As explained, marine litter originates from various sources, and despite international regulations, monitoring and controlling waste discharge into the ocean is challenging. Most waste entering the sea is plastic, including microplastics, which constitute an increasing proportion of marine debris (FAO 2017). According to the FAO (2017), they have been documented in all open ocean habitats and enclosed seas. Major regions worldwide, including the East China Sea, the Yellow Sea, and the Mediterranean, have already exceeded levels of plastic contamination, reaching points where significant ecological risks can arise. In the coming years, several other regions will also cross these thresholds. The ubiquity of these contaminants makes their interaction with aquatic organisms across several trophic levels inevitable. Some studies have investigated the mechanisms of biota-microplastics interactions in laboratory settings, such as adhesion to external appendages and microplastic adsorption through ingestion, which is likely the most frequent interaction between organisms and microplastics (FAO 2017; Cole et al. 2015). Organism exposure to plastics and microplastics can occur through different routes; for instance, microplastics present in water cause direct exposure to organisms, while those ingested by other prey organisms become an indirect source of microplastic contamination for predators through the food chain (Kühn and Van Franeker 2020). In other words, direct entry of plastics into organisms can happen through filter feeding, respiratory ingestion, and ingestion while foraging. Indirect consumption pathways include trophic transfer, where a predator eats an organism that has consumed and retained plastic beforehand. Additionally, plastics can adhere to the external surfaces or gills of organisms, or organisms like phytoplankton can attach to the plastic's surface, creating a "plastisphere" that makes the plastic an appealing food source (Tuuri and Leterme 2023). It is also reported that microplastics in the ocean can acquire the smell of dimethyl sulfide, a compound many marine organisms associate with food presence. This odor makes microplastics attractive to these organisms, leading them to ingest them by mistake (FAO 2017; Savoca et al. 2016). Animals have been ingesting plastic for a long time. The first records of these events were made in the 1960s in New Zealand, where pellets were found in the stomachs of seabirds, such as those in the Pachyptila genus (Weideman et al. 2020; Harper and Fowler 1985). Additionally, plastic container caps and toys were discovered in deceased Laysan albatross chicks. These discoveries occurred over two decades before the Great Pacific Garbage Patch was identified (NOAA 2023). Laboratory studies have demonstrated that many marine organisms, spanning all trophic levels, can absorb or ingest microplastics. Controlled laboratory exposures enable the tracking of the absorption, movement, and distribution of these synthetic particles within entire organisms and specific tissues such as gills, intestines, and liver (FAO 2017; Kühn and Van Franeker 2020; NOAA 2023). It is essential to understand the degree of plastic ingestion by organisms to monitor and determine the damage suffered, both at the individual and population levels, and ultimately its impact on the ecological functioning of the species (Alomar et al. 2020).

New species that not only ingest plastics but also become trapped when entangled in them are frequently discovered. Recently, Kühn and Van Franeker (2020) conducted a review that identified over 700 species, including seabirds, fish, turtles, and marine mammals, as confirmed plastic consumers. This number is anticipated to rise over time (NOAA 2023).

Likewise, Alomar et al. (2020) reported that plankton, fish, birds, shellfish, sea turtles, and marine mammals from various ocean depths and regions worldwide have been confirmed to ingest plastic debris. Therefore, a few of the findings on some species and groups of marine animals that have ingested plastic waste, including microplastics, are mentioned below.

5.3.1 Seabirds

Most publications have focused on plastic ingestion by seabirds. Kühn and Van Franeker (2020)) published a review in which information was compiled from studies involving 226 species of seabirds of the 409 known species, of which plastic ingestion was reported in 180 species. In total, 43,525 individual samples of seabirds were analyzed, and plastic waste was found in 27.7% of the samples (12,065 individuals). Among seabird species, fulmars, albatrosses, and shearwaters are significantly impacted by marine litter that can become lodged in their gizzards, which are specialized stomach compartments that grind food, thus hindering their passage through the digestive system.

Nono Almeida et al. (2023) investigated the presence of plastics in 143 regurgitated pellets (balls composed of indigestible materials that certain birds regurgitate after feeding) by Larus michahellis, 284 plastic items were found in total. With an average of 2.01 ± 1.68 per pellet.

The species *Fulmarus glacialis* has also been studied. Collard et al. (2022a) in 2022 researched the plastic content ingested by these seabirds. The birds were obtained in Kongsfjorden, Svalbard, in 1997. They were kept frozen until April 2021. The stomach contents for each bird were analyzed, finding an average value of 0.070 g with a standard error of 0.013 per individual. In addition, an examination of 107 Procellariiformes stomachs from the Uruguayan coast revealed that 4.17% of albatrosses and 62.86% of petrels had ingested plastics (Muñoz et al. 2023).

Nam et al. (2021) studied 387 individuals of 11 species of seabirds from the Korean Peninsula. *Swinhoe's petrels* from the family Procellariiformes were found to have ingested 18.73 ± 13.41 pieces of plastic on average, while most other species had less than one piece per individual. Classifying by size, microplastics (1–5 mm) were the most prevalent, accounting for 87.2% (2615 pieces), followed by mesoplastics (5–20 mm) at 12.1% (364 pieces), and macroplastics (20–100 mm) at 0.6% (19 pieces). It is also noted that the size of plastic fragments ingested by seabirds, particularly petrels, correlates with body size, with smaller species tending to consume larger quantities. The significant presence of plastics in this study, especially among petrels, underscores the critical role of surface waste availability in the ocean.

In the cited articles, the morphology of the plastics reported were fragments, thread-like type, foam, industrial pellets and sheets, and the polymers polyethylene (PE), polypropylene (PP), polystyrene (PS), polyethylene terephthalate (PET), polyvinyl chloride (PVC), polyamide (PA) and in general, user plastics (Kühn and Van Franeker 2020; Nono Almeida et al. 2023; Collard et al. 2022a; Muñoz et al. 2023; Nam et al. 2021).

5.3.2 Fish

Despite several studies on the ingestion of plastics in fish, out of the 20,000 marine fish species, it is estimated that only about 2% of them have been examined for plastic ingestion so far (Savoca et al. 2016). Some findings are mentioned below.

Chenet et al. (2021) assessed the presence of plastics in Atlantic horse mackerel (*Trachurus trachurus*), a commercial fish species that feeds primarily on euphausiid crustaceans. The contents of the gastrointestinal tract of the collected specimens were analyzed. The *T. trachurus* specimens were collected in the Strait of Sicily in April and May 2018; 53 were selected to evaluate the presence of plastic particles. Microplastics, including fragments and filaments, were detected in 90.6% of the specimens, with fragments present in 73.6% of the fish. Most of the filaments were dark in color. The fragments primarily consisted of PS, PET, and PP, whereas the filaments mainly comprised nylon 6, acrylic, and PEST.

While most research on plastic ingestion has centered on marine teleost fish, there has been less focus on elasmobranchs. A notable study by Trindade et al. examined the white skate *Potamotrygon leopoldi*, a threatened and endemic species of the Xingu River in the Amazon Basin. The gastrointestinal tracts of 24 manta rays were analyzed, revealing that 16 (66.6%) had ingested plastic particles. A total of 81 plastic particles were identified, including microplastics and mesoplastics, which were categorized into fibers and fragments, with blue being the predominant color.

The European catfish is a fish of excellent sporting and nutritional value, very common in the fish fauna of Slovakia and frequently found in plain rivers. Mičušík et al. (2021) found in one Silurian among 17 that were captured in the Bodrog River, a relatively large plastic waste, on which a chemical analysis was carried out with μFTIR spectroscopy. The plastic measured 7 cm in length and 3.5 cm in width, and its match with the LDPE reference database was approximately 98%.

On the other hand, Chen et al. (2021)) investigated the ingestion of plastic debris by marine fish in the coastal waters of the Hengchun Peninsula, Taiwan. Among the 117 fish specimens from 18 families analyzed, 94.87% had ingested microplastics, with an average of 5.6 ± 5.1 pieces per fish. The primary types of microplastics were fibers (96%) and blue-colored fragments (43%). The study found no significant correlation between the amount of ingested microplastics and the fish's trophic level.

In another study (Neto et al. 2020), 965 specimens of eight commercially exploited fish species from the southeast-south coast of Brazil were analyzed to evaluate the ingestion of plastic waste. It was found that all species ingested plastics, with pelagic species exhibiting higher intake and greater diversity, frequency, and sizes of plastic items compared to demersal-pelagic and demersal species. The pelagic skipjack tuna (*Katsuwonus pelamis*) had the highest frequency of plastic ingestion (25.8%), whereas the demersal fish *Prionotus punctatus* showed the lowest frequency (5%). No correlation was observed between the size of the individuals and the number of plastics ingested based on the available data. Additionally, microplastics were the predominant type ingested across all species. The main items found were fibers/filaments and fragments, likely from fishing materials. PA and polyurethane (PU) were the main polymers identified, and transparent, black, and blue were the most abundant colors.

5.3.2.1 Invertebrates

There has been an inclination toward studying microplastic ingestion in fish, while there is less interest and a lack of understanding regarding ingestion by invertebrates. Provided evidence of microplastic contamination in brachyuran crabs from the Thames Estuary, specifically the native shore crab (*Carcinus maenas*) and the Chinese mitten crab (*Eriocheir sinensis*). They examined the gills, gastric mill, and intestines of 94 *C. maenas* and 41 *E. sinensis*, finding 874 plastic particles. The plastics ranged in size from 34 μm to 34 mm. Among the crabs, 71.3% of *C. maenas* and 100% of *E. sinensis* contained at least one plastic particle. Fibers were the most common elements (78.5%).

Alomar et al. (2020)) investigated plastic ingestion in the edible mussel species *Mytilus galloprovincialis* in Mallorca within experimental aquaculture facilities and assessed plastic contamination in surrounding surface waters. They found that 94% of *Mytilus galloprovincialis* samples contained plastics, while all surface water samples (100%) were contaminated. The mussels primarily ingested fibers, with 97% composed of cellulose acetate. Pedà et al. (2022) claim to have conducted the first research on plastic ingestion in *Octopus vulgaris*. Sixty-two plastic fragments were recovered from six octopus specimens, ranging from 1 to 38 elements per individual. The average dimensions of ingested plastics were 1.56 ± 2.26 mm in length and 0.07 ± 0.11 mm in width. The most identified polymers in *O. vulgaris* were PET and PS, which comprised 68% of the plastics. More than 50% of the plastics were small microplastics, but large microplastics (37%) and mesoplastics (6%) were also found. The most common colors of the plastic ingested by the common octopus were transparent (50%) and red (19%). Additionally, some observed elements came from the fishing lines of bottom trap nets. Another species studied is *Bunodactis reynaud*, in which the ingestion of macro-litter by sand anemone captured on Muizenberg Beach in False Bay, South Africa, was investigated (Weideman et al. 2020). Monthly analyses were conducted from May 2015 to August 2019, during which a total of 491 pieces of ingested waste were collected from *B. reynaudi* (averaging 9.4 ± 14.9 items per month and 39.8 ± 71.5 g per month), with over 99% of the items being plastics. A correlation was observed between the number of ingested items and the abundance of stranded garbage. Ingestion rates peaked in the fall, coinciding with increased trash influx into the bay due to seasonal rains. Additionally, a preference for flexible plastics, such as bags and food packaging, was evident. Likewise, experimental feeding tests revealed that *B. reynaudi* preferentially selected pieces of HDPE bags that had been suspended in seawater for 2–20 days. This suggests biofilms may enhance flexible plastic's palatability (Weideman et al. 2020).

5.3.3 Marine Turtles

Regarding marine turtles, it has been confirmed that all seven existing species ingest marine debris, which poses a significant risk, as several of them are endangered. Marine turtles, like jellyfish, often mistake plastic sheets and bags for prey. These debris items get trapped in their digestive system, causing them to feel full, which leads to a loss of appetite, resulting in poor nutrition and hunger (NOAA 2023). Kühn and Franeker (Kühn and Van Franeker 2020) estimated the presence of plastic marine debris in turtles' stomachs, finding that around 32% of them have it. In the Sea of the Balearic Islands, a study investigated plastic ingestion in loggerhead turtles (*Caretta caretta*). The research (Solomando et al. 2022) involved necropsying 45 stranded dead *C. caretta* specimens and analyzing their digestive tract contents for plastic presence. Plastics were detected in 27 of these turtles (60.0%), with an average of 12.7 ± 4.7 plastic objects per individual. Additionally, fecal samples from 67 living turtles were examined, revealing the presence of plastic in 46 (68.7%), with an average of 9.7 ± 3.3 plastic items per individual. In total, 785 plastic items were identified, measured, weighed, and categorized, with plastic sheets being the predominant item found in 65.3% of the turtles analyzed (Solomando et al. 2022). In another research published in 2020 (Choi et al. 2021), plastic ingestion by green turtles (*Chelonia mydas*) was investigated over 33 years along the Texas coast, USA. The study examined the gastrointestinal tracts of 464 green turtles stranded between 1987 and 2019, of which 226 (48.7%) were found to have ingested plastic. The incidence of plastic ingestion doubled from 32.5% in the period 1987–1999 to 65.5% in 2019. Smaller turtles (<25 cm straight shell length) were observed to ingest plastic more frequently and in greater quantities than larger turtles. Despite this, no evidence of death directly related to plastic ingestion was found in any of the turtles. In the study by Moon et al. (2022), plastic ingestion by sea turtles in Korean waters between 2012 and 2018 was investigated. An analysis was conducted on the plastics ingested by a total of 34 turtles, including green turtles, loggerhead turtles, olive ridley turtles, and leatherback turtles. Ingestion frequency was 100% for green turtles, 81% for loggerhead turtles, 50% for olive ridley turtles, and 50% for leatherback turtles. The average amount of plastic ingested was 108 ± 253 mg/kg. The ingested waste was mainly films and fibers (>80%), with a predominance of light colors (white and transparent; 65%) and light polymers such as PE, PP, and PS. Observations revealed that green turtles predominantly consumed fibers (51%), including items like rope, thread, and net, whereas loggerhead turtles primarily ingested films (61%), such as plastic bags and packaging. The variations in the types and quantities of ingested waste between species are likely linked to their unique feeding behaviors and geographic movements.

5.3.4 Mammals

The review carried out by Kühn and Franeker (2020) reported that 69 marine mammals ingest waste. Marine mammals comprise 56% of all marine mammal species, including 44 species of odontocetes (toothed whales), seal species, and manatees. Due to their high level of protection, researching marine mammals is difficult, resulting in much of the research being conducted postmortem.

Another study (Foskolos et al. 2023) documents multiple sightings of sawtooth dolphins in different locations in the Mediterranean Sea and northwest Greece between 2016 and 2021. In addition, information is provided on the objects found in a serrated-toothed dolphin stranded on the island of Salamis, Greece, on July 23, 2021. Seven plastic bags, both whole and fragmented, along with nine squid-like plastic fishing lures, were discovered in this individual's stomach, leading to a total gastric blockage. The lures were an orange-brown color, almost identical to that of real live squid. All of this plastic waste was macroplastics (>20–100 mm), except for one bag of megaplastic (>100 mm). They were shiny and made of PVC. Im et al. (2020) documented the first instance of plastic ingestion by the vulnerable baleen whale species *Balaenoptera physalus* in the East Asian sea, highlighting the potential risks of plastic pollution to these whales. In 2019, a young deceased whale was discovered floating near Jeju Island, Republic of Korea. An examination of the whale's body revealed 45 plastic particles, including fishing lines, plastic filaments, fragments of fishing nets, and expanded polystyrene. Among the findings, the most significant was a fishing line measuring 1180 mm in length and 1.15 mm in thickness.

A review was also carried out by López-Martínez et al. (2023), in which the ingestion of plastic by two families of cetaceans was identified: Ziphiidae and Delphinidae. This study analyzed plastic found in the digestive contents of 623 stranded specimens, including two Cuvier's beaked whales and 14 common dolphins (*Stenella coeruleoalba*) in the Alboran Sea, in the western Mediterranean. Among stranded cetaceans, beaked whales showed the highest macro, meso, and microplastic concentrations in the western Pacific Ocean. Common dolphins had the highest plastic intake in Eastern Spain. Beaked whales, which dive deep, ingested more plastic than common dolphins, probably due to their feeding behavior and habitat. The study highlights conservation issues and demonstrates that cetacean hotspots are significantly affected by the ingestion of marine plastic.

5.4 Impacts on Marine Life

Past examples have made clear that plastic pollution has already infiltrated the marine food chain, severely impacting many of the world's most vital marine ecosystems, especially their productivity, such as mangroves and coral reefs. Additionally, nearly all groups of species in the oceans have been exposed to plastic pollution, and scientists estimate that there are adverse effects on almost 90% of the species studied for their interaction with plastics (Tekman et al. 2022). However, although plastic pollution in the marine environment, especially microplastics, is a current global concern, detailed knowledge about its effects remains unclear (Hodkovicova et al. 2022). There are currently efforts to help clarify the consequences of the widespread problem of marine pollution. According to data from the Litter base project, based on 1956 publications, it has been documented that 4076 species are affected by waste found in marine ecosystems, with different types of litter impacting aquatic life, with plastic being the most abundant at 40.25% of the total quantified waste, 3.36% pellets, 2.87% expanded polystyrene, 11.3% plastic fibers, and 8.05% fishing plastics, with the remaining waste falling into other categories such as textiles, metals, cardboard, among others. The interactions of marine species with all these wastes include ingestion (40.42%), entanglement (17.44%), and colonization (34.29%), and this has diverse effects, such as behavioral changes, physical effects like internal and external injuries, toxicity, affected mobility, population changes, among many others. Most species studied or observed are fish, followed by seabirds and crustaceans (Alfred Wegener Institute Helmholtz Centre for Polar and Marine Research 2024).

Entanglement can occur in various ways and has been documented on multiple sites regarding the most common interactions. However, it is known that most of these cases occur due to the loss or abandonment of fishing tools (Jepsen and de Bruyn 2019). Ingestion is a process that affects more organisms and can happen intentionally or accidentally. The intentional ingestion of plastic waste by organisms depends on various factors that vary between species and interact with each other. Both types of ingestion have been observed in tiny organisms and oysters, as well as fish, seabirds, and large mammals like whales (Kühn et al. 2015).

It is worth mentioning that although the ingestion of plastic waste has been considered a widespread and increasing problem, a study conducted in Brazil showed that, contrary to expectations, the observed fish did not interact with plastic waste, not even when covered with biofilms. From this study, it was inferred that the behavior of these fish was probably affected in such a way that they avoided areas contaminated by plastics (Menezes et al. 2022). However, microplastics should not be considered biologically inert materials as there is sufficient ecotoxicological information to confirm their danger both in physical aspects and for biochemical, molecular, and cellular pathways (Pitacco et al. 2022).

5.4.1 Physiological Effects of Plastic Ingestion on Marine Organisms

The physical impact of microplastic consumption on organisms in their natural habitats, including the consequences for their health and survival, remains largely unknown. Despite numerous controlled laboratory studies, they typically use microplastic concentrations much higher than those found in most polluted marine environments. Nonetheless, these studies have observed that microplastics can have potentially harmful effects on marine organisms (Egbeocha et al. 2018).

5.4.1.1 Physical and Internal Damage

Among the consequences of plastic ingestion, it has been proven that these can cause abrasions and internal injuries (Egbeocha et al. 2018). A noticeable result of plastic ingestion is damage or perforation to the digestive system and obstruction due to the passage of hard fragments in the intestines, often observed in stranded individuals (Nelms et al. 2016; Tourinho et al. 2010). This type of physical harm can lead to direct death, which has been observed in numerous marine vertebrates, such as seabirds and marine mammals (Markic et al. 2020).

Other observed effects include cellular damage and impacts on biological systems. For example, in a study performed on fish, oxidative stress biomarkers were activated after consuming microplastics, demonstrating the inflicted damage. However, the impact of microplastics on each biomarker can differ based on factors such as species, the organ studied, concentration, and the duration of exposure, among others (Compa et al. 2024). A study on adult sea urchins also observed effects on immune cells, indicating a stress syndrome (Murano et al. 2020). Moreover, depending on their size, microplastics can accumulate in the digestive tract, be translocated to other organs or systems, such as muscles or the circulatory system, or be excreted after hours or days (Agathokleous et al. 2021; Duis and Coors 2016). The same study on sea urchins found that ingested microplastics were distributed to different organs after accumulation (Murano et al. 2020).

5.4.1.2 Intestinal Blockages and Nutrient Displacement

Another consequence of plastic ingestion is intestinal blockages, which have been observed in various species, including sea turtles, dolphins, whales, and sharks. Specifically, the digestive tract of sea turtles is susceptible to waste accumulation once ingested due to their inability to regurgitate them; however, the damage largely depends on the size and shape of the debris (Wilcox et al. 2018). In young turtles, it was observed that even a small amount of debris was sufficient to cause an intestinal blockage, which subsequently has negative consequences, potentially leading to death (Santos et al. 2015).

Additionally, the digestive tracts of different shark species were analyzed for microplastics, and it was concluded that they might have different capacities to excrete them due to differences in stomach and intestinal findings. The retention time in each species is presumed to be different, so feeding habits could affect the metabolism and processing of ingested microplastics (Valente et al. 2019).

The consumption of microplastics has provided evidence of causing negative changes in nutrient and energy sources to predators, as they can consume microplastics directly or through their prey, as trophic base-level species like krill and zooplankton also consume microplastics. This ultimately disrupts the energy transfer in food chains (Egbeocha et al. 2018). For example, an adult Risso's dolphin died from ingesting plastic gloves, resulting in malnutrition and starvation (Wan et al. 2023).

5.4.1.3 Reproductive and Growth Impacts

Furthermore, plastic ingestion can deteriorate reproductive functions and affect growth rates and vulnerability to diseases. Earlier research on the effects of virgin microplastics on fish reproduction revealed that these substances impact endocrine glands and reproductive pathways, lower female sexual steroids, impair the quality and quantity of gametes, and delay gametogenesis (Sharifinia et al. 2020). On the other hand, exposure to plastic waste had a significantly negative effect on the growth of cold-water corals because the physical barrier of the plastic changed the hydrodynamic conditions around the corals, thereby reducing the rate of successful encounters of coral polyps with their prey, consequently decreasing food availability, causing energy losses in the corals and compensatory responses, adverse effects on physiological functions such as respiration and calcification (Chapron et al. 2018).

5.4.2 Ecological Implications of Plastic Ingestion

5.4.2.1 Impact on Marine Food Webs and Effects on Biodiversity

The full impact and extent of plastic entering food webs remain unclear. It is known that marine debris can infiltrate food webs both directly and indirectly, affecting organisms at all trophic levels. The exposure of organisms to plastics and, consequently, their entry into food chains varies depending on the marine zone—pelagic, demersal, or benthic—as the physical conditions and accumulated plastics change. Trophic level is another factor tied to each species' degree of exposure to plastics; higher level species tend to accumulate larger amounts of microplastics because they consume other organisms that have ingested them (Sultana et al. 2024).

At the lower levels of the trophic chain, marine microbes, essential for primary production and nutrient cycling, are affected by plastics and their leachates, altering their growth, protein production, phosphorus acquisition mechanisms, nitrogen fixation, genetic transcription, and photosynthesis. These effects vary by microbial species and the characteristics of the plastic (size, shape, charge, and chemical additives). Plastic pollution also facilitates the growth of harmful bacteria and the dispersion of invasive species by adhering to the surface of plastics, transporting microbes across oceans and into the digestive systems of marine organisms. Fundamental differences have been observed in the composition and functionality of microbes associated with plastics compared to planktonic microbes in the open ocean, varying by season and location. Additionally, plastics can alter marine plants' photosynthesis, growth, and physiology (Tuuri and Leterme 2023). In laboratory settings, zooplankton have been observed to reduce their fitness, growth, reproduction, and survival when exposed to microplastics. This plastic contamination is documented at various trophic levels, affecting primary, secondary, and tertiary predators, with potential consequences for humans due to trophic transfer. The effects of plastic pollution can transcend all levels of the marine food chain, increasing with higher contamination (Tuuri and Leterme 2023).

Furthermore, the consumption of microplastics transferred along the chain can trigger enzymatic responses, including the bioaccumulation of mercury and other metals, causing oxidative stress. Therefore, the transfer of these microplastics could lead to diseases in final consumers (Cruz 2023), such as humans, in whom microplastics can also infiltrate the circulatory and lymphatic systems, enter the liver, or be absorbed by the intestines, causing metabolic changes (Arif et al. 2024). Given all the negative consequences previously outlined, plastic ingestion can lead to population declines in certain species, impacting genetic diversity and ecosystem resilience (Compa et al. 2024). For instance, sea turtles face a substantial risk of ingesting plastic debris throughout all stages of their life cycle, which can be fatal. A study found that there is a 50% chance of mortality if a turtle has 14 pieces of plastic in its intestines and a 22% chance of death from consuming just one piece of debris (Wilcox et al. 2018). Comprehending the effects of microplastic exposure on marine species is essential, as alterations in lower trophic level species can impact higher levels. This occurs through the bioaccumulation of harmful chemicals as well as through the reduction of available resources (Pitacco et al. 2022). On the other hand, microplastics can also act as carriers for pathogenic bacteria that cause diseases in animals and humans, potentially causing more significant effects on species and their ecosystems (Abdulaziz et al. 2024).

5.4.2.2 Bioaccumulation and Biomagnification

Bioaccumulation and biomagnification are two different processes that generally occur together and result in the accumulation of toxins (Fig. 5.1) (CIMI - Catalina Island Marine Institute 2020). Bioaccumulation is defined as the process by which an organism accumulates substances (contaminants) over time within itself, sometimes reaching harmful levels, from the surrounding environment. The entry of substances into the organism can occur through different routes, such as respiration and food intake, as well as sources like water, sediments, and prey items (Borgå 2008). In summary, bioaccumulation occurs when the absorption of a contaminant exceeds the organism's ability to eliminate it (Miller et al. 2020). Biomagnification, on the other hand, refers to the increase in the concentration of substances as they move along the food chain. The main route of exposure for this process is through the organism's diet, so its mobility continues at each trophic level, resulting in top predators, such as large fish, marine mammals, and even humans, having higher concentrations than organisms at lower levels of the trophic chain (Miller et al. 2020; Drouillard 2008). Regarding the issue of bioaccumulation and biomagnification of microplastics, it is generally inferred that this occurs in marine and coastal food webs, so studies primarily focus on this field.

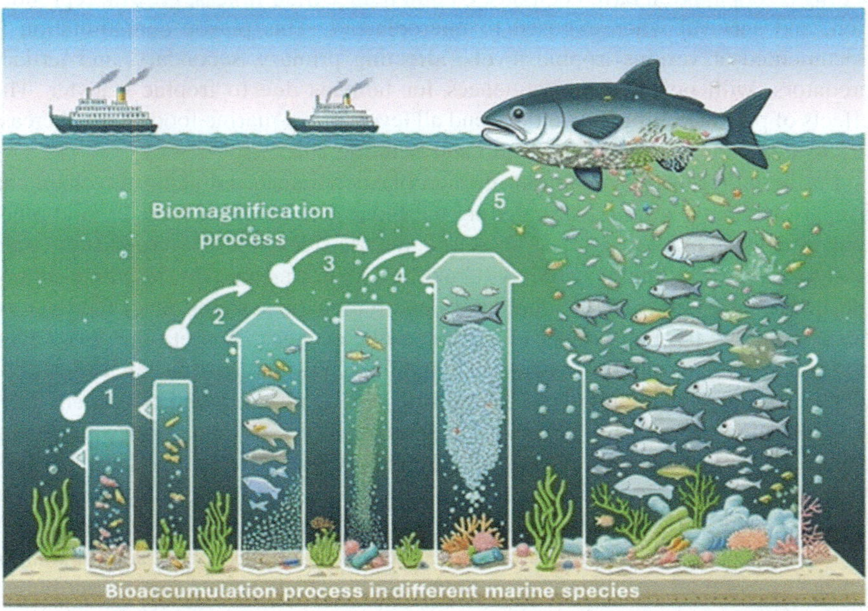

Fig. 5.1 Bioaccumulation and biomagnification processes

Miller, Hamann, and Kroon in 2020 investigated, through a literature review of microplastic studies in different marine species, whether MPs and their chemical additives bioaccumulated and biomagnified at each of the main trophic levels of a marine food web (Miller et al. 2020). The findings indicated that microplastic bioaccumulation occurs in various marine species across four primary trophic levels that represent consumers, such as mollusks, annelids, fish, arthropods, and sea turtles. However, the different exposure pathways contributing to this process are unclear and need more detailed study. Some chemical substances have been found in marine species collected from the field. However, results from laboratory exposure suggest that environmental exposure to chemical additives influences bioaccumulation more significantly than exposure to the additives linked to microplastics. Finally, no evidence was found that biomagnification occurs in the field from lower to higher trophic levels, even though some laboratory studies have reported the trophic transfer of microplastics from *Carcinus maenas* crabs to *Mytilus edulis* mussels (Watts et al. 2014; Farrell and Nelson 2013).

Parloni and collaborators in 2023 conducted a literature review on the bioaccumulation of microplastics in marine organisms from different taxonomic groups and sampled from various geographical locations. The results indicated that microplastic concentrations in organisms varied across taxonomic groups and trophic levels; for instance, mammals, birds, and reptiles had the highest levels of microplastics, while zooplankton exhibited the lowest levels. Regarding seasons or geographical areas, there were no differences in microplastic bioaccumulation, regardless of taxonomic groups. Therefore, it is concluded that all marine organisms can ingest and accumulate microplastics; still, the significant changes in concentration between taxonomic groups do not allow for the identification of global contamination patterns (Parolini et al. 2023).

5.4.2.3 Exposure to Harmful Chemicals

In addition to the above, plastic waste contamination (including MP) influences toxin accumulation in marine organisms in various ways. Firstly, plastic waste acts as a vector by introducing contaminants and microorganisms attached to its surfaces into aquatic environments and organisms, thus causing ecological risks. Among the pollutants they can introduce are hydrophobic organic compounds present in water due to their high surface-to-volume ratio, increasing their affinity. Examples of these contaminants are alkylphenols, polycyclic aromatic hydrocarbons (PAHs), bisphenol, polychlorinated biphenyls (PCBs), polybrominated diphenyl ethers (PBDEs), and dichloro-diphenyl-trichloroethane and its metabolites (DDTs) (Agboola and Benson 2021). Regarding microorganisms, their accumulation can lead to a layer called biofilm, which can impact chemical contaminants' sorption. In one study, it was found that the biofilm of the alga *Dunaliella salina* formed over 20 days on HDPE particles facilitates the sorption of contaminants such as PBDEs, PCBs, α-hexabromocyclododecane (α-HBCDD), and organophosphate flame retardants (OPFRs), resulting in higher concentrations than in virgin HDPE (Cui et al. 2023). On the other hand, during manufacturing, plastics are combined with chemical compounds called additives to give them specific properties. Some examples of additives are plasticizers, which are used to provide softness and flexibility; flame retardants, which help slow the combustion of plastics; antioxidants, which prevent the degradation of plastics due to contact with oxygen; antistatic agents, which prevent the accumulation of static electricity; and pigments, which provide a particular color (FOW Mould 2024; BPF - British Plastics Federation 2024). However, additives are not bound to the chemical structure of the plastic polymers; they are at the surface level. This situation causes them to desorb and, therefore, leach easily into their environment, including the biota that ingests them, due to factors such as temperature, salinity, and pH, to name a few examples (Benson et al. 2022). The presence and accumulation of toxins in marine species, especially those of commercial importance, can impact human health. One example is fish, a vital human protein source necessary for growth. Therefore, the contamination of fish by microplastics and chemical substances poses a serious risk that warrants careful consideration. Following exposure to microplastics alone or in conjunction with other pollutants, these organisms may experience tissue damage, oxidative stress, alterations in the expression of immune-related genes, neurotoxicity, stunted growth, and behavioral issues. Consequently, there is a possibility that humans may also experience cytotoxicity, oxidative stress, immune system alteration, neurotoxicity, and the transfer of microplastics to other tissues after exposure to them (Bhuyan 2022). Another example of commercially important seafood is bivalves, which are eaten whole, increasing the likelihood of humans ingesting microplastics through their consumption (Unuofin and Igwaran 2023).

5.5 Methodology

To obtain information about recent studies on the subject, a search was performed on Science Direct and Springer libraries in June 2024 using the following keywords: Plastic ingestion + marine species. Since the results were far too many, the filter was adjusted to only display research articles from 2021 forward. Reviews or book chapters were excluded.

The first 100 of each library were reviewed from all the articles found to select those that met the criteria needed for this work. Only research articles that studied plastic ingestion in marine species were selected. In total, 114 research articles were chosen and analyzed, and their information was registered in a database later analyzed.

5.6 Results

Plastic ingestion by marine species and its effects are of increasing interest. Numerous studies have been published in the past years since marine litter was recognized as a severe and aggravating problem worldwide. Notably, the number of articles found and selected for this work increased with every year (Fig. 5.2), considering the search was conducted in June of 2024.

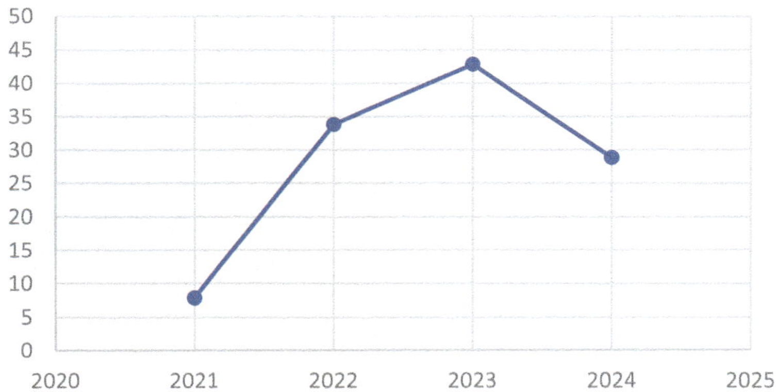

Fig. 5.2 Publication years of the selected articles

Most of the research in the analyzed articles was conducted in locations within Europe, with 38% of the total, followed by Asia with 28%, and America with 25%. Africa and Oceania correspond to 4% and 3%, respectively. Lastly, 2% of the studies were conducted in the Antarctic (Fig. 5.3). Italy, Brazil, and Spain were the countries where most of the studies were carried out, with 14, 13, and 10 studies, respectively, followed by India and China. In total, locations in 35 different countries were registered. The broad area of the places where the research took place is shown in Fig. 5.4.

5.6.1 Key Discoveries, Trends, or Patterns Observed in Different Marine Ecosystems

Most of the research articles reviewed were about species captured or found in their natural habitats. Then, they were transported to a laboratory, where samples were taken and processed. These will be referred to as in situ throughout this work. Only 12 articles were experiments performed inside a laboratory (10.5%), where the organisms were fed and then analyzed.

5.6.2 Organisms

In studies conducted within laboratories (Fig. 5.5), the most widely studied organisms corresponded to fish (52%). Among these organisms, species that stand out as mainly sampled are *Sparus aurata* (Alomar et al. 2022; Capó et al. 2022) and *Oreochromis niloticus* (Feitosa et al. 2024; do Nascimento et al. 2023), the first of which is a carnivore that is commercially important within the Mediterranean (Ortega 2008). The second one is an omnivore, popular in aquaculture globally due to its high protein content (FAO - Food and Agriculture Organization 2009).

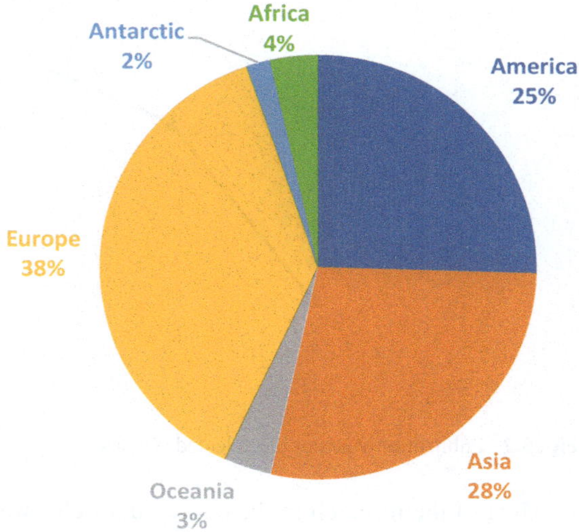

Fig. 5.3 The continent where the studies were conducted

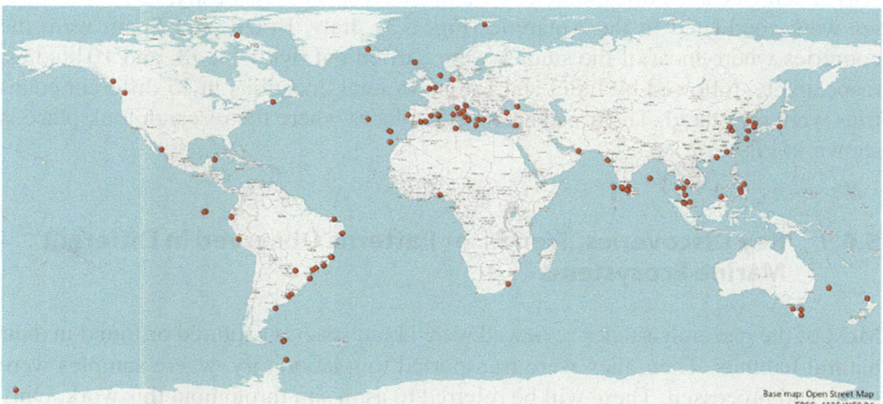

Fig. 5.4 Locations where studies were conducted in the selected articles

The second most studied organism was zooplankton (16%), mainly samples of rotifers of the Brachionus genre, employing species like *B. koreanus* (Yoo et al. 2024) and *B. plicatilis* (Zhao et al. 2024), which are frequently used in aquaculture for fish production (Contreras-Sillero et al. 2019) and as a model for ecotoxicity assessments (Park et al. 2020), respectively. In studies carried out in situ (Fig. 5.6), the organisms that were most sampled were fishes of species *Mullus barbatus* (Giani et al. 2023; Koraltan et al. 2022; Valente et al. 2022; Aytan et al. 2023; Miccoli et al. 2022; Atamanalp et al. 2021), *Merluccius merluccius* (Giani et al. 2023; Valente et al. 2022; Miccoli et al. 2022; Carrillo-Barragán et al. 2024), and *Merlangius merlangus* (Aytan et al. 2023; Carrillo-Barragán et al. 2024; Horton et al. 2024),

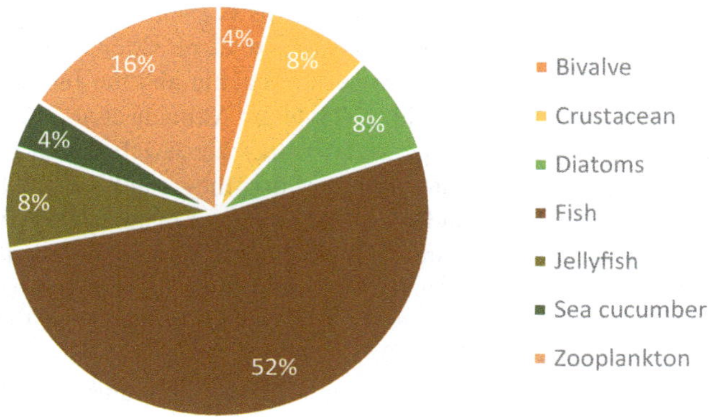

Fig. 5.5 Most studied organisms in laboratory conditions

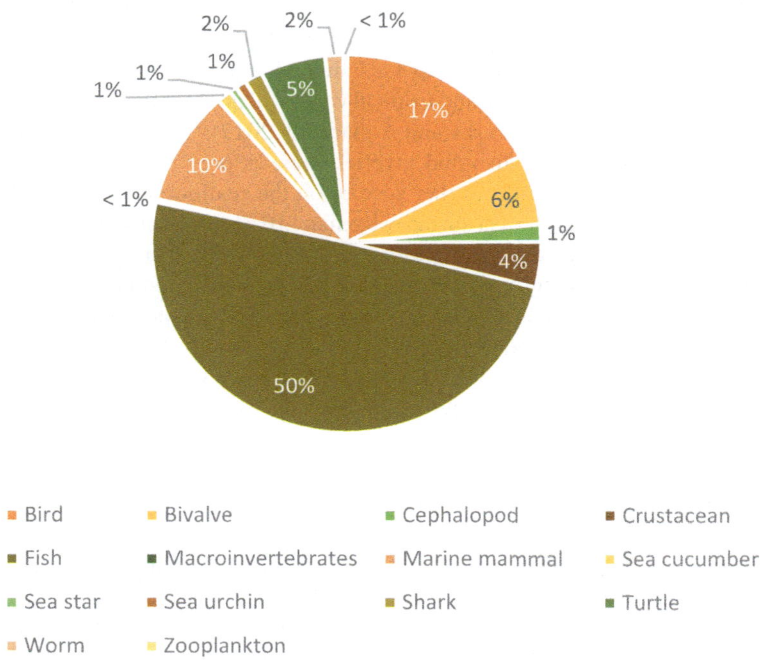

Fig. 5.6 Most in situ studied organisms

which are all carnivores, demersal, of commercial interest for human consumption, and with distribution across the Northeast Atlantic Ocean, the Mediterranean Sea, and the Black Sea (MITECO - Ministerio para la Transición Ecológica y el Reto Demográfico 2008; FAO - Food and Agriculture Organization 2003).

The second most common organism was birds (17%), in which case the main studied species were *Ardenna carneipes* (Muñoz et al. 2023; Lavers and Bond 2023), distributed in the subtropical area of the Pacific and the Indian Ocean, and *Larus michahellis* (Nono Almeida et al. 2023; Nono Almeida et al. 2024), present in the Mediterranean and within Europe. *A. carneipes* has a rich diet of intertidal and terrestrial invertebrates, crustaceans, reptiles, fish they catch on their own or as bycatch from boats, other birds, and even organic matter in landfills (Arizaga 2023). Therefore, their interaction with humans is much more significant than for *L. michahellis*, which feeds almost exclusively on offshore fish, crustaceans, cephalopods, and invertebrates caught by themselves (DCCEEW - Department of Climate Change Energy the Environment and Water 2024).

The third type of most studied organisms were marine mammals (10%), among which cetaceans studied by Merrill et al. (2023) stand out. Some of these species were *Eschrichtius robustus*, which is distributed along the North Pacific (CONANP - Comisión Nacional de Áreas Naturales Protegidas 2013), and *Delphinapterus leucas*, which inhabits arctic and subarctic waters of the northern hemisphere (AMMPA - Alliance of Marine Mammal Parks and Aquariums 2019), in addition to dolphins of the *Tursiops truncantus* species, that were also studied by Padula et al. (2023), and that live in temperate and tropical waters all around the world (AMMPA - Alliance of Marine Mammal Parks and Aquariums 2017). These organisms have a relevant role as predators of fish and invertebrates, as prey for animals like killer whales, as carcasses that nourish other species on the seafloor, and as nutrient conveyers from the depths to the sea surface (Roman 2017). Another relevant organism is seals, with species such as the *Erignathus barbatus* (Merrill et al. 2023), which is distributed in the Arctic region of the Atlantic sector, feeding on benthonic invertebrate species and fish and serving as food for predators like polar bears and killer whales, as well as humans (Nelson 2008).

It should be noted that Costa et al. (2023) studied macroinvertebrates, but did not specify the studied species within the main article.

5.6.3 Plastic Sizes

The types of plastics to be analyzed within the studies were defined according to their sizes. In the case of research carried out within laboratory conditions, it focused exclusively on microplastics (Capó et al. 2022; do Nascimento et al. 2023; Yoo et al. 2024; Zhao et al. 2024; Miccoli et al. 2022; Iwalaye et al. 2021; Köster and Paffenhöfer 2022; Bonifacio et al. 2022; Martellini et al. 2023; Chen et al. 2023; Kelly et al. 2023a; Ge et al. 2023), while in situ studies reported a variety of sizes (Fig. 5.7), with the presence of microplastics in all types of organisms, and their prevalence in relatively small animals like sea cucumbers (Rios-Fuster et al. 2022), macroinvertebrates (Costa et al. 2023), sea stars (Cossi et al. 2021), sea urchins (Fagiano et al. 2023; Digka et al. 2023), cephalopods (Pedà et al. 2022; Alfaro-Núñez et al. 2021; Sambolino et al. 2023), and zooplankton (Alfonso et al. 2024). This size was followed by macroplastics, which were predominant in turtles (65%)

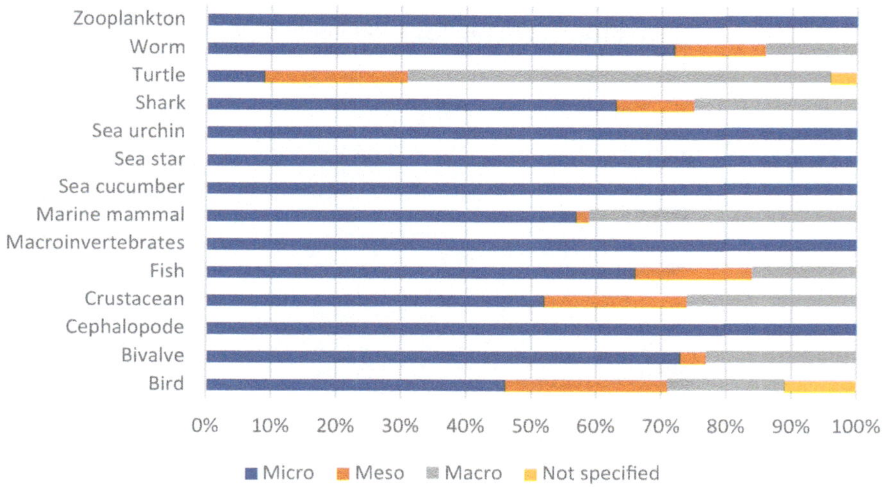

Fig. 5.7 Types of plastic sizes studied in situ

(Solomando et al. 2022; Moon et al. 2022; Feitosa et al. 2024; Duncan et al. 2024; Bruno et al. 2022; Palmer et al. 2021; Ramon-Gomez et al. 2024; Yaghmour et al. 2022), marine mammals (41%) (Padula et al. 2023; Kelly et al. 2023b; da Silva et al. 2024), and crustaceans (26%) (Leila et al. 2023; Soe et al. 2024; Cáceres-Farias et al. 2023). Lastly, mesoplastics were present mainly within birds (25%) (Nono Almeida et al. 2023; Muñoz et al. 2023; Lavers and Bond 2023; Schutten et al. 2023; Benjaminsen et al. 2022; Bange et al. 2023), turtles (22%) (Solomando et al. 2022; Feitosa et al. 2024; Bruno et al. 2022; Rodríguez et al. 2022) and sharks (12%) (Mancuso et al. 2022).

In the case of studies carried out in the laboratory, the most reported size was <1.00 mm with 41.38%, while in the in situ studies, the size with the highest proportion was 1.01–5.00 mm (19.51%) (Fig. 5.8). Knowing the particle size is necessary to determine what plastics are being studied. Although there is still no clear consensus on the size limits to establish where nanoplastics are located, it is known among the scientific community that, in general, microplastics are those particles with a size less than or equal to 5 mm, and that plastics larger than that size can be mesoplastics or macroplastics. In the case of particles <1.0 mm, these include both nanoplastics and tiny microplastics; however, none of the research analyzed here specifically studied nanoplastics. On the other hand, it is also relevant to mention that there are still studies that do not report this data, with percentages of 32.51% and 51.72% for work carried out in situ and in the laboratory, respectively. Knowing this characteristic of the particle, as well as its color, polymer, and type of plastic, can also be relevant to know other important data, for example, which plastics are the ones they ingest the most, if they prefer a particular color or polymer and from this establish specific actions related to the management of plastic waste or the development of activities in the plastic industry.

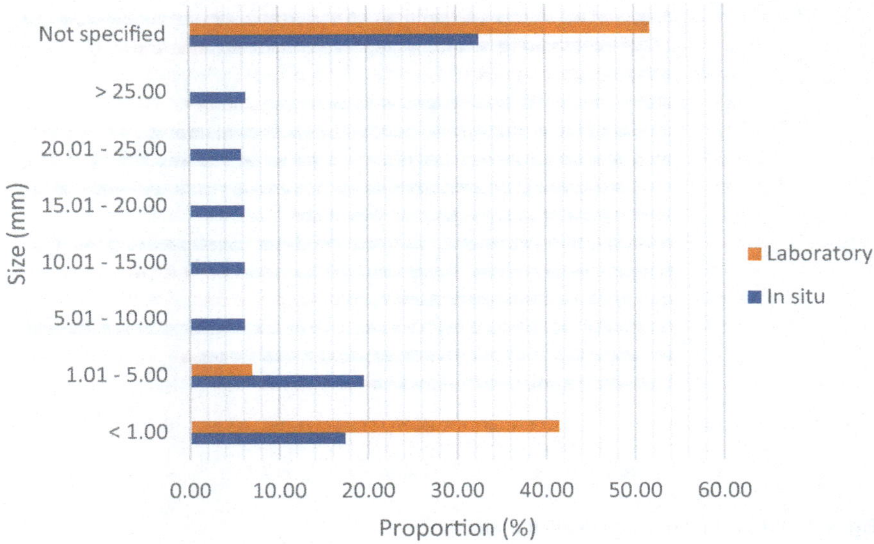

Fig. 5.8 Size of ingested plastics

5.6.4 Duration of Research Conducted

The duration of the research conducted in laboratory conditions varied between organisms and authors, as observed in Table 5.1. Zooplankton of the *Brachionus plicatilis* species was subjected to the shortest exposure time (approximately 4 h) (Zhao et al. 2024), while *Sparus aurata* fish have been studied for up to 120 days (Capó et al. 2022). Some studies did not include information about exposition time (Yoo et al. 2024; Chen et al. 2023; Kelly et al. 2023a).

On the other hand, sampling for in situ research can be carried out on a single day (Miccoli et al. 2022; Gong et al. 2023; Ahuir-Baraja 2022; Yu et al. 2024), or samples can be collected throughout the years, be it through the capture of live specimens, like fish and crustaceans sampled by Carrillo-Barragán et al. (Carrillo-Barragán et al. 2024) during 49 years—the lengthiest period covered within studies—or collecting carcasses and incidentally captured specimens, like the marine mammals gathered by Padula et al. (2023)) during 32 years (Table 5.2).

5.6.5 Types of Samples

As presented in Fig. 5.9, in general, for laboratory studies, the samples were obtained mainly from the gastrointestinal section of the organisms, whether it be the whole digestive tract (Capó et al. 2022; Kelly et al. 2023a; Horie et al. 2024) or parts of it, such as the guts (Iwalaye et al. 2021), stomach content (Miccoli et al. 2022), or feces, (Köster and Paffenhöfer 2022; Martellini et al. 2023), or specialized organs,

Table 5.1 Study duration in laboratory

Type of organism	Laboratory study duration	Reference
Bivalve	17 days	Bonifacio et al. (2022)
Crustacean	28 days	Martellini et al. (2023)
Diatoms	Not specified	Chen et al. (2023)
Fish	3–120 days	Capó et al. (2022), do Nascimento et al. (2023), Miccoli et al. (2022), Kelly et al. (2023a)
Jellyfish	0.5–0.63 days	Ge et al. (2023)
Sea cucumber	2–3 days	Iwalaye et al. (2021)
Zooplankton	0.16–0.25 days	Zhao et al. (2024), Köster and Paffenhöfer (2022)

like gastric pouches, the manubrium (the central structure that contains the mouth and stomach) and oral arms (appendages for prey capture) of jellyfish (Ge et al. 2023). Since one of the routes of exposure to plastics is ingestion, studying different parts of the digestive process can be very useful to explore their effects.

According to Miccoli et al. (2022), other parts of fish have been studied, such as their gills, organs that allow oxygen withdrawal from water (Sam 2023), and that, therefore, represent a crucial exposure pathway; the head kidney, a structure where immune cells are produced and matured, aside from being responsible for the initial immunity response (Soulliere and Dixon 2017); and the spleen, a secondary lymphatic organ that induces adaptive immunity responses, apart from eliminating immune complexes (Sayed et al. 2022). The most commonly studied samples for bivalves were soft tissue (Bonifacio et al. 2022). Some studies for diatoms (Chen et al. 2023), jellyfish (Ge et al. 2023), fish (do Nascimento et al. 2023), and zooplankton (Zhao et al. 2024) did not specify the types of samples collected. The most frequently studied samples within in situ research (Fig. 5.10) also correspond to the whole digestive tract (Muñoz et al. 2023; Pedà et al. 2022; Solomando et al. 2022; Moon et al. 2022; Alomar et al. 2022; Yoo et al. 2024; Giani et al. 2023; Koraltan et al. 2022; Valente et al. 2022; Aytan et al. 2023; Atamanalp et al. 2021; Horton et al. 2024; Nono Almeida et al. 2024; Padula et al. 2023; Costa et al. 2023; Rios-Fuster et al. 2022; Fagiano et al. 2023; Digka et al. 2023; Alfaro-Núñez et al. 2021; Duncan et al. 2024; da Silva et al. 2024; Cáceres-Farias et al. 2023; Rodríguez et al. 2022; Mancuso et al. 2022; Ahuir-Baraja 2022; Baes et al. 2024; Puskic et al. 2024; Mills et al. 2024; Navarro et al. 2023; Collard et al. 2022b; Pollet et al. 2024; Cohen-Sánchez et al. 2022; Sbrana et al. 2022; Brawn et al. 2023; Patsiou et al. 2024; Batalha et al. 2024; Rodrigues et al. 2023; Fatema et al. 2024; Lim et al. 2023; Ragesh et al. 2024; Sciutteri et al. 2023; Zhang et al. 2022; Mboglen et al. 2024; Pinho et al. 2022; Ojeda et al. 2024; Devi et al. 2024; Bottari et al. 2022; Alves et al. 2024; Rivera-Garibay et al. 2024; Kalaiselvan et al. 2022) or parts of it, such as the upper digestive tract (Kim et al. 2023a), digestive glands (Ciocan et al. 2024), esophagus (Yaghmour et al. 2022), stomach (Alomar et al. 2022; Sambolino et al. 2023; Yaghmour et al. 2022; Kelly et al. 2023b; Soe et al. 2024; Benjaminsen et al. 2022; Bange et al. 2023; Gong et al. 2023; Pon et al. 2023; Puskic et al. 2024; Scotti et al. 2023; de Lemos Santana et al. 2022; de Almeida et al. 2024; Hajisamae et al.

Table 5.2 In situ sample collection duration

Type of organism	Sample collection duration	Reference
Bird	1 day to 15 years	Nono Almeida et al. (2023), Lavers and Bond (2023), Nono Almeida et al. (2024), Schutten et al. (2023), Benjaminsen et al. (2022), Bange et al. (2023), Lavers et al. (2022), Pon et al. (2023), Keerthika et al. (2024), Baes et al. (2024), Puskic et al. (2024), Mills et al. (2024), Navarro et al. (2023), Matos et al. (2024), Collard et al. (2022b), Pollet et al. (2024)
Bivalve	4 months to 4 years	Alomar et al. (2022), Rios-Fuster et al. (2022), Alfaro-Núñez et al. (2021), Yu et al. (2024), Ciocan et al. (2024), Diaz-Santibañez et al. (2023), Joyce and Falkenberg (2023)
Cephalopod	25 days to 1 month	Pedà et al. (2022), Alfaro-Núñez et al. (2021), Sambolino et al. (2023)
Crustacean	25 days to 49 years	Carrillo-Barragán et al. (2024), Alfaro-Núñez et al. (2021), Leila et al. (2023), Soe et al. (2024), Cáceres-Farias et al. (2023), Compa et al. (2022), Scotti et al. (2023), de Lemos Santana et al. (2022)
Fish	1 day to 49 years	Giani et al. (2023), Koraltan et al. (2022), Valente et al. (2022), Miccoli et al. (2022), Atamanalp et al. (2021), Carrillo-Barragán et al. (2024), Horton et al. (2024), Costa et al. (2023), Rios-Fuster et al. (2022), Alfaro-Núñez et al. (2021), Sambolino et al. (2023), Cáceres-Farias et al. (2023), Ahuir-Baraja (2022), de Almeida et al. (2024), Cohen-Sánchez et al. (2022), Sbrana et al. (2022), Cardozo et al. (2023), Brawn et al. (2023), Patsiou et al. (2024), Batalha et al. (2024), Dantas et al. (2024), Morgan et al. (2021), Rodrigues et al. (2023), Fatema et al. (2024), Lim et al. (2023), Ragesh et al. (2024), Hajisamae et al. (2022), Sciutteri et al. (2023), Alves and Figueiredo (2023), Motivarash et al. (2024), Selden and Baker (2023), Kuroda et al. (2022), Zhang et al. (2022)
Macroinvertebrates	6 months	Costa et al. (2023)
Marine mammal	1 day to 32 years	Merrill et al. (2023), Padula et al. (2023), Kelly et al. (2023b), Pereira et al. (2023), Hernandez-Milian et al. (2023)
Sea cucumber	1 year	Rios-Fuster et al. (2022)
Sea star	Not specified	Cossi et al. (2021)
Sea urchin	Not specified	Fagiano et al. (2023), Digka et al. (2023)
Shark	4 months to 7 years	Mancuso et al. (2022), Gong et al. (2023), Mboglen et al. (2024), Yong et al. (2021)

(continued)

Table 5.2 (continued)

Type of organism	Sample collection duration	Reference
Turtle	5 months to 11 years	Solomando et al. (2022), Moon et al. (2022), Feitosa et al. (2024), Duncan et al. (2024), Bruno et al. (2022), Palmer et al. (2021), Ramon-Gomez et al. (2024), Yaghmour et al. (2022), Rodríguez et al. (2022), Camedda et al. (2022)
Worm	6 months	Keerthika et al. (2024), James et al. (2023), Vecchi et al. (2021)
Zooplankton	3 days	Alfonso et al. (2024)

Some of the in situ studies did not specify the duration of sample collection Muñoz et al. (2023), Aytan et al. (2023), Cossi et al. (2021), Fagiano et al. (2023), Digka et al. (2023), da Silva et al. (2024), Mancuso et al. (2022), Keerthika et al. (2024), Vecchi et al. (2021), Fagiano et al. (2024), Pinho et al. (2022), Rajendran et al. (2022), Ojeda et al. (2024), Yücel (2023), Abd Rahman et al. (2024), Dada and Bello (2023), Devi et al. (2024), Bottari et al. (2022), Kim et al. (2023a), Alves et al. (2024)

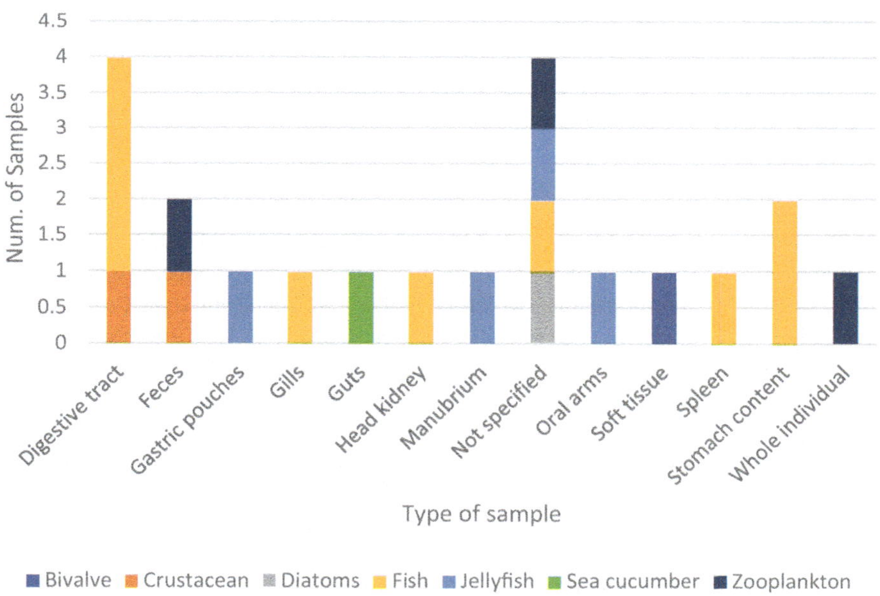

Fig. 5.9 Types of samples studied in laboratory conditions

2022; Alves and Figueiredo 2023; Selden and Baker 2023; Kim et al. 2023a; Compa et al. 2023), intestines (Sambolino et al. 2023; Yaghmour et al. 2022; Scotti et al. 2023), the upper part of the small intestine (Kim et al. 2023a), guts (Palmer et al. 2021; Motivarash et al. 2024; Rajendran et al. 2022; Dada and Bello 2023; Keerthika et al. 2023), gastrointestinal (Feitosa et al. 2024; Dantas et al. 2024; Camedda et al. 2022; Vecchi et al. 2021) or stomach content (Lavers and Bond 2023; Leila et al.

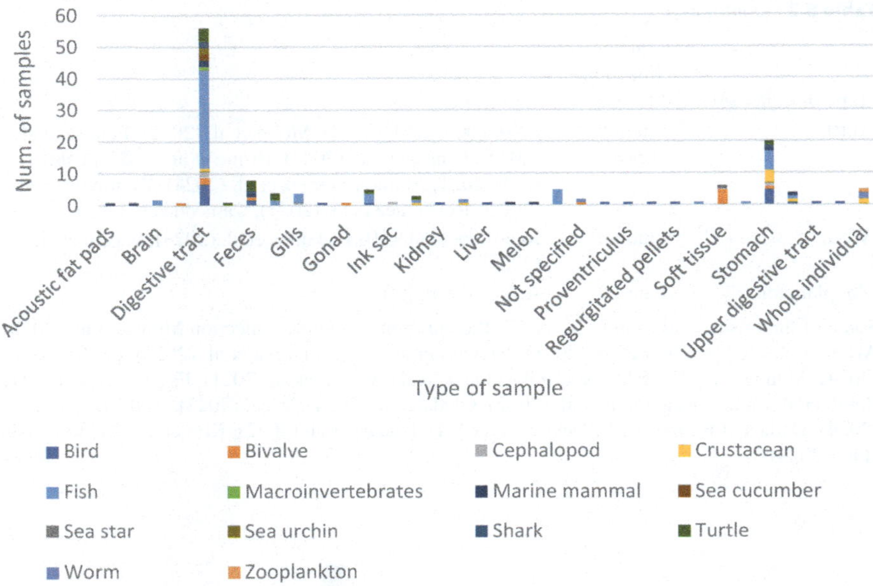

Fig. 5.10 Types of samples mostly studied in situ

2023; Kuroda et al. 2022; Pereira et al. 2023), or regurgitated pellets (Nono Almeida et al. 2023) or feces (Solomando et al. 2022; Bruno et al. 2022; Ramon-Gomez et al. 2024; Bange et al. 2023; Matos et al. 2024; Diaz-Santibañez et al. 2023; Hernandez-Milian et al. 2023; Yong et al. 2021). Specific digestive organs have also been sampled, such as the proventriculus, which is in charge of preparing digestion, and the gizzard of birds like *Pachyptila turtur* (Lavers et al. 2022), as well as the spiral valve that covers the intestine of fish like *Scyliorhinus canicula* (Morgan et al. 2021).

Samples of pumice stone, pieces of rock stored in the gizzard and used to facilitate digestion, have also been taken from birds to analyze for microplastic content in species of *A. carneipes* and *A. pacífica* (Lavers and Bond 2023). To perform histopathological studies to asses microplastic effects on birds, Puskic et al. (2024) also analyzed the pectoralis major muscle, the liver, and kidneys of *A. carneipes* and *A. tenuirostris*.

Furthermore, lipid-rich organs in marine mammals have also been studied. Merrill et al. (2023) studied the potential translocation of microplastics in organs like the blubber, which is composed of several layers of adipose and vascularized tissue important for energy storage and movement, acoustic fat pads, vascularized organs that provide oxygen through the circulatory system, and the melon, which is located on the forehead of odontocetes and is used for echolocation (Harper et al. 2008). Lungs were also included in the samples. Occasionally, whole individuals are taken as samples, such is the case of zooplankton (Alfonso et al. 2024) and some small crustaceans (Fagiano et al. 2024; Yücel 2023) and worms (James et al. 2023; Yücel 2023), in addition to crustacean and fish larvae (Carrillo-Barragán et al.

2024). Other types of samples include fish (Atamanalp et al. 2021; Motivarash et al. 2024) and cephalopod (Sambolino et al. 2023) gills; fish skin (Motivarash et al. 2024), muscular tissue (Atamanalp et al. 2021; Alfaro-Núñez et al. 2021; Cáceres-Farias et al. 2023; Motivarash et al. 2024) and brains (Atamanalp et al. 2021); ink sacs from cephalopods (Sambolino et al. 2023), and soft tissue from bivalves (Alomar et al. 2022; Fagiano et al. 2023; Yu et al. 2024; Ciocan et al. 2024; Joyce and Falkenberg 2023) and sea stars (Cossi et al. 2021). A couple of articles did not specify the type of samples taken and only mentioned the organisms they studied, which were worms (Vecchi et al. 2021) and bivalves (Abd Rahman et al. 2024).

5.6.6 Number of Samples Analyzed

The graph in Fig. 5.11 presents the number of samples analyzed in the different studies. This is generally very varied since the value is between 1 and 10,537 samples (Mills et al. 2024) for in situ studies, while for laboratory studies, the range is from 1 to 3000 samples (Yoo et al. 2024). However, it must be considered that the number of samples to be analyzed may be related to the species to be examined and the study's objective. For example, suppose a research project focuses on studying the ingestion of plastics in marine mammals, such as toothed whales, for years. In that case, this makes it challenging to obtain a considerable number of samples since they cannot be easily captured due to their size, and it also becomes complex when it comes to species with a risk category. In this case, the best way is to collect the dead mammals and analyze the different organs of interest (Padula et al. 2023). On the contrary, if the organisms to be studied are small in size and are not in any risk category or do not live in a protected area, it is feasible to obtain a more significant number of samples, such is the case of crabs (Soe et al. 2024), zooplankton (Alfonso et al. 2024), and fish (Carrillo-Barragán et al. 2024; Hajisamae et al. 2022).

5.6.7 Quantity Ingested per Individual

It is worth noting that not all studies report the feeding rate in laboratory experiments, and those that do use varying units, even when studying the same type of organism. For example, the feeding rate for fish is reported in units of pieces/volume/time (Ge et al. 2023), mass of particles/cage/time (Capó et al. 2022), and mass of particles/mass of the specimen/time (do Nascimento et al. 2023). This lack of uniformity in reporting feeding rates across studies underscores the need for standardization in research protocols to facilitate comparisons. Not all studies reported data on the presence of plastic particles in terms of pieces per individual (Fig. 5.12). In in situ investigations, the percentage was 29.74%, while in laboratory work, it rose to 44.44%. However, it is essential to clarify that within these percentages are those cases where the presence of plastics was reported in other units, such as pieces/mass of the individual (Ramon-Gomez et al. 2024; Ciocan et al. 2024) and pieces/mass of the organ analyzed (Iwalaye et al. 2021), whose percentages had low

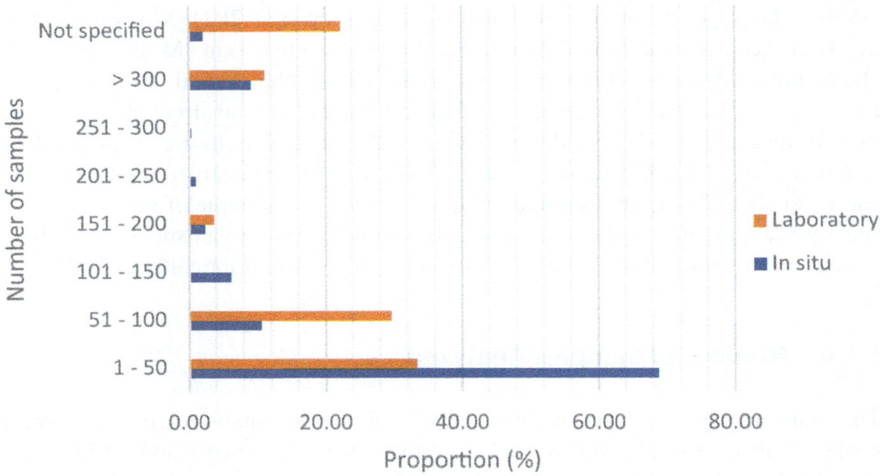

Fig. 5.11 Proportion of analyzed samples

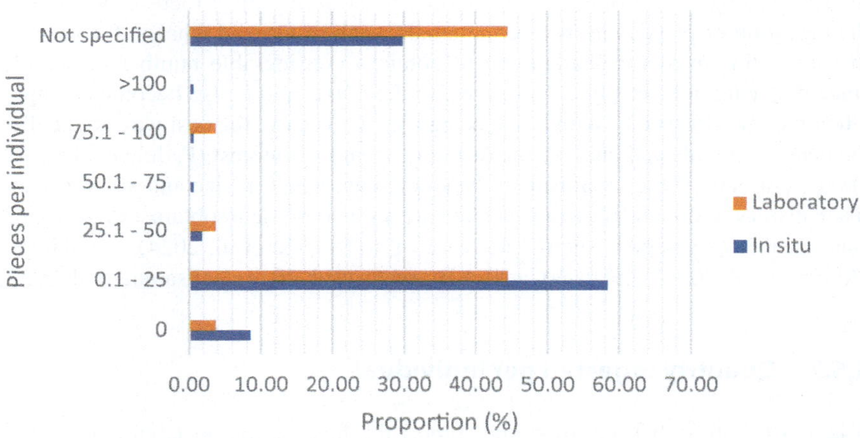

Fig. 5.12 Number of pieces ingested per individual

values (2.92% and 3.70%, respectively). On the other hand, it is also important to highlight that there were investigations where no plastics were found in some of the organisms analyzed, both in the in situ studies (Koraltan et al. 2022; Cáceres-Farias et al. 2023; Cardozo et al. 2023) and in the laboratory (Horie et al. 2024). These percentages corresponded to 8.45% and 3.70%, respectively. On the other hand, the interval where the highest presence of plastics was reported in the various organisms was from 0.1 to 25.0 pieces/individual, with proportions of 58.31% and 4.44% for in situ and laboratory studies, respectively. The highest number of plastic pieces per individual was reported for bivalves *Venerupis* spp. from the Philippines, where the mean was 195 ± 0.17 pieces of fibers/individual (Yu et al. 2024).

5.6.8 Classification by Color

The review distinguished plastics and microplastics ingested by marine animals based on color. Figure 5.13 illustrates the colors reported in situ and laboratory studies and the percentage of studies that did not provide this information. For in situ studies, the color of plastics was not reported in 5.54% of cases (Padula et al. 2023; Pollet et al. 2024; Ciocan et al. 2024). Among the remaining studies, the most common colors were: blue (16.17%), black (13.16%), red (12.01%), white (11.09%), transparent (9.24%), green (9.24%), yellow (5.31%), brown (3.46%), purple (2.31%), grey (2.08%), orange (2.54%), pink (2.77%), silver (1.15%), a mix of colors (0.91%), cream (0.23%), and other colors (2.77%).

In laboratory studies (Fig. 5.14), the colors of plastics ingested by marine animals were reported as blue (14.29%), yellow (14.29%), red (14.29%), green (7.14%), black (7.14%), and transparent (7.14%). The color specification was not provided in 35.71% of investigations (Capó et al. 2022; do Nascimento et al. 2023; Zhao et al. 2024; Bonifacio et al. 2022; Chen et al. 2023). The category "blue" includes shades such as dark blue, light blue, and turquoise; "green" includes light green; "purple" encompasses violet and lilac; "yellow" includes fluorescent yellow; "white" includes clear and off-white shades; "transparent" includes translucent and colorless descriptions; "mix" includes multicolored and various colored items; and "others" refers to opaque materials. Green and blue particles likely originate from fishing nets and related materials (Soe et al. 2024). Additionally, certain species exhibit preferences for specific plastic colors (Horie et al. 2024). For instance, some fish mistake blue, black, and white microplastics for prey due to their color

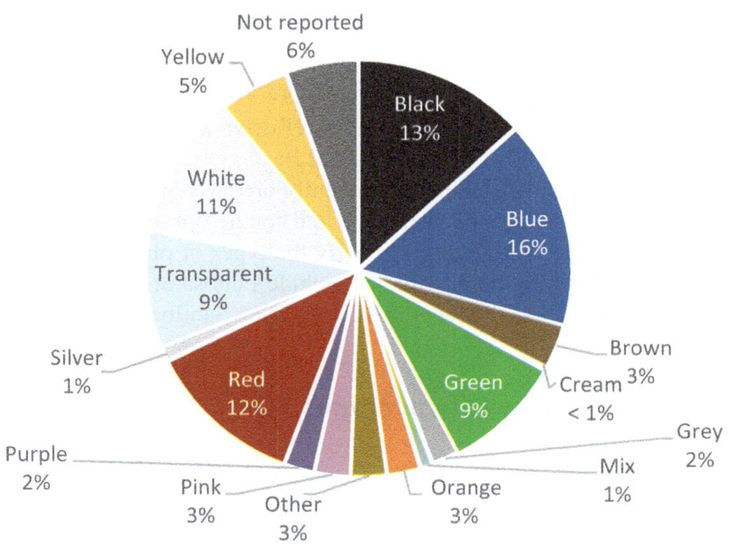

Fig. 5.13 Percentage of colors in situ

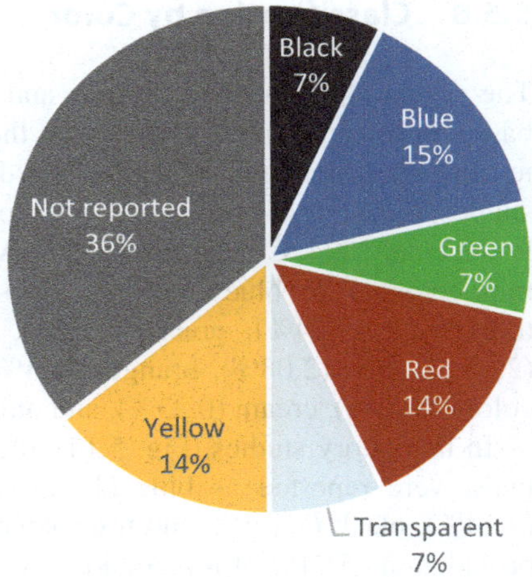

Fig. 5.14 Percentage of polymer colors reported in Laboratory studies

similarity (Soe et al. 2024; Ojeda et al. 2024; Horie et al. 2024). In (Ramon-Gomez et al. 2024), tortoises (*Chelonoidis porteri*) were observed ingesting light blue, green, and white plastics, mistaking them for vegetation.

5.6.9 Classification by Type and Morphology

Plastics were categorized based on type and morphology, though the lack of a standardized classification method resulted in a wide variety of shapes grouped into 13 categories. Table 5.3 provides detailed information on each category.

In laboratory studies (Fig. 5.15), four main types of microplastics were identified. The most prevalent were spherical particles (40%) (Zhao et al. 2024; Ge et al. 2023; Horie et al. 2024), followed by fragments (30%) (Yoo et al. 2024; Iwalaye et al. 2021; Bonifacio et al. 2022), fibers (20%) (Köster and Paffenhöfer 2022; Chen et al. 2023), and pellets (10%) (Capó et al. 2022).

In situ studies (Fig. 5.16), the most frequently observed plastics were fragments (26.02%), fibers (17.10%), films (14.87%), and foamed plastics (1.86%). A small percentage of studies (7.43%) did not specify the types of plastics ingested by the animals under study. Contributing studies include Feitosa et al. (2024), Soe et al. (2024), Puskic et al. (2024), Pollet et al. (2024), Diaz-Santibañez et al. (2023).

5.6.10 Classification by Polymer Types

The types of polymers ingested by organisms were identified in both in situ and laboratory studies. Given the wide variety reported, some were grouped as shown in the first column of Tables 5.4 and 5.5. Among in situ studies, the most frequently mentioned polymers were PE (27.59%), PA (10.61%), PS (7.96%), with lesser

Table 5.3 Plastics classification

Category of plastics	Plastics included
Elastic plastics	Elastic plastics
Fishery-related	Fishing buoys, fishing equipment, fragments of fishing nets
Foamed plastics	Styrofoam
Fragments	Irregular, hard fragments, hard plastics
Fibers	Glass fiber reinforced plastic (GRP) fibers
Sphere shaped	Microspheres, microbeads, beads
Miscellaneous	Other
Pellets	Industrial granules, commercial dry granules
Filaments	Rope, ropes monofilaments
Film	Sheet plastic, sheet-like plastic, plastic wrapping sheets
Single use plastics	Bags, balloons, food packages, food wrappers, containers, lollipop packaging, complete plastics
Flexible plastic	Soft plastics
Thread-like	Threads, line

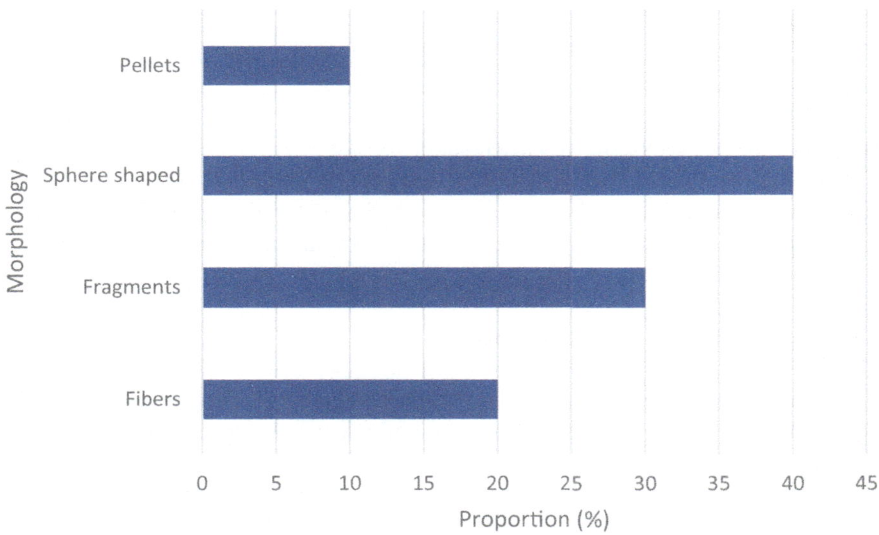

Fig. 5.15 Types of microplastics reported in laboratory studies

occurrences of PVC (2.92%), PEST (2.92%), HDPE (2.65%), and LDPE (2.39%). In 12.2% of in situ studies, polymers were classified as "other" and in 9.02% the polymers found were unidentified. A study by (Sambolino et al. 2023) identified four phthalate esters (PAEs): diethyl phthalate (DEP), diisobutylphthalate (DIBP), benzyl-butyl phthalate (BBP), and Bis (2–ethylhexyl) phthalate (DEHP), found across all analyzed fish and cephalopod species.

Laboratory studies identified five polymer types: PS (30.77%), PE (23.08%), PET (7.69%), LDPE (7.69%), and two biopolymers, BIO1 and BIO_2 (7.69%),

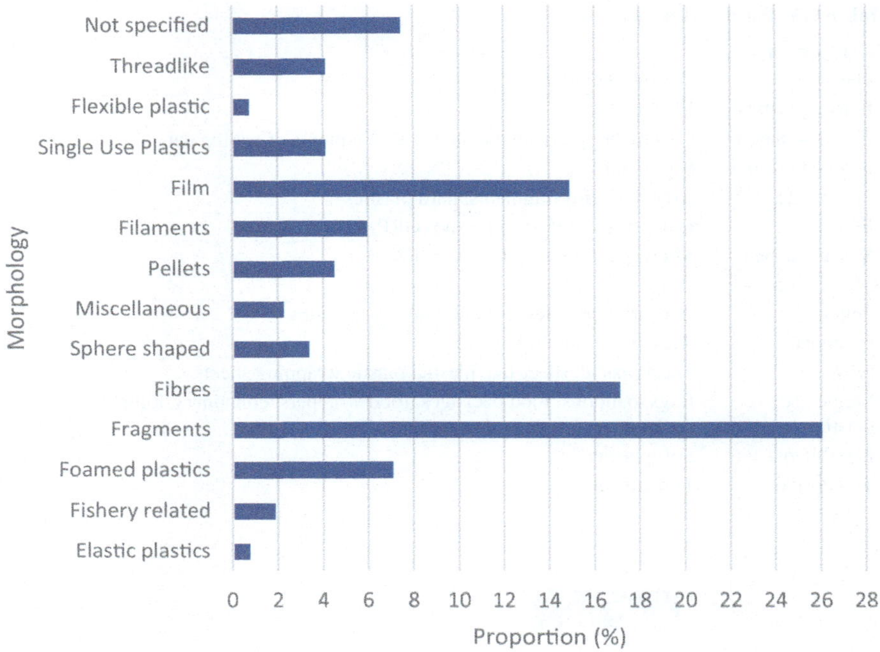

Fig. 5.16 Types of microplastics reported in situ

commonly used in supermarket bags, as specified by (Martellini et al. 2023). Studies that did not specify the polymer type ingested by organisms represented 23.08%.

5.6.11 Frequency of Occurrence

The occurrence percentage, indicating the proportion of specimens ingesting plastics and microplastics, was also documented in reviewed studies. Tables 5.6 and 5.7 present the minimum and maximum occurrence percentages observed across studies by organism type. Cephalopods exhibited the highest minimum ingestion percentage among all animal types, ranging from 93% to 100% (Pedà et al. 2022). Holothurians followed, with a minimum of 73% (Rios-Fuster et al. 2022) and a maximum of 100% (Iwalaye et al. 2021; Rios-Fuster et al. 2022) in in situ studies. In contrast, echinoids showed the lowest ingestion percentage, reaching a maximum of 39% (Digka et al. 2023). No ingestion percentage was reported for jellyfish (Ge et al. 2023). For birds, fish, and marine mammals, ingestion percentages ranged from 0% to 100%. Birds such as guillemots (Pollet et al. 2024), the great horned owl (*Bubo virginianus*), the barn owl (*Tyto alba*), the red-tailed hawk (*Buteo jamaicensis*), and the Cooper's hawk (*Accipiter cooperii*) (Muñoz et al. 2023) showed 0% ingestion. Species like *Brachionus plicatilis*, *Ardenna gravis* (Muñoz et al. 2023), and the gentoo penguin (*Pygoscelis papua*) exhibited 100% occurrence. Conversely,

Table 5.4 Polymer types reported in studies "in situ"

Polymer type reported in situ	Percentage	References
ABS	0.53	Sambolino et al. (2023), Palmer et al. (2021)
Acrylic	1.33	Aytan et al. (2023), Soe et al. (2024), Zhang et al. (2022), Pinho et al. (2022)
Alkyd resin	0.27	James et al. (2023)
Cellophane	0.80	Carrillo-Barragán et al. (2024), Brawn et al. (2023)
Cellulose	1.59	Alomar et al. (2022), Aytan et al. (2023), Atamanalp et al. (2021), Horton et al. (2024), Ojeda et al. (2024)
CPE	0.27	Carrillo-Barragán et al. (2024)
DEP, DIBP, DEHP, BBP	0.27	Sambolino et al. (2023)
EPDM	0.27	Alfonso et al. (2024)
EVA	1.33	Aytan et al. (2023), Horton et al. (2024), Yaghmour et al. (2022), Pereira et al. (2023), Camedda et al. (2022)
PVDF	0.27	Yu et al. (2024)
HDPE	2.65	Solomando et al. (2022), Alomar et al. (2022), Miccoli et al. (2022), Costa et al. (2023), Yu et al. (2024), Motivarash et al. (2024)
LDPE	2.39	Solomando et al. (2022), Alomar et al. (2022), Brawn et al. (2023), Rodrigues et al. (2023)
n-Hexadecanoic acid and methyl ester	0.27	Rajendran et al. (2022)
Nitrile	0.53	Yong et al. (2021), Abd Rahman et al. (2024)
Nova-thene polyolefine	0.27	Devi et al. (2024)
Other	2.12	Nono Almeida et al. (2023), Moon et al. (2022), Nono Almeida et al. (2024), Rios-Fuster et al. (2022), Kuroda et al. (2022)
PA	10.61	Moon et al. (2022), Duncan et al. (2024), Pon et al. (2023), Patsiou et al. (2024), Sciutteri et al. (2023), Hernandez-Milian et al. (2023)
PANI	0.27	Devi et al. (2024)
PB	0.53	Alfonso et al. (2024), Dantas et al. (2024)
PBT	0.27	Alfonso et al. (2024)
PC	0.80	Pollet et al. (2024), Hernandez-Milian et al. (2023), Abd Rahman et al. (2024)
PE	27.59	Alomar et al. (2022), Nono Almeida et al. (2024), Yaghmour et al. (2022), Kelly et al. (2023b), Soe et al. (2024), Ojeda et al. (2024)
PES	0.80	Devi et al. (2024)
PEST	0.27	Merrill et al. (2023)
PET/PETE	7.96	Pedà et al. (2022), Nono Almeida et al. (2024), Yaghmour et al. (2022), Patsiou et al. (2024), Sciutteri et al. (2023), Ojeda et al. (2024)
PEV/PEVA	0.80	Nono Almeida et al. (2023), Giani et al. (2023)

(continued)

Table 5.4 (continued)

Polymer type reported in situ	Percentage	References
PMMA	0.80	Batalha et al. (2024), Abd Rahman et al. (2024), Devi et al. (2024)
Poly (butyl acrylate)	0.27	Brawn et al. (2023)
Poly(isobutylene)	0.27	Moon et al. (2022)
Polyacrylamide	0.80	Atamanalp et al. (2021), Sciutteri et al. (2023), Pinho et al. (2022)
Polyacrylate	0.80	Valente et al. (2022), Rodrigues et al. (2023)
Polyacrylonitrile	0.27	Giani et al. (2023)
Polychloroprene	0.27	Atamanalp et al. (2021)
PEST	2.92	Aytan et al. (2023), Atamanalp et al. (2021), Merrill et al. (2023), Matos et al. (2024), Yong et al. (2021), Bottari et al. (2022)
Polyhexamethylene	0.27	Duncan et al. (2024)
Polyisoprene	0.53	Moon et al. (2022), Dantas et al. (2024)
Polysulfide	0.27	Atamanalp et al. (2021)
Mix	1.33	Nono Almeida et al. (2023), Alfonso et al. (2024), Yaghmour et al. (2022), Sciutteri et al. (2023), Vecchi et al. (2021)
PTFE	0.80	Atamanalp et al. (2021), Vecchi et al. (2021), Devi et al. (2024)
PU	1.06	Pereira et al. (2023), Camedda et al. (2022), Devi et al. (2024), Bottari et al. (2022)
PVA	1.06	Nono Almeida et al. (2023), Giani et al. (2023)
PS	7.96	Nono Almeida et al. (2023), Solomando et al. (2022), Bruno et al. (2022), Collard et al. (2022b)
PVC:	2.92	Moon et al. (2022), Duncan et al. (2024), Bruno et al. (2022)
PVDC	0.27	Kuroda et al. (2022)
Rubber: natural or synthetic rubber	0.80	Pedà et al. (2022), Horton et al. (2024)
Semisynthetic	0.27	Aytan et al. (2023)
Silicone/polysiloxane	0.53	Koraltan et al. (2022)
SAN	0.53	Alomar et al. (2022), Cohen-Sánchez et al. (2022)
Unidentified	1.33	Nono Almeida et al. (2024), Yaghmour et al. (2022), Sciutteri et al. (2023), Lavers et al. (2021)
VAAE	0.27	Merrill et al. (2023)
Varnish	0.27	Matos et al. (2024)
Viscosa	0.53	Giani et al. (2023), Ahuir-Baraja (2022), Lim et al. (2023)
Not specified	7.69	Muñoz et al. (2023), Ramon-Gomez et al. (2024), Cáceres-Farias et al. (2023), Baes et al. (2024), de Almeida et al. (2024)
Total	100.00	

ABS acrylonitrile butadiene styrene, *CPE* chlorinated polyethylene, *EPDM* ethylene propylene diene monomer, *EVA* ethylene vinyl acetate, *PVDF* polyvinylidene fluoride, *PANI* polyaniline, *PB* polybutylene, *PBT* polybutylene terephthalate, *PC* polycarbonate, *PES* polyethersulfone, *PET/PETE* polyethylene terephthalate, *PEV/PEVA* polyethylene vinyl acetate, *PMMA* polymethyl methacrylate, *PTFE* polytetrafluoroethylene, *PU* polyurethane, *PVA* polyvinyl alcohol, *PVDC* polyvinylidene chloride, *SAN* styrene acrylonitrile, *VAAE* vinyl acetate ethylene copolymer

Table 5.5 Polymer types reported in laboratory studies

Polymer type reported in laboratory	Percentage	References
LDPE	7.69	Bonifacio et al. (2022)
BIO1 and BIO_2	7.69	Martellini et al. (2023)
PE	23.08	Iwalaye et al. (2021), Kelly et al. (2023a), Horie et al. (2024)
PET	7.69	Yoo et al. (2024)
PS	30.77	Zhao et al. (2024), Kelly et al. (2023a), Ge et al. (2023)
Not specified	23.08	Capó et al. (2022), Miccoli et al. (2022), Navarro et al. (2023)
Total	100	

fish species such as *Mullus surmuletus*, *Nemipterus randalli*, *Oblada melanura* (Koraltan et al. 2022), and *Sparus aurata* (Capó et al. 2022) showed no presence of plastics. Species including *Mola mola* (Ahuir-Baraja 2022), *Merluccius merluccius* (Miccoli et al. 2022), and *Katsuwonus pelamis* (Ragesh et al. 2024) had plastics in all examined specimens. Marine mammal species like *Globicephala melas*, *Grampus griseus*, *Lagenorhynchus obscurus*, *Orcinus orca*, *Pseudorca crassidens*, *Stenella clymene*, *Stenella coeruleoalba*, *Tursiops truncatus*, *Kogia breviceps*, and *Phocoena dioptrica* showed no ingestion of plastics, representing 0% ingestion (Padula et al. 2023). In contrast, species such as the bottlenose dolphin (*Tursiops truncatus*), beluga whale (Delphinapterus leucas), gray whale (*Eschrichtius robustus*), bearded seal (*Erignathus barbatus*), and the Guiana dolphin (*Sotalia guianensis*) exhibited a 100% occurrence of plastic ingestion (da Silva et al. 2024).

In laboratory studies, the minimum occurrence frequency was observed in crustaceans (Martellini et al. 2023) and fish (Capó et al. 2022), while 100% occurrence frequencies were reported in birds (Navarro et al. 2023), bivalves (Bonifacio et al. 2022), holothurians (Iwalaye et al. 2021), and zooplankton (Zhao et al. 2024).

5.6.12 Effects of Plastic Ingestion

Most of the articles did not report on the effects of plastic ingestion. Six articles expressly reported that no effect was observed on the organisms due to plastic ingestion under certain conditions. The organisms in which no effect was observed after microplastic ingestion were fish *Micropogonias furnieri* (Alves and Figueiredo 2023), Zooplankton *Eucalanus pileatus* and *Dolioletta gegenbauri* (Köster and Paffenhöfer 2022), seabird *Ardenna carneipes* (Lavers and Bond 2023), shark *Scyliorhinus canicular* (Mancuso et al. 2022), sea urchin *Paracentrotus lividus* (Digka et al. 2023), and jellyfish *Rhopilema esculentum* (Ge et al. 2023). All the studies that reported on found effects were conducted in a laboratory, meaning the conditions of microplastic ingestion were controlled inside the laboratory. The reported effects are shown in Table 5.8.

All the above studies focused on one specific subject without considering if damage was inflicted in other tissues or systems. Thus, it must be considered that the effects reported could be one of many consequences occasioned by exposure to

Table 5.6 Percentage of ingestion occurrence reported in studies "in situ"

Type of organism "in situ"	Minimum percentage of occurrence	Maximum percentage of occurrence	References
Bird	0	100	Muñoz et al. (2023), Schutten et al. (2023), Pollet et al. (2024), Kim et al. (2023b)
Bivalve	33	100	Alomar et al. (2022), Abd Rahman et al. (2024)
Cephalopod	93	100	Pedà et al. (2022), Sambolino et al. (2023)
Crustacean	11	100	Fagiano et al. (2024), Yücel (2023)
Echinoidea	26	39	Digka et al. (2023)
Fish	0	100	Koraltan et al. (2022), Duncan et al. (2024), Patsiou et al. (2024)
Holothurian	73	100	Rios-Fuster et al. (2022)
Macroinvertebrates	55	NA	UN Habitat - United Nations Human Settlement Programme and WHO - World Health Organization (2021), Costa et al. (2023)
Marine mammal	0	100	Padula et al. (2023), da Silva et al. (2024)
Polychaete	3	91	James et al. (2023), Vecchi et al. (2021)
Sea star	56	70	Cossi et al. (2021)
Shark	36	100	Gong et al. (2023), Mboglen et al. (2024)
Turtle	5,7	100	Moon et al. (2022), Palmer et al. (2021)
Zooplankton	Not specified	Not specified	Alfonso et al. (2024)

Table 5.7 Percentage of ingestion occurrence reported in Laboratory studies

Type of organism in "laboratory"	Minimum percentage of occurrence	Maximum percentage of occurrence	References
Bird	2.8	100	Navarro et al. (2023), Matos et al. (2024)
Bivalve	–	100	Bonifacio et al. (2022)
Crustacean	0	0	Martellini et al. (2023)
Fish	0	66	Capó et al. (2022), Aytan et al. (2023)
Holothurian	–	100	Iwalaye et al. (2021)
Zooplankton	Not specified	100	Zhao et al. (2024)

microplastics. However, it is important to note that all reported effects are related to cellular and internal damage.

5.6.13 Mitigation of Plastic Waste Pollution in the Oceans

Due to the increasing and notable problem of marine litter, action must be taken to protect the environment and marine ecosystems along with all the species that inhabit them, which represent an essential part of the biodiversity and ecosystems worldwide. However, it must be considered that the ocean can be regarded as a single entity, which covers 71% of the planet's surface and is in continuous movement (NOAA 2021), so the measures to mitigate plastic pollution are complex; these range from local to global actions, including political, technological, economic, and social approaches.

5.6.14 Regulations

Bans and Restrictions: Implement bans on single-use plastics, such as plastic bags, straws, utensils, and food containers (Bharadwaj et al. 2023). For example, many Caribbean countries have adopted legislative bans on these products, which include fines and sanctions for noncompliance (Clayton et al. 2021). Extended Producer Responsibility (EPR): Forcing producers to be responsible for the generation, marketing, and disposal of plastics, this type of system has contributed to reducing their income in many countries (Tumu et al. 2023; Harris et al. 2021b). Improving Waste Management: Strengthen waste management systems to ensure proper plastic waste collection, recycling, and disposal; this includes the development of efficient recycling infrastructure and the promotion of sustainable waste management practices (Ugwu et al. 2021b), in addition to new approaches such as the adoption of a simple and informative packaging labeling system for correct classification, more awareness, and motivation toward the population by companies and governments (Sandhi and Rosenlund 2024).

5.6.15 Technological Innovations

Bioplastics and Alternative Materials: Promote the development and use of bioplastics that biodegrade faster in the environment. For example, gillnets, rope, and monofilament nets with biodegradable polybutylene succinate co-adipate-co-terephthalate (PBSAT), Polybutylene succinate (PBS), and polybutylene adipate succinate (PBAT) resins are being investigated to reduce contamination from lost fishing gear at sea known as ghost nets (Grimaldo et al. 2019; Le Gu et al. 2023).

Detection and Mitigation Technologies: Advanced technologies such as satellite image processing and aerial systems (Topouzelis et al. 2019), artificial intelligence systems to monitor and predict the flow of plastics in the ocean in real-time (Lebreton

Table 5.8 Effects reported on organisms after plastic ingestion

	Species	Effect reported	
Jellyfish	*Rhopilema esculentum*	• At environmentally relevant concentrations, no effect was observed • At high concentrations, there was a reduction in the lipase enzyme activity in the juveniles after 15 days of exposure, and possibly the epithelium of the digestive filament was abraded. There was also an increase in the respiration rate and a possible reduced energy uptake caused by the microplastics	Kelly et al. (2023a))
Fish	*Sparus aurata*	After 60 days • Antioxidant and proinflammatory enzyme activities in the liver significantly increased • BPA levels were significantly higher in muscle tissue	Alomar et al. (2022)
		After 120 days • Antioxidant and proinflammatory enzyme activities in the liver significantly increased • SOD activity in blood cells significantly decreased • MDA plasma levels were significantly lower • ROS production in the liver was significantly higher • BPA and BPS levels were significantly higher in muscle tissue	
Crustacean	*Talitrus saltator*	• Survival rate improved with bioplastic one compared to bioplastic two and the control group • Plasticizing agents were found in all samples	Martellini et al. (2023)
Fish	*Merluccius merlucciu* and *Mullus barbatus*	• Exposure to micro litter induced a decrease in median cell viability after incubating 72 h • Median intracellular ATP levels were lower	Lim et al. (2023)

Fish	*Oreochromis niloticus*	Fish fed for 30 days with 100 μg of PP • Leukocyte, lymphocyte, and thrombocyte counts increased significantly, suggesting inflammatory effects on the digestive tract • Dose-respondent serum enzymatic activity of ALT and AST significantly increased, indicating liver disorders • Possible congestive circulatory alterations	do Nascimento et al. (2023)
		Fish fed for 30 days with 500 μg of PP • Leukocyte, lymphocyte, and thrombocyte counts increased significantly, suggesting inflammatory effects on the digestive tract • Dose-respondent serum enzymatic activity of ALT and AST significantly increased, indicating liver disorders • Significant increase in cholesterol levels and decrease in creatinine levels • Possible congestive circulatory alterations • Significant change in intestinal microbiota: increase in gram-negative microorganisms and decrease in mesophilic aerobic microorganisms	

BPA plasticizer: Bisphenol A, *SOD* antioxidant enzyme: superoxide dismutase, *MDA* oxidative damage marker: malondialdehyde, *ROS* reactive oxygen species, *BPS* plasticizer: bisphenol S, *ATP* adenosine triphosphate, *ALT* enzyme: alanine transaminase, *AST* enzyme: aspartate transferase, *PP* polypropylene et al. 2019), and aerial remote sensing through spectral reflectance (Garaba and Dierssen 2018) are used.

5.6.16 Education and Public Awareness

Awareness Campaigns: Develop public education campaigns to raise awareness among the population about the impacts of plastic pollution and encourage more sustainable behaviors. These campaigns can include information about reducing plastic use, proper recycling, and keeping oceans clean (Ocean-Conservancy 2024).

Citizen Science Programs: Involve the community in data collection and marine pollution monitoring. Citizen science programs can provide valuable data to evaluate the effectiveness of mitigation policies and strategies (Harris et al. 2021b).

5.6.17 Collaboration and International Alliances

Regional and International Cooperation: Help foster collaboration between regions and countries to address plastic pollution; this includes sharing technologies, knowledge, and best practices and working together on global initiatives to reduce the entry of plastic waste into the environment (Wittstock and Wessolowski 2022).

5.6.18 Economic and Financial Incentives

Grants and Funds: Establish funds and grants to support innovative projects and technologies that seek to minimize plastic pollution. This may include funding research and development initiatives for new materials and recycling methods (Mushtaq et al. 2024). Taxes and Fees: Implement taxes and fees on plastic products to discourage their use and generate funds for reinvestment in waste management and marine cleanup programs (Abate and Elofsson 2024). In summary, mitigating plastic pollution in the oceans requires a comprehensive approach that combines effective policies, technological innovations, public education, international collaboration, and financial mechanisms. These measures must be coordinated to address the growing threat of plastics in marine ecosystems effectively.

5.6.19 Conclusion

It is clear and well-documented that plastic waste, in various sizes, is present in virtually all ecosystems on the planet, with potential consequences for many organisms. However, comparing and systematizing all the information generated to date is challenging due to several key issues. One significant problem is the lack of consensus on how to analyze samples and the methodology for determining the number of specimens to study. For future research, establishing an agreement on the minimum number of specimens needed for statistically valid results is crucial. Moreover, there is no standardized definition to clearly distinguish between nanoplastics, microplastics, and mesoplastics, nor are there uniform criteria for reporting the morphology and color of the plastics identified. . Comparing studies becomes challenging without precise specifications on the characteristics of the plastics analyzed. Another issue is the variable duration of studies, which hinders comparisons. Standardizing study durations and the units of plastic concentration in different parts of the world would yield more robust and comparable results. Additionally, although information has been increasingly generated in recent years, certain regions lack sufficient information. These areas need to be characterized and analyzed for the presence of plastics. The scientific community must adopt international agreements on the issues discussed before to make modeling effects and comparing data possible and practical. This comprehensive approach is essential for accurately assessing the global impact of plastic pollution. Moreover, while the effects of microplastics on ecosystems and food chains are not entirely clear, it is evident that they can have negative impacts. However, many laboratory studies use concentrations that do not reflect real-world conditions, underscoring the need for further research focused on the ingestion of microplastics and the characteristics that attract different types of organisms. Understanding these characteristics would help identify the species most affected and the mechanisms driving plastic ingestion.

Finally, addressing plastic pollution in the oceans necessitates a holistic strategy that integrates effective policies, technological advancements, public awareness initiatives, international cooperation, and financial resources. These actions must be synchronized to tackle the escalating threat of plastics in marine ecosystems effectively. Only through a concerted global effort can we significantly reduce the impact of plastics on our oceans and preserve marine biodiversity for future generations.

References

Abate TG, Elofsson K (2024) Environmental taxation of plastic bags and substitutes: balancing marine pollution and climate change. J Environ Manag 359:120868

Abd Rahman NN et al (2024) Evaluation of the microplastics in bivalves and water column at Pantai Teluk Likas, North Borneo, Malaysia. Environ Sci Pollut Res 31(15):23178–23192

Abdulaziz A et al (2024) Plastic litter: a hidden reservoir for antibiotic-resistant pathogens in coastal ecosystems. Reg Stud Mar Sci 76:103616

Agathokleous E, Iavicoli I, Barceló D, Calabrese EJ (2021) Micro/nanoplastics effects on organisms: a review focusing on 'dose'. J Hazard Mater 417:1–2

Agboola OD, Benson NU (2021) Physisorption and chemisorption mechanisms influencing micro (nano) plastics-organic chemical contaminants interactions: a review. Front Environ Sci 9:678574

Ahuir-Baraja AE (2022) First evidence of presence of plastic debris in digestive system of Mola mola (L.) from western Mediterranean Sea. Mar Pollut Bull 175:113326

Alfaro-Núñez A et al (2021) Microplastic pollution in seawater and marine organisms across the tropical eastern Pacific and Galápagos. Sci Rep 11(1):1–8

Alfonso MB et al (2024) Small microplastic ingestion by the calanoid Centropages furcatus in the Gulf of Thailand. Sci Total Environ 930:172837

Alfred Wegener Institute Helmholtz Centre for Polar and Marine Research (2024) Litterbase. https://litterbase.awi.de/. Accessed 6 July 2024
Alomar C, Deudero S, Compa M, Guijarro B (2020) Exploring the relation between plastic ingestion in species and its presence in seafloor bottoms. Mar Pollut Bull 160:111641
Alomar C et al (2022) A holistic approach to plastic pollution in integrated multi-trophic aquaculture facilities: plastic ingestion in Sparus aurata and Mytilus galloprovincialis. Aquaculture 561:738666
Alves VEN, Figueiredo GM (2023) Assessment of microplastic impacts on whitemouth croaker (Micropogonias furnieri) and ecosystem services in Guanabara Bay, Brazil. Environ Biol Fish 106(12):2177–2192
Alves NM et al (2024) Like noodles in a soup: anthropogenic microfibers are being ingested by juvenile fish in nursery grounds of the southwestern Atlantic Ocean. Mar Pollut Bull 202:116368
AMMPA - Alliance of Marine Mammal Parks and Aquariums (2017) Hoja de hechos: delfín nariz de botella
AMMPA - Alliance of Marine Mammal Parks and Aquariums (2019) Información especializada: Belugas
Andrady AL (2011) Microplastics in the marine environment. Mar Pollut Bull 62(8):1596–1605
Andrady AL, Neal MA (2009) Applications and societal benefits of plastics. Philos Trans R Soc B Biol Sci 364(1526):1977–1984
Angiolillo M et al (2021) Distribution of seafloor litter and its interaction with benthic organisms in deep waters of the Ligurian Sea (northwestern Mediterranean). Sci Total Environ 788:147745
Arif Y, Mir AR, Zieliński P, Hayat S, Bajguz A (2024) Microplastics and nanoplastics: source, behavior, remediation, and multi-level environmental impact. J Environ Manag 356:120618
Arizaga J (2023) Gaviota patiamarilla—Larus michahellis Naumann, 1840. Museo Nacional de Ciencias Naturales, Madrid
Atamanalp M et al (2021) Microplastics in tissues (brain, gill, muscle and gastrointestinal) of Mullus barbatus and Alosa immaculata. Arch Environ Contam Toxicol 81(3):460–469
Aytan Ü, Başak Esensoy F, Şentürk Y, Güven O, Karaoğlu K, Erbay M (2023) Plastic occurrence in fish caught in the highly industrialized gulf of İzmit (Eastern Sea of Marmara, Türkiye). Chemosphere 324:138317
Baes L, Santiago CD, Roman L, dos Santos Costa PC, Pugliesi É, Reigada C (2024) Beached seabirds as plastic biomonitors in Brazil from the beach monitoring project of the Santos Basin (PMP-BS). Mar Pollut Bull 199:115847
Bange A, Backes A, Garthe S, Schwemmer P (2023) Prey choice and ingestion of microplastics by common shelducks and common eiders in the Wadden Sea world heritage site. Mar Biol 170(5):54
Batalha DC, Vaz-dos-Santos AM, da Silveira EL, Ávila-da-Silva AO, dos Santos LD (2024) How are plastic debris affecting the diet of the whitemouth croaker in the southeastern Brazilian bight? Reg Stud Mar Sci 75:103567
Benjaminsen SC, Bourgeon S, Herzke D, Ask A, Collard F, Gabrielsen GW (2022) First documentation of plastic ingestion in the arctic glaucous gull (Larus hyperboreus). Sci Total Environ 834:155340
Benson NU, Agboola OD, Fred-Ahmadu OH, De-la-Torre GE, Oluwalana A, Williams AB (2022) Micro(nano)plastics prevalence, food web interactions, and toxicity assessment in aquatic organisms: a review. Front Mar Sci 9:851281
Bergmann M, Gutow L, Klages M (2015) Marine anthropogenic litter. Springer, Cham
Bharadwaj B, Subedi MN, Rai RK (2023) Retailer's characteristics and compliance with the single-use plastic bag ban. Sustain Anal Model 3:100019
Bhuyan MS (2022) Effects of microplastics on fish and in human health. Front Environ Sci 10:827289
Bonifacio PSP, Metillo EB, Romano EF (2022) Microplastic in sediments and ingestion rates in three edible bivalve mollusc species in a southern Philippine estuary. Water Air Soil Pollut 233(11):455

Borgå K (2008) Bioaccumulation. In: Jørgensen SE, Fath BD (eds) Encyclopedia of ecology, vol 1. Academic Press, New York, pp 346–348

Bottari T et al (2022) Anthropogenic microparticles in the emerald rockcod Trematomus bernacchii (Nototheniidae) from the Antarctic. Sci Rep 12(1):1–9

BPF - British Plastics Federation (2024) Plastics additives. https://www.bpf.co.uk/plastipedia/additives/Default.aspx. Accessed 6 June 2024

Brawn C, Hamilton BM, Savoca MS, Bardarson B, Vermaire JC, Provencher J (2023) Suspected anthropogenic microparticle ingestion by Icelandic capelin. Mar Pollut Bull 196:115551

Brown E, MacDonald A, Allen S, Allen D (2023) The potential for a plastic recycling facility to release microplastic pollution and possible filtration remediation effectiveness. J Hazard Mater Adv 10:100309

Bruno C et al (2022) Polymer composition analysis of plastic debris ingested by loggerhead turtles (Caretta caretta) in southern Tyrrhenian Sea through ATR-FTIR spectroscopy. Mar Environ Res 179:105676

Bucci K, Tulio M, Rochman CM (2020) What is known and unknown about the effects of plastic pollution: a meta-analysis and systematic review. Ecol Appl 30(2):1–16

Cáceres-Farias L, Espinoza-Vera MM, Orós J, Garcia-Bereguiain MA, Alfaro-Núñez A (2023) Macro and microplastic intake in seafood variates by the marine organism's feeding behaviour: is it a concern to human health? Heliyon 9(5):e16452

Camedda A et al (2022) Polymer composition assessment suggests prevalence of single-use plastics among items ingested by loggerhead sea turtles in the western Mediterranean sub-region. Environ Pollut 292:118274

Cannon SME, Lavers JL, Figueiredo B (2016) Plastic ingestion by fish in the southern hemisphere: a baseline study and review of methods. Mar Pollut Bull 107(1):286–291

Capó X et al (2022) Quantification of differential tissue biomarker responses to microplastic ingestion and plasticizer bioaccumulation in aquaculture reared sea bream Sparus aurata. Environ Res 211:113063

Cardozo ALP, Yofukuji KY, da Silva Júnior RC, de Castro-Hoshino LV, Fugi R (2023) Plastic ingestion by carnivore fish in a neotropical floodplain: seasonal and interspecific variations. Environ Sci Pollut Res 30(14):40712–40723

Carrillo-Barragán P, Fitzsimmons C, Lloyd-Hartley H, Tinlin-Mackenzie A, Scott C, Sugden H (2024) Fifty-year study of microplastics ingested by brachyuran and fish larvae in the central English North Sea. Environ Pollut 342:123060

Castañeta G, Gutiérrez AF, Nacaratte F, Manzano CA (2020) Microplásticos: un contaminante que crece en todas las esferas ambientales, sus características y posibles riesgos para la salud pública por exposición. Rev Boliv Quím 37(3):160–175

Chapron L et al (2018) Macro- and microplastics affect cold-water corals growth, feeding and behaviour. Sci Rep 8(1):1–8

Chen KJ, Chen MC, Chen TH (2021) Plastic ingestion by fish in the coastal waters of the Hengchun peninsula, Taiwan: associated with human activity but no evidence of biomagnification. Ecotoxicol Environ Saf 213:112056

Chen F et al (2023) Impact of facemask debris on marine diatoms: physiology, surface properties, sinking rate, and copepod ingestion. Sci Total Environ 905:167222

Chenet T et al (2021) Plastic ingestion by Atlantic horse mackerel (Trachurus trachurus) from Central Mediterranean Sea: a potential cause for endocrine disruption. Environ Pollut 284:117449

Choi DY, Gredzens C, Shaver DJ (2021) Plastic ingestion by green turtles (Chelonia mydas) over 33 years along the coast of Texas, USA. Mar Pollut Bull 173:113111

CIMI - Catalina Island Marine Institute (2020) Bioaccumulation and biomagnification: increasingly concentrated problems! https://cimi.org/blog/bioaccumulation-and-biomagnification-increasingly-concentrated-problems/

Ciocan C, Annels C, Fitzpatrick M, Couceiro F, Steyl I, Bray S (2024) Glass reinforced plastic (GRP) boats and the impact on coastal environment—evidence of fibreglass ingestion by marine bivalves from natural populations. J Hazard Mater 472:134619

Clayton CA, Walker TR, Carlos J, Adam I (2021) Policy responses to reduce single-use plastic marine pollution in the Caribbean. Mar Pollut Bull 162:111833

Cohen-Sánchez A et al (2022) First detection of microplastics in Xyrichtys novacula (Linnaeus 1758) digestive tract from Eivissa Island (Western Mediterranean). Environ Sci Pollut Res 29:65077–65087

Cole M, Lindeque P, Fileman E, Halsband C, Galloway TS (2015) The impact of polystyrene microplastics on feeding, function and fecundity in the marine copepod Calanus helgolandicus. Environ Sci Technol 49(2):1130–1137

Collard F, Bangjord G, Herzke D, Gabrielsen GW (2022a) Plastic burdens in northern fulmars from Svalbard: looking back 25 years. Mar Pollut Bull 185:114333

Collard F et al (2022b) Plastic ingestion and associated additives in Faroe Islands chicks of the northern fulmar Fulmarus glacialis. Water Biol Secur 1(4):1–9

Compa M, Wilcox C, Hardesty BD, Alomar C, March D, Deudero S (2022) Quantifying the risk of plastic ingestion by ichthyofauna in the Balearic Islands (western Mediterranean Sea). Mar Pollut Bull 183:114075

Compa M, Perelló E, Box A, Colomar V, Pinya S, Sureda A (2023) Ingestion of microplastics and microfibers by the invasive blue crab Callinectes sapidus (Rathbun 1896) in the Balearic Islands, Spain. Environ Sci Pollut Res 30(56):119329–119342

Compa M, Capó X, Alomar C, Deudero S, Sureda A (2024) A meta-analysis of potential biomarkers associated with microplastic ingestion in marine fish. Environ Toxicol Pharmacol 107:104414

CONANP - Comisión Nacional de Áreas Naturales Protegidas (2013) Monitoreo de la Ballena Gris (Eschrichtius robustus) en la Reserva de la Biosfera El Vizcaíno

Contreras-Sillero ME et al (2019) Evaluation of growth and bromatological composition of the rotifer Brachionus plicatilis fed different native microalgae from the Gulf of California. Rev Bio Ciên 6:509

Cossi PF et al (2021) First evidence of microplastics in the marine protected area Namuncurá at Burdwood Bank, Argentina: a study on Henricia obesa and Odontaster penicillatus (Echinodermata: Asteroidea). Polar Biol 44(12):2277–2287

Costa LL, David da Costa I, da Silva Oliveira A, Zalmon IR (2023) 'Microplastic ecology': testing the influence of ecological traits and urbanization in microplastic ingestion by sandy beach fauna. Estuar Coast Shelf Sci 290:108406

Coyle R, Hardiman G, Driscoll KO (2020) Microplastics in the marine environment: a review of their sources, distribution processes, uptake and exchange in ecosystems. Case Stud Chem Environ Eng 2:100010

Cruz AH (2023) Impact of plastic waste ingestion by fish. Circ Econ Sustain 3(1):607–616

Cui W et al (2023) Sorption of representative organic contaminants on microplastics: effects of chemical physicochemical properties, particle size, and biofilm presence. Ecotoxicol Environ Saf 251:114533

da Silva LVG et al (2024) Identification and characterization of plastic debris in the gastrointestinal tract of Guiana dolphins (Sotalia guianensis) from Espírito Santo coast, Brazil. Mar Pollut Bull 200:116076

Dada OA, Bello JO (2023) Microplastics in carnivorous fish species, water and sediments of a coastal urban lagoon in Nigeria. Environ Sci Pollut Res 30(19):55948–55957

Dantas DV et al (2024) Influence of trophic overlaps and trophic niche amplitude on microplastic intake of fish species in shallow areas of a neotropical coastal lagoon. Sci Total Environ 927:172235

DCCEEW - Department of Climate Change Energy the Environment and Water (2024) Ardenna carneipes—flesh-footed shearwater, fleshy-footed shearwater. Species profile and threats database. http://www.environment.gov.au/cgi-bin/sprat/public/publicspecies.pl?taxon_id=82404. Accessed 3 July 2024

de Almeida AB, Henrique BVR, de Souza Azevedo J (2024) Stomach content analysis in ariids: insights of three catfish species from a marine protected area. Reg Stud Mar Sci 75:103573 de Lemos Santana J, dos Santos Calado TC, de Souza-Filho JF (2022) Feeding of Eriphia gonagra (Crustacea: Eriphiidae) in two polluted reef areas in tropical Brazil with records of ingestion of microplastics. Thalass An Int J Mar Sci 38(1):431–443

Devi SS, Saifudeen N, Kumar KS, Kumar AB (2024) Does the microplastics ingestion patterns and polymer composition vary across the oceanic zones? A case study from the Indian coast. Mar Pollut Bull 204:116532

Diaz-Santibañez I, Clark BL, Zavalaga CB (2023) Guanay cormorant (Leucocarbo bougainvilliorum) pellets as an indicator of marine plastic pollution along the Peruvian coast. Mar Pollut Bull 192:115104

Digka N, Patsiou D, Kaberi H, Krasakopoulou E, Tsangaris C (2023) Microplastic ingestion and its effects on sea urchin Paracentrotus lividus: a field study in a coastal East Mediterranean environment. Mar Pollut Bull 196:115613

do Nascimento LS et al (2023) Deleterious effects of polypropylene microplastic ingestion in nile tilapia (Oreochromis niloticus). Bull Environ Contam Toxicol 111(1):13

Drouillard K (2008) Biomagnification. In: Jørgensen SE, Fath BD (eds) Encyclopedia of ecology. Academic Press, New York, pp 441–448

Duis K, Coors A (2016) Microplastics in the aquatic and terrestrial environment: sources (with a specific focus on personal care products), fate and effects. Environ Sci Eur 28(1):1–25

Duncan EM et al (2024) Marine turtles as bio-indicators of plastic pollution in the eastern Mediterranean. Mar Pollut Bull 201:116141

Egbeocha CO, Malek S, Emenike CU, Milow P (2018) Feasting on microplastics: ingestion by and effects on marine organisms. Aquat Biol 27:93–106

Fagiano V et al (2023) Breaking the paradigm: marine sediments hold two-fold microplastics than sea surface waters and are dominated by fibers. Sci Total Environ 858:159722

Fagiano V, Alomar C, Ventero A, de Puelles MLF, Iglesias M, Deudero S (2024) First assessment of anthropogenic particle ingestion in Pontellid copepods: Pontella mediterranea as a potential microplastic reservoir in the Neuston. Sci Total Environ 908:168480

FAO (2017) Microplastics in fisheries and aquaculture. FAO, Rome

FAO - Food and Agriculture Organization (2003) Species fact sheet: Merlangius merlangus. Cultured Aquatic Species Information Programme

FAO - Food and Agriculture Organization (2009) Oreochromis niloticus (Linnaeus, 1758) [Cichlidae]. Cultured acuatic species fact sheets. https://www.fao.org/fishery/docs/DOCUMENT/aquaculture/CulturedSpecies/file/es/es_niletilapia.htm. Accessed 27 June 2024

Farrell P, Nelson K (2013) Trophic level transfer of microplastic: Mytilus edulis (L.) to Carcinus maenas (L.). Environ Pollut 177:1–3

Fatema K et al (2024) Occurrence of microplastics in fish gastrointestinal tracts belongs to different feeding habits from the Bangladesh coast of the bay of Bengal. Environ Sci Pollut Res 31(16):24329–24343

Feitosa AF et al (2024) The impact of chronic and acute problems on sea turtles: the consequences of the oil spill and ingestion of anthropogenic debris on the tropical semi-arid coast of Ceará, Brazil. Aquat Toxicol 269:106867

Foskolos I et al (2023) Insights into the distribution and ingestion of prey-like plastic fishing lures in Mediterranean rough-toothed dolphins. Mar Pollut Bull 188:114701

FOW Mould (2024) Plastic additives complete guide. https://www.immould.com/plastic-additives/

Galgani F, Hanke G, Werner S, De Vrees L (2013) Marine litter within the European marine strategy framework directive. ICES J Mar Sci 70:1055–1064

Gao Z, Cizdziel JV, Wontor K, Vianello A (2022) Spatiotemporal characteristics of microplastics in a university wastewater treatment plant: influence of sudden on-campus population swings. J Environ Chem Eng 10(6):108834

Garaba SP, Dierssen HM (2018) An airborne remote sensing case study of synthetic hydrocarbon detection using short wave infrared absorption features identified from marine-harvested macro- and microplastics. Remote Sens Environ 205:224–235

Ge J et al (2023) Ingestion, egestion and physiological effects of polystyrene microplastics on the marine jellyfish Rhopilema esculentum. Mar Pollut Bull 187:114609

Geyer R, Jambeck JR, Law KL (2017) Production, use, and fate of all plastics ever made. Sci Adv 3(7):25–29
Giani D et al (2023) Trophic niche influences ingestion of micro- and mesoplastics in pelagic and demersal fish from the Western Mediterranean Sea. Environ Pollut 328:121632
Giesy JP, Feyk LA, Jones PD, Kannan K, Sanderson T (2003) Review of the effects of endocrine-disrupting chemicals in birds. Pure Appl Chem 75(11-12):2287–2303
Golwala H, Zhang X, Iskander SM, Smith AL (2021) Solid waste: an overlooked source of microplastics to the environment. Sci Total Environ 769:144581
Gong Y et al (2023) Plastic ingestion and trophic transfer in an endangered top predator, the longfin mako shark (Isurus paucus), from the tropical western Pacific Ocean. Environ Sci Pollut Res 30(49):107365–107370
Grimaldo E et al (2019) Comparison of fishing efficiency between biodegradable gillnets and conventional nylon gillnets. Fish Res 213:67–74
Hajisamae S, Soe KK, Pradit S, Chaiyvareesajja J, Fazrul H (2022) Feeding habits and microplastic ingestion of short mackerel, Rastrelliger brachysoma, in a tropical estuarine environment. Environ Biol Fish 105(2):289–302
Harper C, Fowler A (1985) Plastic pellets in New Zealand storm-killed prions, Pachyptila
Harper CJ, McLellan WA, Rommel SA, Gay DM, Dillaman RM, Pabst DA (2008) Morphology of the melon and its tendinous connections to the facial muscles in bottlenose dolphins (Tursiops truncatus). J Morphol 269(7):820–839
Harris PT, Westerveld L, Nyberg B, Maes T, Macmillan-Lawler M, Appelquist LR (2021a) Exposure of coastal environments to river-sourced plastic pollution. Sci Total Environ 769:145222
Harris L, Liboiron M, Charron L, Mather C (2021b) Using citizen science to evaluate extended producer responsibility policy to reduce marine plastic debris shows no reduction in pollution levels. Mar Policy 123:104319
Hernandez-Milian G et al (2023) Monk seal faeces as a non-invasive technique to monitor the incidence of ingested microplastics and potential presence of plastic additives. Mar Pollut Bull 193:115227
Hodkovicova N, Hollerova A, Svobodova Z, Faldyna M, Faggio C (2022) Effects of plastic particles on aquatic invertebrates and fish—a review. Environ Toxicol Pharmacol 96:104013
Hofmann T et al (2023) Plastics can be used more sustainably in agriculture. Commun Earth Environ 4(1):332
Horie Y et al (2024) Variability in microplastic color preference and intake among selected marine and freshwater fish and crustaceans. Discov Ocean 11(1):1–12
Horton AA, Weerasinghe KDI, Mayor DJ, Lampitt R (2024) Microplastics in commercial marine fish species in the UK—a case study in the river Thames and the river Stour (East Anglia) estuaries. Sci Total Environ 915:170170
Im J, Joo S, Lee Y, Kim BY, Kim T (2020) First record of plastic debris ingestion by a fin whale (Balaenoptera physalus) in the sea off East Asia. Mar Pollut Bull 159:111514
Iwalaye OA, Moodley GK, Robertson-Andersson DV (2021) Water temperature and microplastic concentration influenced microplastic ingestion and retention rates in sea cucumber (Holothuria cinerascens Brandt, 1835). Ocean Sci J 56(2):141–155
Jackson GD, Buxton NG, George MJA (2000) Diet of the southern opah Lampris immaculatus on the Patagonian shelf. Mar Ecol Prog Ser 206:261–271
James K et al (2023) Microplastic ingestion by the polychaete community in the coastal waters of Kochi, southwest coast of India. Reg Stud Mar Sci 62:102948
Jepsen EM, de Bruyn PJN (2019) Pinniped entanglement in oceanic plastic pollution: a global review. Mar Pollut Bull 145:295–305
Joyce PWS, Falkenberg LJ (2023) Microplastic abundances in co-occurring marine mussels: species and spatial differences. Reg Stud Mar Sci 57:102730
Kalaiselvan K, Pandurangan P, Velu R, Robinson J (2022) Occurrence of microplastics in gastrointestinal tracts of planktivorous fish from the Thoothukudi region. Environ Sci Pollut Res 29(29):44723–44731

Kalogerakis N et al (2017) Microplastics generation: onset of fragmentation of polyethylene films in marine environment mesocosms. Front Mar Sci 4:1–15

Kantharajan G, Pandey PK, Krishnan P, Bharti VS, Samuel VD (2018) Plastics: a menace to the mangrove ecosystems of megacity Mumbai, India. ISME/GLOMIS Electron J 16(1):1–5

Keerthika K et al (2023) Ingestion of microplastics in commercially important species along Thoothukudi coast, south East India. Environ Monit Assess 195(4):511

Keerthika K, Padmavathy P, Rani V, Jeyashakila R, Aanand S, Kutty R (2024) Evidence of microplastics in the polychaete worm (capitellids—Capitella capitata) (Fabricicus, 1780) along Thoothukudi region. Environ Monit Assess 196(6):1–13

Kelly ERM, Trujillo JE, Setiawan A, Pether S, Burritt D, Allan BJM (2023a) Investigating the impacts of biofouled marine plastic debris on the olfactory behaviour of juvenile yellowtail kingfish (Seriola lalandi). Mar Pollut Bull 192:115079

Kelly NE et al (2023b) Long term trends in floating plastic pollution within a marine protected area identifies threats for endangered northern bottlenose whales. Environ Res 227:115686

Kim Y et al (2023a) Microplastics in gastrointestinal tracts of gentoo penguin (Pygoscelis papua) chicks on King George Island, Antarctica. Sci Rep 13(1):13016

Kim Y et al (2023b) Microplastics in gastrointestinal tracts of gentoo penguin (Pygoscelis papua) chicks on King George Island, Antarctica. Sci Rep 13(1):1–11

Koraltan İ, Mavruk S, Güven O (2022) Effect of biological and environmental factors on microplastic ingestion of commercial fish species. Chemosphere 303:135101

Köster M, Paffenhöfer GA (2022) Feeding of marine zooplankton on microplastic fibers. Arch Environ Contam Toxicol 83(2):129–141

Kotrikla AM, Zavantias A, Kaloupi M (2021) Waste generation and management onboard a cruise ship: a case study. Ocean Coast Manag 212:105850

Kühn S, Van Franeker JA (2020) Quantitative overview of marine debris ingested by marine megafauna. Mar Pollut Bull 151:110858

Kühn S, Bravo Rebolledo EL, van Franeker JA (2015) Deleterious effects of litter on marine life. In: Marine anthropogenic litter. Springer, Cham, pp 75–116

Kuroda M, Uchida K, Kitakado T (2022) Relationship between ocean area and incidence of anthropogenic debris ingested by longnose lancetfish (Alepisaurus ferox). Reg Stud Mar Sci 55:102476

Lai H, Liu X, Qu M (2022) Nanoplastics and human health: hazard identification and biointerface. Nano 12(8):1298

Laist DW (1997) Impacts of marine debris: entanglement of marine life in marine debris including a comprehensive list of species with entanglement and ingestion records. In: Marine debris. Springer, Cham

Lavers JL, Bond AL (2023) Pumice ingestion in seabirds: interannual variation, and relationships with chick growth and plastic ingestion. Mar Biol 170(5):1–9

Lavers JL, Hutton I, Bond AL (2021) Temporal trends and interannual variation in plastic ingestion by flesh-footed shearwaters (Ardenna carneipes) using different sampling strategies. Environ Pollut 290:118086

Lavers JL et al (2022) Ingested plastics in beach-washed fairy prions Pachyptila turtur from Tasmania. Mar Pollut Bull 184:114096

Le Gu L, Davies P, Arhant M, Vincent B, Tanguy E (2023) Mitigating plastic pollution at sea: natural seawater degradation of a sustainable PBS/PBAT marine rope. Mar Pollut Bull 193:115216

Lebreton L, Andrady A (2019) Future scenarios of global plastic waste generation and disposal. Palgrave Commun 5(1):1–11

Lebreton LCM, Van Der Zwet J, Damsteeg JW, Slat B, Andrady A, Reisser J (2017) River plastic emissions to the world's oceans. Nat Commun 8:1–10

Lebreton L, Egger M, Slat B (2019) A global mass budget for positively buoyant macroplastic debris in the ocean. Sci Rep 9(1):12922

Leila B, Sedláček P, Anastasopoulou A (2023) Plastic pollution in the deep-sea Giant red shrimp, Aristaeomorpha foliacea, in the eastern Ionian Sea; an alarm point on stock and human health safety. Sci Total Environ 877:162783

Li WC, Tse HF, Fok L (2016) Plastic waste in the marine environment: a review of sources, occurrence and effects. Sci Total Environ 566:333–349

Lim KP et al (2023) First evidence of microplastic ingestion by crescent perch (Terapon jarbua) in Malaysia. Reg Stud Mar Sci 67:103202

López-Martínez S et al (2023) Plastic ingestion by two cetacean groups: Ziphiidae and Delphinidae. Environ Pollut 333:121932

MacLeod M, Arp HPH, Tekman MB, Jahnke A (2021) The global threat from plastic pollution. Science 373(6550):61–65

Mancuso M et al (2022) Investigating the effects of microplastic ingestion in Scyliorhinus canicula from the south of Sicily. Sci Total Environ 850:157875

Markic A, Gaertner JC, Gaertner-Mazouni N, Koelmans AA (2020) Plastic ingestion by marine fish in the wild. Crit Rev Environ Sci Technol 50(7):657–697

Martellini T et al (2023) Bioplastics on marine sandy shores: effects on the key species Talitrus saltator (Montagu, 1808). Sci Total Environ 876:162811

Martin C, Almahasheer H, Duarte CM (2019) Mangrove forests as traps for marine litter. Environ Pollut 247:499–508

Matos DM et al (2024) Microplastics ingestion and endocrine disrupting chemicals (EDCs) by breeding seabirds in the east tropical Atlantic: associations with trophic and foraging proxies (δ15N and δ13C). Sci Total Environ 912:168664

Mboglen D, Gong Y, Guo Z, Ngo Nola D, Li Y (2024) First report of plastic and non-plastic microparticles in stomach of slandertail lanternshark and shortspine spurdog from the edge of East China Sea. Mar Pollut Bull 204:116531

Meaza I, Toyoda JH, Wise JP (2021) Microplastics in sea turtles, marine mammals and humans: a one environmental health perspective. Front Environ Sci 8:1–16

Meijer LJJ, van Emmerik T, van der Ent R, Schmidt C, Lebreton L (2021) More than 1000 rivers account for 80% of global riverine plastic emissions into the ocean. Sci Adv 7(18):1–14

Menezes M, Dias JD, Longo GO (2022) Plastic debris decrease fish feeding pressure on tropical reefs. Mar Pollut Bull 185:114330

Merrill GB, Hermabessiere L, Rochman CM, Nowacek DP (2023) Microplastics in marine mammal blubber, melon, & other tissues: evidence of translocation. Environ Pollut 335:122252

Miccoli A, Mancini E, Saraceni PR, Della Ventura G, Scapigliati G, Picchietti S (2022) First evidence of in vitro cytotoxic effects of marine microlitter on Merluccius merluccius and Mullus barbatus, two Mediterranean commercial fish species. Sci Total Environ 813:152618

Mičušík M et al (2021) Plastic ingestion by the Wels catfish (Silurus glanis L.): detailed chemical analysis and degradation state evaluation. Toxicol Rep 8:1869–1876

Miller ME, Hamann M, Kroon FJ (2020) Bioaccumulation and biomagnification of microplastics in marine organisms: a review and meta-analysis of current data. PLoS One 15:e0240792

Mills WF et al (2024) Plastic ingestion and trace element contamination of Manx shearwaters Puffinus puffinus on The Faroe Islands. Mar Pollut Bull 204:116537

MITECO - Ministerio para la Transición Ecológica y el Reto Demográfico (2008) Guía de Especies Marinas Desembarcadas en Puertos Españoles, Madrid

Mohamed Nor NH, Obbard JP (2014) Microplastics in Singapore's coastal mangrove ecosystems. Mar Pollut Bull 79(1-2):278–283

Monira S, Bhuiyan MA, Haque N, Pramanik BK (2022) Road dust-associated microplastics from vehicle traffics and weathering. In: Plastic waste for sustainable asphalt roads. Elsevier, Amsterdam, pp 257–271

Moon Y et al (2022) What type of plastic do sea turtles in Korean waters mainly ingest? Quantity, shape, color, size, polymer composition, and original usage. Environ Pollut 298:118849

Morales-Caselles C et al (2021) An inshore–offshore sorting system revealed from global classification of ocean litter. Nat Sustain 4(6):484–493

Morgan E, Hutchinson D, Gaion A (2021) Plastic ingestion by the small-spotted catshark (Scyliorhinus canicula) from the south west coast of the United Kingdom. Bull Environ Contam Toxicol 106(6):910–915

Motivarash Y, Bhatt A, Kardani H (2024) Microplastic (MP) occurrence in pelagic and demersal fishes of Gujarat, northwest coast of India. Environ Sci Pollut Res 31(11):17239–17255
Muñoz J, Forselledo R, Domingo A, Jiménez S (2023) Interspecific variability in plastic ingested by Procellariiformes off the Uruguayan coast. Mar Pollut Bull 197:115725
Murano C et al (2020) How sea urchins face microplastics: uptake, tissue distribution and immune system response. Environ Pollut 264:114685
Mushtaq S, Jamil F, Inayat A, Ghenai C, Shanableh A (2024) The green revolution in plastics: unveiling breakthroughs, applications, and the path forward. Curr Opin Green Sustain Chem 49:100950
Myszka R, Enfrin M, Giustozzi F (2023) Microplastics in road dust: a practical guide for identification and characterisation. Chemosphere 315:137757
Nabi G et al (2022) The adverse health effects of increasing microplastic pollution on aquatic mammals. J King Saud Univ Sci 34(4):102006
Nafea TH, Al-Maliki AJ, Al-Tameemi IM (2023) Sources, fate, effects, and analysis of microplastic in wastewater treatment plants: a review. Environ Eng Res 29(1):230040
Nam KB, Kim M, Hong MJ, Kwon YS (2021) Plastic debris ingestion by seabirds on the Korean peninsula. Mar Pollut Bull 166:112240
Nava V, Leoni B (2021) A critical review of interactions between microplastics, microalgae and aquatic ecosystem function. Water Res 188:116476
Navarro A et al (2023) Microplastics ingestion and chemical pollutants in seabirds of gran Canaria (Canary Islands, Spain). Mar Pollut Bull 186:114434
Nelms SE et al (2016) Plastic and marine turtles: a review and call for research. ICES J Marine Sci 73:165–181
Nelson M (2008) Bearded seal. Alaska Department of Fish & Game
Neto JGB et al (2020) Ingestion of plastic debris by commercially important marine fish in southeast-South Brazil. Environ Pollut 267:115508
NOAA (2021) Ocean and coasts. NOAA-National Oceanic and Atmospheric Administration Education
NOAA (2023) Why is Marine Debris a problem? https://marinedebris.noaa.gov/why-marine-debris-problem/ingestion. Accessed 7 Jun 2024
Nono Almeida F et al (2023) Changes in plastic ingestion by yellow-legged gulls (Larus michahellis) over the breeding season. Mar Pollut Bull 187:114483
Nono Almeida F et al (2024) Among-colony variation in plastic ingestion by yellow-legged gulls (Larus michahellis) across the western Mediterranean basin. Mar Pollut Bull 204:116508
Nukpezah D, Quarshie JT, Nyarko E, Hogarh JN (2022) Characterisation of litter and their deposition at the banks of coastal lagoons in Ghana. Heliyon 8(3):e08997
Nyberg B, Harris PT, Kane I, Maes T (2023) Leaving a plastic legacy: current and future scenarios for mismanaged plastic waste in rivers. Sci Total Environ 869:161821
Oceana (2023) Plastic-free paradise: tackling marine plastic pollution in the tourism industry. Oceana, Washington
Ocean-Conservancy (2024) Fighting for trash free seas. Trash Free Seas
Ojeda M, Rimondino GN, Fraysse CP, Cossi PF, Boy CC, Pérez AF (2024) Microplastic ingestion in key fish species of food webs in the Southwest Atlantic (marine protected area Namuncurá/ Burdwood Bank). Aquat Toxicol 267:106827
Okoffo ED, Rauert C, Thomas KV (2023) Mass quantification of microplastic at wastewater treatment plants by pyrolysis-gas chromatography–mass spectrometry. Sci Total Environ 856:159251
Oluwoye I et al (2023) Degradation and lifetime prediction of plastics in subsea and offshore infrastructures. Sci Total Environ 904:166719
Ortega A (2008) Cuadernos de acuicultura: Cultivo de Dorada. Fundación Observatorio Español de Acuicultura, Madrid
Padula AD et al (2023) Marine debris ingestion by odontocete species from the Southwest Atlantic Ocean: absence also matter. Mar Pollut Bull 186:114486

Palmer JL et al (2021) Dietary analysis of two sympatric marine turtle species in the eastern Mediterranean. Mar Biol 168(6):94

Park JC et al (2020) The genome of the marine rotifer Brachionus koreanus sheds light on the antioxidative defense system in response to 2-ethyl-phenanthrene and piperonyl butoxide. Aquat Toxicol 221:105443

Parolini M, Stucchi M, Ambrosini R, Romano A (2023) A global perspective on microplastic bioaccumulation in marine organisms. Ecol Indic 149:110179

Pasqualini V et al (2023) Harmful algae and pathogens on plastics in three mediterranean coastal lagoons. Heliyon 9(3):e13654

Patsiou D, Digka N, Galli M, Baini M, Fossi MC, Tsangaris C (2024) Assessment of the impact of microplastic ingestion in striped red mullets from an eastern Mediterranean coastal area (Zakynthos Island, Ionian Sea). Mar Environ Res 196:106438

Pawar PR, Shirgaonkar SS, Patil RB (2016) Plastic marine debris: sources, distribution and impacts on coastal and ocean biodiversity. Pencil Publ Biol Sci 3(1):40–54

Pedà C et al (2022) The waste collector: information from a pilot study on the interaction between the common octopus (Octopus vulgaris, Cuvier, 1797) and marine litter in bottom traps fishing and first evidence of plastic ingestion. Mar Pollut Bull 174:113185

Pereira LG et al (2023) Exploring microplastic contamination in Guiana dolphins (Sotalia guianensis): insights into plastic pollution in the southwestern tropical Atlantic. Mar Pollut Bull 194:115407

Pierdomenico M et al (2018) Effects of trawling activity on the bamboo-coral Isidella elongata and the sea pen Funiculina quadrangularis along the Gioia canyon (Western Mediterranean, southern Tyrrhenian Sea). Prog Oceanogr 169:214–226

Pinho I, Amezcua F, Rivera JM, Green-Ruiz C, Piñón-Colin TJ, Wakida F (2022) First report of plastic contamination in batoids: plastic ingestion by Haller's round ray (Urobatis halleri) in the Gulf of California. Environ Res 211:113077

Pitacco V, Orlando-Bonaca M, Avio CG (2022) Plastic impact on marine benthic organisms and food webs. In: Plastic pollution and marine conservation. Elsevier, Amsterdam

Pollet IL et al (2024) The relationship between plastic ingestion and trace element concentrations in Arctic seabirds. Mar Pollut Bull 203:116509

Pon JPS, Álvarez VA, Nicolini AT, Rosenthal AF, García GO (2023) Ingestion of marine debris by juvenile Magellanic penguins (Spheniscus magellanicus) in wintering grounds of coastal Argentina. Mar Pollut Bull 193:115247

Porterfield KK, Hobson SA, Neher DA, Niles MT, Roy ED (2023) Microplastics in composts, digestates, and food wastes: a review. J Environ Qual 52(2):225–240

Puskic PS et al (2024) Exploring the pathology of liver, kidney, muscle, and stomach of fledgling seabirds associated with plastic ingestion. J Hazard Mater 465:133306

Radford F, Horton A, Hudson M, Shaw P, Williams I (2023) Agricultural soils and microplastics: are biosolids the problem? Front Soil Sci 2:941837

Ragesh S et al (2024) Environmental and ecological risk of microplastics in the surface waters and gastrointestinal tract of skipjack tuna (Katsuwonus pelamis) around the Lakshadweep Islands, India. Environ Sci Pollut Res 31(15):22715–22735

Rajendran K, Rajendiran R, Ravichandran R, Velu RK (2022) Investigation of microplastic accumulation in Rastrelliger kanagurta fish gut and microplastic degradation behaviour of existing gut bacteria pseudomonas sp. Arch Microbiol 204(10):1–10

Raju M, Gandhimathi R, Nidheesh PV (2023) The cause, fate and effect of microplastics in freshwater ecosystem: ways to overcome the challenge. J Water Process Eng 55:104199

Ramon-Gomez K et al (2024) Plastic ingestion in giant tortoises: an example of a novel anthropogenic impact for Galapagos wildlife. Environ Pollut 340(P2):122780

Rios-Fuster B et al (2022) Assessing microplastic ingestion and occurrence of bisphenols and phthalates in bivalves, fish and holothurians from a Mediterranean marine protected area. Environ Res 214:114034

Rivera-Garibay O, Méndez-López ME, Torres-Irineo E, Rivas M, Santillo D, Álvarez-Filip L (2024) Presence of microplastic in target species of small scale fisheries and possible social implications on the local communities. Mar Biol 171(4):78
Rodrigues D, Pequeno J, Pais J, Antunes J, Sobral P, Costa MH (2023) Comparing microplastics ingested by the Bogue Boops boops (Teleostei: Sparidae) to those available in its feeding areas: a case study on the Portuguese west coast. Reg Stud Mar Sci 65:103093
Rodríguez Y et al (2022) Litter ingestion and entanglement in green turtles: an analysis of two decades of stranding events in the NE Atlantic. Environ Pollut 298:118796
Roman J (2017) Reporte del Taller 'Rol de los Cetáceos en el funcionamiento de los ecosistemas: Definiendo las políticas de conservación marina en el Siglo 21'. In 28°Congreso Internacional de Biología de la Conservación (ICCB), Cartagena
Roman L et al (2019) Is plastic ingestion in birds as toxic as we think? Insights from a plastic feeding experiment. Sci Total Environ 665:660–667
Sahu BK, Pati P, Panigrahy RC (2013) Plastic litters: a major environmental issue in chilika lagoon. Curr Sci 104(9):1133–1134
Sam L (2023) Structure and important functions of fish gills. Fish Aquac J 14(3):339
Sambolino A, Iniguez E, Herrera I, Kaufmann M, Dinis A, Cordeiro N (2023) Microplastic ingestion and plastic additive detection in pelagic squid and fish: implications for bioindicators and plastic tracers in open oceanic food webs. Sci Total Environ 894:164952
Sandhi A, Rosenlund J (2024) Municipal solid waste management in Scandinavia and key factors for improved waste segregation: a review. Clean Waste Syst 8:100144
Santos RG, Andrades R, Boldrini MA, Martins AS (2015) Debris ingestion by juvenile marine turtles: an underestimated problem. Mar Pollut Bull 93(1-2):37–43
Savoca MS, Wohlfeil ME, Ebeler SE, Nevitt GA (2016) Marine plastic debris emits a keystone infochemical for olfactory foraging seabirds. Sci Adv 2(11):1–9
Savoca MS, McInturf AG, Hazen EL (2021) Plastic ingestion by marine fish is widespread and increasing. Glob Chang Biol 27(10):2188–2199
Sayed RKA, Zaccone G, Capillo G, Albano M, Mokhtar DM (2022) Structural and functional aspects of the spleen in molly fish Poecilia sphenops (Valenciennes, 1846): synergistic interactions of stem cells, neurons, and immune cells. Biology 11(5):779
Sbrana A et al (2022) Ask the shark: blackmouth catshark (Galeus melastomus) as a sentinel of plastic waste on the seabed. Mar Biol 169(7):98
Schmaltz E et al (2020) Plastic pollution solutions: emerging technologies to prevent and collect marine plastic pollution. Environ Int 144:106067
Schmidt C, Krauth T, Wagner S (2017) Export of plastic debris by Rivers into the sea. Environ Sci Technol 51(21):12246–12253
Schutten K et al (2023) Assessing plastic ingestion in birds of prey from British Columbia, Canada. Environ Sci Pollut Res 30(31):76631–76639
Schuyler Q, Hardesty BD, Wilcox C, Townsend K (2014) Global analysis of anthropogenic debris ingestion by sea turtles. Conserv Biol 28(1):129–139
Sciutteri V et al (2023) Integrated approach for marine litter pollution assessment in the southern Tyrrhenian Sea: information from bottom-trawl fishing and plastic ingestion in deep-sea fish. Mar Pollut Bull 188:1–10
Scotti G, D'Alessandro M, Esposito V, Vivona P, Panti C (2023) Anthropogenic fibers and microplastics in the pelagic gooseneck barnacle Lepas (Lepas) anatifera in capo Milazzo marine protected area (Tyrrhenian Sea): a first characterization. Ecol Indic 152:110368
Selden KR, Baker MR (2023) Influence of marine habitat on microplastic prevalence in forage fish and salmon in the Salish Sea. Mar Pollut Bull 197:115748
Sharifinia M, Bahmanbeigloo ZA, Keshavarzifard M, Khanjani MH, Lyons BP (2020) Microplastic pollution as a grand challenge in marine research: a closer look at their adverse impacts on the immune and reproductive systems. Ecotoxicol Environ Saf 204:111109
Sheavly SB, Register KM (2007) Marine debris & plastics: environmental concerns, sources, impacts and solutions. J Polym Environ 15(4):301–305

Shen M et al (2021) Can incineration completely eliminate plastic wastes? An investigation of microplastics and heavy metals in the bottom ash and fly ash from an incineration plant. Sci Total Environ 779:146528

Sheriff I, Yusoff MS, Halim HB (2023) Microplastics in wastewater treatment plants: a review of the occurrence, removal, impact on ecosystem, and abatement measures. J Water Process Eng 54:104039

Soe KK, Maae S, Jaafar Z, Chuaduangpui P, Jantarat S, Hajisamae S (2024) Plastic ingestion by three species of Scylla (Brachyura) from the coastal areas of Thailand. Mar Pollut Bull 198:115914

Solomando A, Pujol F, Sureda A, Pinya S (2022) Ingestion and characterization of plastic debris by loggerhead sea turtle, Caretta caretta, in the Balearic Islands. Sci Total Environ 826:154159

Soulliere C, Dixon B (2017) Immune system organs of bony fishes. In: Reference module in life sciences. Elsevier, Amsterdam

Sühring R et al (2022) Co-contaminants of microplastics in two seabird species from the Canadian Arctic. Environ Sci Ecotechnol 12:100189

Sultana S et al (2024) Ecological risk assessment of microplastics and mesoplastics in six common fishes from the bay of Bengal coast. Mar Pollut Bull 204:116544

Tekman MB, Walther BA, Peter C, Gutow L, Bergmann M (2022) Impacts of plastic pollution in the oceans on marine species, biodiversity and ecosystems, Berlin

Thushari GGN, Senevirathna JDM (2020) Plastic pollution in the marine environment. Heliyon 6(8):e04709

Topouzelis K, Papakonstantinou A, Garaba SP (2019) Detection of floating plastics from satellite and unmanned aerial systems (plastic litter project 2018). Int J Appl Earth Obs Geoinf 79:175–183

Tourinho PS, do Sul JAI, Fillmann G (2010) Is marine debris ingestion still a problem for the coastal marine biota of southern Brazil? Mar Pollut Bull 60(3):396–401

Tumu K, Vorst K, Curtzwiler G (2023) Global plastic waste recycling and extended producer responsibility laws. J Environ Manag 348:119242

Tuuri EM, Leterme SC (2023) How plastic debris and associated chemicals impact the marine food web: a review. Environ Pollut 321:121156

Ugwu K, Herrera A, Gómez M (2021a) Microplastics in marine biota: a review. Mar Pollut Bull 169:112540

Ugwu CO, Ozoegwu CG, Ozor PA, Agwu N, Mbohwa C (2021b) Waste reduction and utilization strategies to improve municipal solid waste management on Nigerian campuses. Fuel Commun 9:100025

UN Habitat - United Nations Human Settlement Programme and WHO - World Health Organization (2021) Progress on wastewater treatment—global status and acceleration needs for SDG indicator 6.3.1. UN-Habitat & WHO, Geneva

Unuofin JO, Igwaran A (2023) Microplastics in seafood: implications for food security, safety, and human health. J Sea Res 194:102410

Valente T et al (2019) Exploring microplastic ingestion by three deep-water elasmobranch species: a case study from the Tyrrhenian Sea. Environ Pollut 253:342–350

Valente T et al (2022) One is not enough: monitoring microplastic ingestion by fish needs a multispecies approach. Mar Pollut Bull 184:114133

van Bijsterveldt CEJ et al (2021) Does plastic waste kill mangroves? A field experiment to assess the impact of macro plastics on mangrove growth, stress response and survival. Sci Total Environ 756:143826

van Emmerik T (2021) Macroplastic research in an era of microplastic. Microplast Nanoplast 1(1):4–5

Vecchi S, Bianchi J, Scalici M, Fabroni F, Tomassetti P (2021) Field evidence for microplastic interactions in marine benthic invertebrates. Sci Rep 11(1):1–12

Waldschläger K et al (2022) Learning from natural sediments to tackle microplastics challenges: a multidisciplinary perspective. Earth Sci Rev 228:104021

Wan X et al (2023) The intestinal microbiota of a Risso's dolphin (Grampus griseus): possible relationships with starvation raised by macro-plastic ingestion. Int Microbiol 26(4):1001–1007

Watts AJR et al (2014) Uptake and retention of microplastics by the shore crab carcinus maenas. Environ Sci Technol 48(15):8823–8830

Webb H, Arnott J, Crawford R, Ivanova E (2012) Plastic degradation and its environmental implications with special reference to poly(ethylene terephthalate). Polymers 5(1):1–18

Weideman EA, Munro C, Perold V, Omardien A, Ryan PG (2020) Ingestion of plastic litter by the sandy anemone Bunodactis reynaudi. Environ Pollut 267:115543

Weiss L et al (2021) The missing ocean plastic sink: gone with the rivers. Science 373(6550):107–111

Wiesinger H, Wang Z, Hellweg S (2021) Deep dive into plastic monomers, additives, and processing aids. Environ Sci Technol 55(13):9339–9351

Wilcox C, Puckridge M, Schuyler QA, Townsend K, Hardesty BD (2018) A quantitative analysis linking sea turtle mortality and plastic debris ingestion. Sci Rep 8(1):1–11

Wittstock K, Wessolowski V (2022) Political boundary conditions—plastics waste management. In Niessner N (ed) Recycling of plastics, p 31

Wojnowska-Baryła I, Bernat K, Zaborowska M (2022) Plastic waste degradation in landfill conditions: the problem with microplastics, and their direct and indirect environmental effects. Int J Environ Res Public Health 19:20

Yaghmour F et al (2022) Junk food: polymer composition of macroplastic marine debris ingested by green and loggerhead sea turtles from the Gulf of Oman. Sci Total Environ 828:154373

Yang Z et al (2021) Is incineration the terminator of plastics and microplastics? J Hazard Mater 401:123429

Yong MMH et al (2021) Microplastics in fecal samples of whale sharks (Rhincodon typus) and from surface water in The Philippines. Microplast Nanoplast 1(1):1–9

Yoo JW et al (2024) Toxic effects of fragmented polyethylene terephthalate particles on the marine rotifer Brachionus koreanus: based on ingestion and egestion assay, in vivo toxicity test, and multi-omics analysis. J Hazard Mater 472:134448

Yose P, Thondhlana G, Fraser G (2023) Conceptualizing the socio-cultural impacts of marine plastic pollution on human Well-being—a perspective. Mar Pollut Bull 194:115285

Yu AJG, Yap-Dejeto LG, Parilla RB, Elizaga NB (2024) Microplastics in Perna viridis and Venerupis species: assessment and impacts of plastic pollution. Int J Environ Sci Technol 21(1):675–686

Yücel N (2023) Detection of microplastic fibers tangle in deep-water rose shrimp (Parapenaeus longirostris, Lucas, 1846) in the northeastern Mediterranean Sea. Environ Sci Pollut Res 30(4):10914–10924

Zhang S, Wang N, Gong S, Gao S (2022) The patterns of trophic transfer of microplastic ingestion by fish in the artificial reef area and adjacent waters of Haizhou Bay. Mar Pollut Bull 177:113565

Zhao J et al (2024) Microplastic fragmentation by rotifers in aquatic ecosystems contributes to global nanoplastic pollution. Nat Nanotechnol 19(3):406–414

Innovative Intervention: Technologies to Tackle the Microplastic Pollution

6

Muhammad Waseem, Qudrat Ullah, Muhammad Qasim, Irfan Haidri, Amna Chaudhary, and Fahad Rahul

Abstract

Microplastics (MPs) are a universal type of pollutants that harm the state of the biosphere and human health. Microplastics are often associated with consumer behavior and commercial activity, so steps must be taken to reduce the amount being released into the environment. In light of such information as the global spread of microplastics in soil and wastewater which has been seen to impact the soil microbial population and characteristics in negative ways, it is crucial now more than ever, to continue reducing the number of plastic inputs in the environment which is a problem requiring collective effort from a cross-section of scholars from across the world. This chapter conducts a critical examination of the

M. Waseem (✉)
Department of Environmental Science, Government College University, Faisalabad, Punjab, Pakistan
e-mail: mw6358673@gmail.com

Q. Ullah · M. Qasim
Department of Environmental Science, Government College University, Faisalabad, Punjab, Pakistan

School of Agricultural Technology and Food Industry, Walailak University, Nakhon Si Thammarat, Thailand

I. Haidri
Department of Environmental Engineering, Mining and Energy University, Gdansk University of Technology, Gdańsk, Poland

A. Chaudhary
Department of Fisheries and Aquaculture, Faculty of Fisheries and Wildlife, University of Veterinary and Animal Sciences, Lahore, Punjab, Pakistan

F. Rahul
Department of Environmental Science and Disaster Management, Gopalganj Science and Technology University, Gopalganj, Bangladesh

M. Oves et al. (eds.), *Marine Microplastics and the Quest for Remediation*,
https://doi.org/10.1007/978-981-96-8979-8_6

current approach to different methods of microplastic removal with a focus on innovative technology. After a critical review of the sources of microplastics, some papers highlighted methods that include physical, chemical, and biological methods for the removal of microplastics. Accordingly, this chapter aims to review new technologies employed to tackle microplastic pollution and reduce its negative effects. This work will identify some of the knowledge gaps in microplastic contamination and compile all available data to provide recommendations based on empirical findings.

Keywords

Microplastics · Pollutants · Microbes · Environment · Consumer

6.1 Introduction

The widespread use of plastics has made them a persistent contaminant in all areas of the environment, with the aquatic ecosystem being especially susceptible. The problem has been made worse by the record-breaking 348 million tons of plastic produced worldwide each year. Plastics' strong molecular structure allows them to last for decades in natural settings, demonstrating their exceptional endurance. However, plastics break down into smaller parts due to weathering, UV radiation exposure, and environmental aging, producing primary and secondary microplastics (MPs), which present different health and environmental hazards (Chamas et al. 2020). The extent of microplastic pollution has made scholars, lawmakers, and environmental activists respond to the situation and recognize the urgent need for sound and, more importantly, holistic measures to address the harm caused by it. Of the nanoplastic forms, microplastics are particularly popular because they are widespread in ecosystems and can interact with biological environments. The impacts of microplastics on abiotic and biotic components are still under study for long-term effects (Bhat et al. 2024; Bhat 2024; Thacharodi et al. 2023).

Microplastics have been reported in various environments, including freshwater, soil, marine environments, and the air (Bhat et al. 2023; Bordós et al. 2019; Guo et al. 2020). Suggestively, the ingestion of microplastics may elicit toxic symptoms, actual body damage, hormonal imbalance, and death in species in varied studies. As carriers of other substances, the multiple effects of microplastics on the ecological system can be aggravated. Some of the recent studies, such as Eraslan et al. (2023) and Fang et al. (2022), have revised the effects on human health and the environment. Fears are focused on threats associated with the ingestion of chemicals and additives present in plastics contaminating the human food chain through diets of seafood or agricultural products. Also, product manufacturing, transportation, and disposal contribute emissions of microplastics in the environment. Use of cosmetic products that have microbeads like body washes and facial scrubs risks introducing microplastics onto the skin (Bashir et al. 2021; Wang et al. 2022a). This further points to the need for more elaborate validations on the effects of microplastic pollution on the sustainability of the environment and its effects on human health.

Current information disproves the concept: microplastics exert high levels of threat to keystone species in world ecosystems, reducing their stability and efficiency. Negative impacts of microplastic pollution have been credited to reduced efficiency and nitrogen turnover, reduced carbon storage, lowered oxygen yield, and lowered plant and animal density (Sridharan et al. 2021). Development of an appropriate and effective risk reduction strategy requires disentanglement of the numerous effects of microplastic pollution. Traditional waste management concepts fail, to some extent, to address microplastic pollution because these particles are difficult to capture and originate from multiple sources (Vanapalli et al. 2021; Zhou et al. 2020). The removal and containment techniques of microplastics in the environment have been developed by various novel methodologies and technologies. These remedies include a range of measures such as enhanced filtration measures, biodegradation processes, and other new-age treatment mechanisms (Gao et al. 2022; Moharir and Kumar 2019). Evidence from the literature shows that to successfully address the issue of microplastic pollution, technical developments have to be complemented by robust and coherent interdisciplinary research, awareness creation, and policy improvements.

One way to avoid the problem of microplastic pollution is to use them as valuable resources to result in a closed-loop system. One of the approaches can be the transformation of microplastics into lightweight components in manufacturing or even high-performance construction materials (Veropalumbo et al. 2023). Moreover, the results of several studies indicate that adding microplastics improves the mechanical characteristics and the durability of concrete or asphalt structures, which might be useful in the possible long-term positive effect on the environment and economies (Veropalumbo et al. 2023). However, new tech in microplastic recycling will help decentralize the industry and break from virgin plastic dependency, in addition to minimizing unfavorable externalities (Walker and Fequet 2023). Exploring new ways can contribute to creating a cycle of materials, which effectively circulate through people's lifecycles again with less fragmentation of resources and less generation of waste.

The specific goal of this chapter is to assess the remediation techniques of microplastics.

6.1.1 Microplastics

Microplastics, which are particles smaller than 5 mm, have in the recent past been established to be hazardous to the environment and human health. Pollutants at submicron levels have been detected in the marine water, freshwater, soil, and air environment (Habib et al. 2022). Because these environments contain microplastics, it is crucial to define them and understand their composition and effects. Recent discoveries have considered particles having a size of 5 mm or less to be a menace to the environment and human health. Hazards have been detected in various sample media, such as the aquatic systems including marine water, freshwater, soil, and atmospheric systems (Habib et al. 2022). A clear understanding of the meaning, characteristics, and consequences of microplastics is critical because they are present in these habitats. This definition has been criticized for excluding substances that should be put under the category of microplastics such as rubber derivatives, inorganic polymers, and PVC with more than 50% additives, along the crystalline fibers, which despite having a similar chemical constitution with other μP to the extent that can be identified through flow test, they do not flow (Hartmann et al. 2019). This brings the question of the need to have a more holistic and broader definition that tends to cater to the different kinds of microplastics.

6.1.2 Source of Microplastics

Such particles of plastic such as bags, bottles, and packaging materials contribute to the distribution of microplastics through disposal. These products biodegrade, biodeteriorate, and biodisintegrate, thereby liberating microparticles in soil and water systems (Habib et al. 2022). In addition, fibers from polyester, nylon, and acrylic clothes have been found to contribute to microplastic pollution during the washing and drying of the garments (Al Mamun et al. 2023; Gavigan et al. 2020; Lellis et al. 2019; Pinlova and Nowack 2023). Microplastic is formed when larger plastics disintegrate and synthetic fibers from clothing and other sources are released (Habib et al. 2022). This exemplifies the importance of careful consumption and manufacturing of textiles in the continuous appearance of microplastic pollution (Gavigan et al. 2020; Lellis et al. 2019). However, more extensive research has to be done to gain a better understanding of the transportation, distribution, and effects of microplastic pollution, which is being derived from the textile industry.

They conducted a study that shows how synthetic clothes release microfibers and then release microplastics into the environment whenever they are washed. Also, tire dust, synthetic paints, lost and discarded fishing gears, ropes, and the wearing of all forms of aquaculture apparatus contribute to synthetic polymers introducing significant levels of microplastics through degradation (Leistenschneider et al. 2021). The marine environment is most susceptible to microplastic pollution because plastic garbage is ubiquitous in the world's environment. Bottles, bags, wrapping materials, and other plastics progressively break up from bottles and other sources of plastic debris by exposure to sunlight and other physical and chemical conditions such as weathering and photochemical changes to generate microplastics (Thompson et al. 2009; Ziani et al. 2023). Also, not to mention, industrial processes are also responsible for the emission of these microplastics. Microplastic pollution is widely generated from the use of the smallest plastic components, such as plastic pellets used in the production of various plastic products (Thompson et al. 2009; Zhang et al. 2022a). This calls for further assessments of the probable source and pathways of microplastics to develop the right strategies to address the issues.

6.1.3 Microplastics in Soil

These physical movement mechanisms: gravitationally, chemically, and biologically enable microplastics (MPs) on soil surfaces to move upward and vertically settle deeper into the soil horizons (Ihenetu et al. 2024; Liu et al. 2023d). They end up as plastic waste dumped in the terrestrial ecosystem due to this process, supported by the fact that plastic production and utilization are largely associated with the terrestrial environment (Su et al. 2022). Recent approximations show that yearly, plastics are emitted to terrestrial environments 4–23 times more than to aquatic ecosystems (Horton et al. 2017). Various actions like using Plastic mulch films, the use of sewage sludge, and plastic-coated fertilizers enhance the entry of microplastics into the soils (Huang et al. 2020; Lian et al. 2021; Rolsky et al. 2020). Additionally, landfill leachate, tire abrasion particles, and atmospheric particulate matter are the secondary terrestrial sources of microplastic (Leifheit et al. 2022; Wan et al. 2022; Zhu et al. 2019). These results highlight the need for a better understanding and characterization of source transport and archives, as well as the complexity of microplastics in terrestrial habitats.

The previous works suggest that the most common microplastic types found in agricultural soils are polyester and polypropylene (Yang et al. 2021). In particular, practices involving mulching are reported on this background to increase the content of microplastics in agricultural soils due to increased plastic debris (Bai et al. 2024; Liu et al. 2023a). Microplastics may accumulate in soils through two mechanisms: direct and indirect deposition. Direct depositions may include the use of plastic mulch and compost made from plastic incorporated in fertilizers, while indirect depositions include flooding of soil with water from rivers and lakes containing microplastics (Zhao et al. 2024). Soil pollution by MP is traced back to road dust, industrial processes, domestic plastic waste, sewage sludge, sedimentation, deposited air, and wastewater spraying (Tariq et al. 2024). These sources show that microplastic pollution affects terrestrial ecosystems through complex pathways as they enter soils directly and indirectly.

6.1.4 Impact of Microplastics on Soil Characteristics and Microbial Communities

It has been substantiated that pollution, particularly that of a microplastic nature, leads to changes in the main soil characteristics including structure, porosity, the pH of the medium, the composition of microbial communities, bulk density, as well as enzymatic activity levels (Nazir et al. 2024; Wan et al. 2019; Yu et al. 2020). It has been shown in prior research that microplastics degrade the soil structure, organic materials, microbial communities, and nutrient accessibility (de Souza Machado et al. 2019; Iqbal et al. 2023b; Shi et al. 2023). In addition, it has been established that microplastics alter the pore size distribution in the soil, and this may affect the mechanical stability, water, and nutrient-holding ability of the soil. This is a paradox of interaction between microplastics and soil, showing that microplastic-generated soil cracks open the soil structure to porosity, where the very same cracks spearhead the entry of microplastics (Iqbal et al. 2023a; Wan et al. 2019; Zhang et al. 2019). As such, these findings underscore the need for a thorough study of the complex interactions between soil characteristics, ecosystem performance, and microplastics.

Polyester microplastics cause some extent of alteration in the availability of nutrients for crop growth by altering the biogeochemical cycle and do not affect the bulk density of soil (Bosker et al. 2019). On the other hand, the efflux of water and fertilizers by the sandy and porous soils diminishes the ability of these soils to hold water and fertilizers. Thus, the vertical movement of microplastics occurs more easily through the sand columns of the soil matrix. Negative correlations between sand content and microplastic concentrations should be exemplified to explain such a phenomenon and elucidate how particle size also influences microplastic transport (Liu et al. 2023b; Ran et al. 2023a, b). Moreover, the presence of microplastics in the soil hampers the soil's biological, chemical and physical characteristics, which in turn declines the plant yield and growth and development (Ali et al. 2024; Lozano et al. 2021). This complex effect opens up the argument for extensive analysis of the extent of microplastic pollution in agricultural soils, given the interaction between physical and chemical attributes of the soil, the presence of microplastics, and the overall functionality of ecosystems.

MPs consequently impact the structure and function of microbial biofilm communities in soil ecosystems and implicitly affect the health of soil ecosystems. However, the complex relationships between state variables, MPs, soil microbiota, and the related biogeochemical processes continue to receive minimal research attention. However, there has been an increasing interest in MP-mediated effects on soil microbial ecology and ecosystem functionality, and yet the basic processes involved are still not well understood. More studies should be conducted to establish the impacts of MPs on individual microbial groups and enzyme activities as they relate to the dose-response model to fill this gap. Thus, understanding the specific forms of microplastics that are detrimental to soil microbial community composition will be crucial to unraveling the dynamics of microplastics and soil interactions. As it has been realized that microplastics are prevalent in various terrestrial ecosystems, developing suitable tactics that can curb the ecological effects of this pollution demands the acquisition of deeper insights into these kinds of associations (Wan et al. 2023).

6.2 Impact of Microplastics on Human Health

Given their versatile use in most human activities, the use of plastics has significantly contributed to the increase in environmental microplastic pollution (Eze et al. 2024). Microplastics are part and parcel of human existence; they are present in contaminated seafood, drinking water, processed foods, food-ware packaging, and plastic cutlery. Ingested microplastics may cause harm to the health of consumers because they are trapped in the gastric cavity and other organs (Al Mamun et al.

2023; Campanale et al. 2020). Despite the current active studies directed at the effects that human exposure to microplastics has on health, some negative effects have been revealed. There is existing research that emphasizes that microplastics are toxic, and may cause genotoxicity, cellular stress, and inflammation in human cells (Caputi et al. 2022; Palaniappan et al. 2022; Roursgaard et al. 2022). Putative pathways that microplastics can be transferred from lower animals to humans have been described, but the overall effect of microplastics on human health is still far from well understood despite documented cases of microplastics in vital internal organs, including the liver, spleen, lungs, heart, kidneys, and brain of lower animals (Çıtar Dazıroğlu and Bilici 2023; Ziani et al. 2023). Additional studies are needed to explain the correlation between microplastics and their effects on individual health, as well as to provide guidelines to control microplastic pollution.

Research over the last years has shown the negative impact of ingested microplastics on the intestinal tissue, its permeability, and inflammation, as well as changes in the gut microbial community (Niu et al. 2023; Osman et al. 2023; Qiao et al. 2019). Furthermore, dermal absorption of microplastic has an influence on the skin and causes irritation, which may lead to reproductive toxicity and endocrine disruption (Campanale et al. 2020; Dubey et al. 2022). Further, changes in the daily rhythms and melatonin regulation have been linked to the contamination of microplastics in the environment and have effects likely to result in sleep disturbances and an increase in susceptibility to sleep pathology (Campanale et al. 2020; Meléndez-Fernández et al. 2023; Moralia et al. 2022; Sinha et al. 2022). However, more evidence conducted in previous studies has implied that microplastic exposure contributes to obesity by influencing metabolic disruptors that include hormones and metabolism (Huang et al. 2022a; Landrigan et al. 2023; Sinha et al. 2014). The current data also capture an indication that increased microplastic exposure might cause an increased intensity of diabetes (Huang et al. 2022b). Taken together, these findings point to the need for an enhanced understanding of the multifaceted link between vulnerability to diseases, human health, and microplastic pollution (Fig. 6.1).

6.3 Strategies to Tackle Microplastics

Microplastic remediation approaches can be categorized into three primary processes: physical, chemical, and biological. Awareness and environmental education have both been looked at as additional types of campaigns. Chemical/biological technologies seek to convert microplastics to CO_2 and H_2O or dem but into easier organic compounds by chemical/biological interactions, while physical removal technologies focus on the transport of microplastics. The above-formulated chemical coagulant or algae extracellular substance assists in the removal of microplastic from water bodies by agglomeration. Furthermore, to decrease the additional input of microplastic waste to the environmental compartments and to provide the upper public with relevant information, the introduction of environmental education and awareness programs is inevitable. Through the approach of developing an

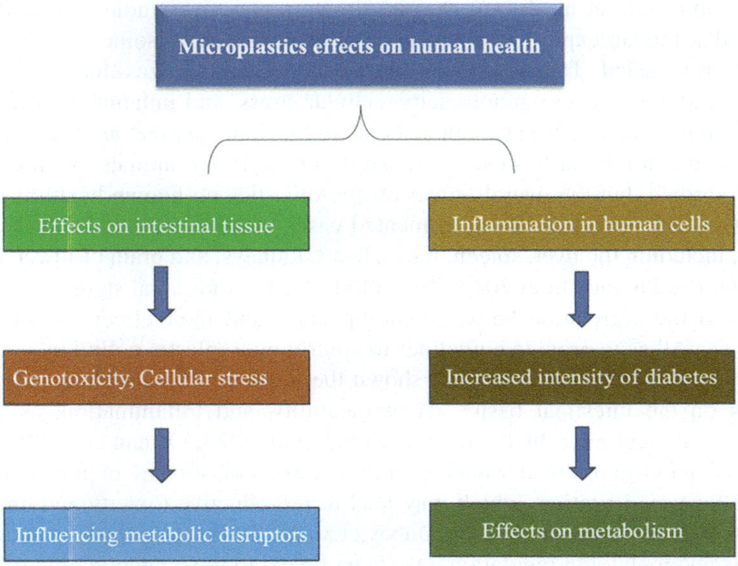

Fig. 6.1 Effects of microplastics on human health

interdisciplinary solution plan, a holistic framework for microplastic remediation can be developed within a vertical application level to provide the societal fabric and environmental multiplicity required in addressing microplastic pollution.

6.3.1 Physical Process Absorption

Nohara et al. (2024) consider adsorption as a surface process explaining how atoms and/or molecules (adsorbates) adhere to the surface of an adsorbent through chemical (chemisorption) or physical (physisorption) actions for retention. New absorbent materials have recently shown that they can efficiently remove microplastics and nanoplastics from water bodies (Huang et al. 2022a; Oliveira et al. 2023; Peng et al. 2022; Zheng et al. 2022). There is a low operating expense, high effectiveness of the adsorption method for the removal of microplastics and nanoplastics, and the possibility of using renewable materials as adsorbents. Still, impediments were noted; for example, the sorption capacity in aqueous conditions remains somewhat limited, post-process filtration provides some inconvenience, and competition for adsorption with other ions is another issue (Huang et al. 2022b; Zheng et al. 2022). The need to overcome these limitations is what contributes to the need to work on advancing adsorption-based microplastic removal technology.

6.3.2 Filtration

Dispersed phase particles and microorganisms in aqueous media possibly get retained during filtration by membranes or granular materials through adhesion forces and transport mechanisms, inclusive of diffusion, interception, and sedimentation. New studies have successfully explained the basic transport properties and their relationship with the filter media efficiency (Chabi et al. 2024). The removal process, which involves adhesion forces and transport mechanisms (diffusion, interception, and sedimentation), involves the use of membranes or granular media that trap particles/microorganisms present in fluid media. Some of the current works have also explained the basic transportation processes, which have also been correlated with filter material effectiveness (Chabi et al. 2024). One of the biggest challenges of filtration technology is the efficiency of microfiber or microplastic filtration, where most particles below 20 μm filter through materials and reduce effectiveness (Hanif et al. 2022). This underlines the need to develop new filter designs and materials that will efficiently trap and hold even the smallest microplastics.

6.3.3 Flotation

Like flocculation, flotation is a physicochemical separation method in which air bubbles are introduced into an aqueous solution to help contaminants aggregate into floccules. Recent investigations have demonstrated the effectiveness of this technique in removing microplastics from liquid media. Improving the flocculant characteristics, rate density, and molecular weight, in particular, is important for improving the efficiency of microplastic removal through flotation-based systems, as the performance of these systems is dependent on the ability of the flocculant and the microplastic. However, the size, shape, and surface chemistry of microplastics are other physicochemical characteristics that determine flocculant affinity and separation. (Jiang et al. 2022; Swart et al. 2022; Zheng et al. 2022).

Hydrophilic debris that has weathered is moved through the surface flow around a bubble and therefore carrying capacity is reduced, while bubbles adhere more effectively to clean hydrophobic surfaces, which ensures efficient transport (Lehmann et al. 2023). New field works carried out in polar zones have shown a direct link between microplastic concentration and bubble attachment probabilities in the case of flotation processes in which low-density microplastics stayed in surface water until biofouling or sedimentation, while high-density microplastics were more likely to adhere to bubbles (Chubarenko et al. 2023). Such conclusions revive interest in the relationship between particle density and the bubble surface, which determines the interaction between microplastics and bubbles and the efficiency of the separation of flotation processes.

6.4 Chemical Methodologies

6.4.1 Coagulation, Flocculation, and Sedimentation (CFS)

A critically analyzed treatment technique that can remove microplastics from water is coagulation, which is often integrated with other techniques like flocculation and sedimentation (Xia et al. 2020; Yao et al. 2023). This work employs the use of flocculation and coagulants together to improve the advancement of microplastic congregation and separation by sedimentation.

6.4.2 Electrocoagulation

Electrocoagulation (EC) uses electrodes to continuously produce coagulants and usually works hand in hand with electrical processes (Akarsu and Deniz 2021; Senathirajah et al. 2023; Shen et al. 2022). The use of this type of approach also has multiple advantages: optimization of process parameters, CDOM reduction, and a decrease in pollutants in wastewater treatment (Elkhatib et al. 2021). However, the requirement for electrical power in the production of coagulants in EC increases the operational cost of the process. The constant demand for the replacement of sacrificial agents presents a significant problem, and problems associated with residuals of coagulants after microplastic removal. These trade-offs between sustainability and efficacy in the electrochemical microplastic remediation show that findings support the utilization of overall life cycle assessments and optimization solutions.

Advanced Oxidation Processes (AOPs) themselves use superoxide radical anion ($O_2•^-$) and hydroxyl radical (–OH) oxidizing agents to break down resistant compounds through oxidative chain scission and end up mineralizing the treated water to CO_2 and H_2O. Conventional treatments have been proven to be unsuccessful in the degradation of microplastics, and in the last few years AOPs have successfully degraded resistant microplastics (Cedillo-González and Siligardi 2022; Nohara et al. 2023; Zheng et al. 2022). Chlorine-based Fenton processes for the degradation of micro-plastic have recently been considered as one of the significant advanced oxidation processes. These processes include the reactions and formation of hydrogen peroxide-catalytic use of H_2O_2 with ferrous ions Fe^{2+} at acidic media. Microplastics dissolve under the action of Fenton reactions, which produce highly reactive hydroxyl radicals (•OH) supplied by this mechanism (Hu et al. 2021; Piazza et al. 2021). One of the good examples of AOPs is the Fenton reaction, whose effectiveness is shown by its ability to break down microplastics to minimize their environmental carry-over.

6.5 Biological Methodologies

6.5.1 Degradation of Microplastics by Bacteria and Fungi

Diversity and importance of microorganisms, such as bacteria, fungi, and other microbes, are essential to keeping nutrient balance and biogeochemical cycling in many ecosystems across the earth (Li et al. 2023). Bacteria that are the primary nutrient transformers can break the synthetic polymers by shaping their usefulness for being able to metabolize long-chain fatty acids (Li et al. 2023). Soil, landfill soil, seawater, sediments, and contaminated water can support bacterial degradation of microplastics in different ecosystems (Auta et al. 2022; Dehghanian et al. 2023; Nanthini Devi et al. 2021; Park and Kim 2019; Yuan et al. 2022).

Fungi are receivers of nutrients and are eukaryotic sporulating cell organisms that do not possess chloroplasts and lack photosynthetic pigments. Several studies have demonstrated that microplastic degradation by fungi is possible through the use of different enzymes produced by these fungi, making fungal-mediated process a potential solution to the problem (Liu et al. 2023b; Moyses et al. 2021).

6.5.2 Algal Degradation

It was established that algae have several mechanisms of reducing the presence of microplastic pollution through direct interception, capture, and entanglement (adsorption), as well as indirect formation of hetero-aggregates through the secretion of extracellular polymeric substances (EPS) from algal cells (Cheng and Wang 2022; Li et al. 2023; Peller et al. 2021). These processes, as identified by Li et al. (2023), are divided into direct and indirect channels. Nanoplastics move to various tissues of the organisms by endocytosis, while micro-plastics adhere to algal surfaces through the direct route. Cell walls and epidermis also act as a barrier to Physical layer size-dependent penetration of plastics in algae tissues, allowing nanoplastics of smaller size to enter the tissues easily compared to micro-plastics. Moreover, studies show that microalgal adhesion increases with increased plastic surface roughness; surfaces with high energies are ideal for biofilm formation (Li et al. 2023). The main predictors of microplastic–algae interactions are the roughness and hydrophilicity of the microplastic surface, of which higher values are expected to enhance microalgal adhesion.

Extracellular polymeric substances (EPS), complex macromolecules with diverse structural, physical, and chemical characteristics, are involved in the indirect microplastic uptake mechanism supported by algae (Li et al. 2023). EPS release and secretion by algae can be upregulated to counter the effects of microplastic exposure and form heterogeneous aggregates that selectively absorb microplastics (Cheng and Wang 2022; Cunha et al. 2020). Perhaps, this EPS-mediated process reduces the harm of microplastics on the environment through their immobilization. The indirect pathway is intricate, and the dynamic interactions between the algal cells, EPS, and microplastics necessarily suggest that the kinetics and mechanisms merit further research.

6.5.3 Remediation Through Nanotechnology

Nanotechnology has now offered the possibility of a new paradigm for soil remediation where conventional technology is combined with nanobiotechnological approaches. Present studies also emphasized the extensive future applications of nanotechnology with the help of combined streams in improving environmental pollution. Remedial activities and uses of nanoparticles and their related technologies include solid waste management, hydrocarbon occurrence management, water treatment, heavy metal pollution control, and radioactive material remediation, based on their unique characteristics (Ajith et al. 2021). In this context, nanotechnology offers a relatively unexplored approach for determining suitable ways to reduce MP pollution of the environment. Different approaches to increasing the biodegradation capacity of plastic materials are being investigated; one of which is the use of nanoparticles in microbial culture. It has also been revealed by prior studies that bacterial proliferation affecting the degradation of plastics can be controlled by altering the concentrations of SiO_2 nanoparticles (Tiwari et al. 2020). There are high prospects for the research of new ecologically friendly methods in the area of nanotechnology as applied to the biodegradation of microplastics. More research has to be conducted to understand the basic concepts and improve the effectiveness of nanotechnology-mediated remediation methods.

6.5.4 Bioremediation Utilizing Biochar

About 70% of above-ground biomass is left as straw after crop harvest, thus resulting in a huge amount of waste biomass in the case of crops like oilseed rape. Waste wood or wood chips are one form of biomass feed, which is widely available biomass that exists in the softwood zone. Residues include waste biomasses and animal excreta, which can be converted to biochar (BC) to address waste management issues and have a wide range of prospects across environmental management (Palansooriya et al. 2022; Tomczyk et al. 2020). The pyrolysis conditions affect the price, degradation profile, and efficiency of BC in the absorptive removal of MPs (Elbasiouny and Elbehiry 2023). Existing comparative analysis has shown that biochar (BC) performs far superior to other carbonaceous materials in batch adsorption in different, dynamic ecological systems like soil and water (Dong et al. 2023; Ganie et al. 2021; Jaafarzadeh Haghighi Fard et al. 2023). Furthermore, in the case of MP, this study also supports the findings of Badola et al. (2022), which shows that BC filters provide efficiency in eliminating MPs. The large surface area and irregular morphology of the BC particles promote the simultaneous adsorption and filtration of particles that are the basis of the BC filters. This provides the means for the physical absorption of MPs, and thus adequate capture is enhanced. As evidence, there are multiple opportunities to reduce or eliminate microplastic pollution in various settings by creating efficient and effective BC filters.

Biochar (BC) is a skillful way of eliminating the negative effects of MPs in both water and land environments. This study also suggests that low-cost solutions exist for addressing MPs, especially when using BC (Badola et al. 2022; Elbasiouny et al. 2023; Tursi et al. 2022). Also, the characteristics of biochar (BC) as well as the pyrolysis temperature improve the soil condition when MP-contaminated soil is mixed with BC (Roy et al. 2023). Notably, magnetic biochar especially Fe_2O_3/Fe_3O_4 biochar has been successful in capturing microplastic particles in the soil and aquifer (Tong et al. 2020). Also, the study recommends that the use of magnetic BC enhances the rate of oxidation, a process that helps separate MPs from other unwanted materials such as heavy metals (Ye et al. 2020). This synergistic effect points out the possibility of using the magnetic BC as a multifunctional restorative material and approach (Table 6.1).

Biochar (BC) is defined as a low-cost and highly efficient adsorbent for the removal of various contaminants, such as inorganic heavy metals and organic MPs in soil and water systems (Kumar et al. 2021; Tong et al. 2020). Previous studies have revealed that the use of BC in treating saturated column porous media limits the flow of various MPs. Its existence can be attributed to the following factors, including a decrease in steric hindrance, repulsion between charged moieties, competition between humic acids and cations with competing species, and the effect of ionic strength. Interactions such as these show that improving BC's adsorptive capacity decreases MP transport in porous surfaces. However, more efforts are needed to understand how BC works to bring clinic-enhancing outcomes for MP

Table 6.1 Microplastic tackle through biochar

Media	Biochar type	Response	References
Sandy porous media	Peanut shells and Fe_3O_4-BC	Polyester transport in porous media was further impeded.	Wang et al. (2022b)
Porous media (quartz sand)	Cellulose-BC and Fe_3O_4-BC	Decreasing the transport of MPs and increasing their retention in the media	Tong et al. (2020)
Soil	Wood chips	BC could accelerate the removal of MPs	Ren et al. (2021)
Soil	Date nuclei	BC mitigated the negative effects of MPs on soil, plants, and microorganisms	Iqbal et al. (2023c)
Soil	Corn straw	BC amendment enhanced bacterial community species evenness and richness, and facilitated N and P metabolism cycle of MPs contaminated soil plants	Ran et al. (2023a, b)
Water	Rice straw	BC adsorbed 99.96% of MPs	Wu et al. (2023)
Water	Palm kernel and coconut shells	Palm kernel shell BC removed a higher percentage of MPs (96.65%) than the coconut shell	Hanif et al. (2023)

repair, as Kumar et al. (2023) mentioned. Studying the effectiveness of the BC amendments to the MP fate and transport in different environmental matrices will help improve the long-lasting strategies for the remediation processes.

6.6 Conclusion

Therefore, this chapter concludes that the increased global population puts into motion an increased daily use of microplastics. As we pointed above, we need sustainable techniques for managing microplastics in our environment. There are several methods used at the present moment to combat microplastics; however, we only discuss the biological ways along with the environmentally friendly ones. In addition, people should be somehow informed that, thanks to their increased level of consciousness, they also use fewer microplastics and know the impact that the presence of these products can have on people and other organisms. Those three R's: as in reduce, reuse, and recycle, do quite a good job in handling microplastic pollution. Furthermore, there must be the development of cost-effective solutions for microplastics, namely BC and other green solutions, as against absorbing or consuming microplastics by physiological and chemical means. Thus, in the future, governments should focus on sectors that produce microplastics and increase people's awareness of the negative effects of microplastic existence.

References

Ajith M, Aswathi M, Priyadarshini E, Rajamani P (2021) Recent innovations of nanotechnology in water treatment: a comprehensive review. Bioresour Technol 342:126000

Akarsu C, Deniz F (2021) Electrocoagulation/electroflotation process for removal of organics and microplastics in laundry wastewater. Clean Soil Air Water 49(1):2000146

Al Mamun A, Prasetya TAE, Dewi IR, Ahmad M (2023) Microplastics in human food chains: food becoming a threat to health safety. Sci Total Environ 858:159834

Ali N, Liu W, Zeb A, Shi R, Lian Y, Wang Q, Liu J (2024) Environmental fate, aging, toxicity and potential remediation strategies of microplastics in soil environment: current progress and future perspectives. Sci Total Environ 906:167785

Auta H, Abioye O, Aransiola S, Bala J, Chukwuemeka V, Hassan A, Fauziah SH (2022) Enhanced microbial degradation of PET and PS microplastics under natural conditions in mangrove environment. J Environ Manag 304:114273

Badola N, Bahuguna A, Sasson Y, Chauhan JS (2022) Microplastics removal strategies: a step toward finding the solution. Front Environ Sci Eng 16:1–18

Bai R, Li Z, Liu Q, Liu Q, Cui J, He W (2024) The reciprocity principle in mulch film deterioration and microplastic generation. Environ Sci Processes Impacts 26(1):8–15

Bashir SM, Kimiko S, Mak C-W, Fang JK-H, Gonçalves D (2021) Personal care and cosmetic products as a potential source of environmental contamination by microplastics in a densely populated Asian city. Front Mar Sci 8:683482

Bhat MA (2024) Unveiling the overlooked threat: macroplastic pollution in indoor markets in an urban city. Case Stud Chem Environ Eng 9:100558

Bhat MA, Gedik K, Gaga EO (2023) Atmospheric micro (nano) plastics: future growing concerns for human health. Air Qual Atmos Health 16(2):233–262

Bhat M, Gedik K, Gaga E (2024) A preliminary study on the natural aging behavior of microplastics in indoor and outdoor environments. Int J Environ Sci Technol 21(2):1923–1936

Bordós G, Urbányi B, Micsinai A, Kriszt B, Palotai Z, Szabó I, Szoboszlay S (2019) Identification of microplastics in fish ponds and natural freshwater environments of the Carpathian basin, Europe. Chemosphere 216:110–116

Bosker T, Bouwman LJ, Brun NR, Behrens P, Vijver MG (2019) Microplastics accumulate on pores in seed capsule and delay germination and root growth of the terrestrial vascular plant Lepidium sativum. Chemosphere 226:774–781

Campanale C, Massarelli C, Savino I, Locaputo V, Uricchio VF (2020) A detailed review study on potential effects of microplastics and additives of concern on human health. Int J Environ Res Public Health 17(4):1212

Caputi S, Diomede F, Lanuti P, Marconi GD, Di Carlo P, Sinjari B, Trubiani O (2022) Microplastics affect the inflammation pathway in human gingival fibroblasts: a study in the Adriatic Sea. Int J Environ Res Public Health 19(13):7782

Cedillo-González EI, Siligardi C (2022) Towards a microplastic-free ocean: green photocatalysis for mitigation of micro-and nanoplastic marine pollution. Paper presented at the International Conference on Microplastic Pollution in the Mediterranean Sea

Chabi K, Li J, Ye C, Kiki C, Xiao X, Li X, Yu X (2024) Rapid sand filtration for <10 μm-sized microplastic removal in tap water treatment: efficiency and adsorption mechanisms. Sci Total Environ 912:169074

Chamas A, Moon H, Zheng J, Qiu Y, Tabassum T, Jang JH, Suh S (2020) Degradation rates of plastics in the environment. ACS Sustain Chem Eng 8(9):3494–3511

Cheng Y-R, Wang H-Y (2022) Highly effective removal of microplastics by microalgae Scenedesmus abundans. Chem Eng J 435:135079

Chubarenko I, Bocherikova I, Esiukova E, Isachenko I, Kupriyanova A, Lobchuk O, Fetisov S (2023) Microplastics in sea ice: a fingerprint of bubble flotation. Sci Total Environ 892:164611

Çıtar Dazıroğlu ME, Bilici S (2023) The hidden threat to food safety and human health: microplastics. Environ Dev Sustain 26(9):1–23

Cunha C, Silva L, Paulo J, Faria M, Nogueira N, Cordeiro N (2020) Microalgal-based biopolymer for nano-and microplastic removal: a possible biosolution for wastewater treatment. Environ Pollut 263:114385

de Souza Machado AA, Lau CW, Kloas W, Bergmann J, Bachelier JB, Faltin E, Rillig MC (2019) Microplastics can change soil properties and affect plant performance. Environ Sci Technol 53(10):6044–6052

Dehghanian Z, Lajayer BA, Atigh ZBQ, Nayeri S, Ahmadabadi M, Taghipour L, Price G (2023) Micro (nano) plastics uptake, toxicity and detoxification in plants: challenges and prospects. Ecotoxicol Environ Saf 268:115676

Dong M, He L, Jiang M, Zhu Y, Wang J, Gustave W, Wang Z (2023) Biochar for the removal of emerging pollutants from aquatic systems: a review. Int J Environ Res Public Health 20(3):1679

Dubey I, Khan S, Kushwaha S (2022) Developmental and reproductive toxic effects of exposure to microplastics: a review of associated signaling pathways. Front Toxicol 4:901798

Elbasiouny H, Elbehiry F (2023) Addressing the microplastic dilemma in soil and sediment with focus on biochar-based remediation techniques. Soil Syst 7(4):110

Elbasiouny H, Mostafa AA, Zedan A, Elbltagy HM, Dawoud SF, Elbanna BA, Darweesh M (2023) Potential effect of biochar on soil properties, microbial activity and vicia faba properties affected by microplastics contamination. Agronomy 13(1):149

Elkhatib D, Oyanedel-Craver V, Carissimi E (2021) Electrocoagulation applied for the removal of microplastics from wastewater treatment facilities. Sep Purif Technol 276:118877

Eraslan FN, Bhat MA, Gedik K, Gaga EO (2023) The single-use plastic pandemic in the COVID-19 era. In: Microplastics in the ecosphere: air, water, soil, and food. Wiley, Hoboken, pp 65–75

Eze CG, Nwankwo CE, Dey S, Sundaramurthy S, Okeke ES (2024) Food chain microplastics contamination and impact on human health: a review. Environ Chem Lett 22(4):1889–1927

Fang M, Liao Z, Ji X, Zhu X, Wang Z, Lu C, Zhang M (2022) Microplastic ingestion from atmospheric deposition during dining/drinking activities. J Hazard Mater 432:128674

Ganie ZA, Khandelwal N, Tiwari E, Singh N, Darbha GK (2021) Biochar-facilitated remediation of nanoplastic contaminated water: effect of pyrolysis temperature induced surface modifications. J Hazard Mater 417:126096

Gao W, Zhang Y, Mo A, Jiang J, Liang Y, Cao X, He D (2022) Removal of microplastics in water: technology progress and green strategies. Green Anal Chem 3:100042

Gavigan J, Kefela T, Macadam-Somer I, Suh S, Geyer R (2020) Synthetic microfiber emissions to land rival those to waterbodies and are growing. PLoS One 15(9):e0237839

Guo J-J, Huang X-P, Xiang L, Wang Y-Z, Li Y-W, Li H, Wong M-H (2020) Source, migration and toxicology of microplastics in soil. Environ Int 137:105263

Habib RZ, Aldhanhani JA, Ali AH, Ghebremedhin F, Elkashlan M, Mesfun M, Thiemann T (2022) Trends of microplastic abundance in personal care products in The United Arab Emirates over the period of 3 years (2018–2020). Environ Sci Pollut Res 29(59):89614–89624

Hanif MA, Ibrahim N, Dahalan FA, Ali UFM, Hasan M, Jalil AA (2022) Microplastics and nanoplastics: recent literature studies and patents on their removal from aqueous environment. Sci Total Environ 810:152115

Hanif MA, Ibrahim N, Dahalan FA, Ali UF, Hasan M, Azhari AW, Jalil AA (2023) Microplastics in facial cleanser: extraction, identification, potential toxicity, and continuous-flow removal using agricultural waste–based biochar. Environ Sci Pollut Res 30(21):60106–60120

Hartmann NB, Huffer T, Thompson RC, Hassellov M, Verschoor A, Daugaard AE, Cole M (2019) Are we speaking the same language? Recommendations for a definition and categorization framework for plastic debris. Environ Sci Technol 53(3):1039–1047

Horton AA, Walton A, Spurgeon DJ, Lahive E, Svendsen C (2017) Microplastics in freshwater and terrestrial environments: evaluating the current understanding to identify the knowledge gaps and future research priorities. Sci Total Environ 586:127–141

Hu K, Zhou P, Yang Y, Hall T, Nie G, Yao Y, Wang S (2021) Degradation of microplastics by a thermal Fenton reaction. ACS ES&T Eng 2(1):110–120

Huang Y, Liu Q, Jia W, Yan C, Wang J (2020) Agricultural plastic mulching as a source of microplastics in the terrestrial environment. Environ Pollut 260:114096

Huang Z, Bu J, Wang H (2022a) Application of two modified kaolin materials in removing microplastics from water. J Mater Cycl Waste Manage 24(4):1460–1475

Huang D, Zhang Y, Long J, Yang X, Bao L, Yang Z, Peng C (2022b) Polystyrene microplastic exposure induces insulin resistance in mice via dysbacteriosis and pro-inflammation. Sci Total Environ 838:155937

Ihenetu S, Li G, Mo Y, Kubwimana J (2024) Impacts of microplastics and urbanization on soil health: an urgent concern for sustainable development. Green Anal Chem 8:100095

Iqbal B, Javed Q, Khan I, Tariq M, Ahmad N, Elansary HO, Du D (2023a) Influence of soil microplastic contamination and cadmium toxicity on the growth, physiology, and root growth traits of Triticum aestivum L. S Afr J Bot 160:369–375

Iqbal B, Li G, Alabbosh KF, Hussain H, Khan I, Tariq M, Ahmad N (2023b) Advancing environmental sustainability through microbial reprogramming in growth improvement, stress alleviation, and phytoremediation. Plant Stress 10:100283

Iqbal B, Zhao T, Yin W, Zhao X, Xie Q, Khan KY, Du D (2023c) Impacts of soil microplastics on crops: a review. Appl Soil Ecol 181:104680

Jaafarzadeh Haghighi Fard N, Jahedi F, Dehdarirad H (2023) The possibility of microplastic removal by earthworms and comparing with conventional chemical removal methods (a global and deeply systematic review). J Polym Environ 31(12):5050–5064

Jiang H, Zhang Y, Bian K, Wang C, Xie X, Wang H, Zhao H (2022) Is it possible to efficiently and sustainably remove microplastics from sediments using froth flotation? Chem Eng J 448:137692

Kumar NS, Shaikh HM, Asif M, Al-Ghurabi EH (2021) Engineered biochar from wood apple shell waste for high-efficient removal of toxic phenolic compounds in wastewater. Sci Rep 11(1):2586

Kumar R, Verma A, Rakib MRJ, Gupta PK, Sharma P, Garg A, Aminabhavi TM (2023) Adsorptive behavior of micro (nano) plastics through biochar: co-existence, consequences, and challenges in contaminated ecosystems. Sci Total Environ 856:159097

Landrigan PJ, Raps H, Cropper M, Bald C, Brunner M, Canonizado EM, Enck J (2023) The Minderoo-Monaco commission on plastics and human health. Ann Glob Health 89(1):23

Lehmann M, Häusl FP, Gekle S (2023) Modeling of vertical microplastic transport by rising bubbles. Microplast Nanoplast 3(1):4

Leifheit EF, Kissener HL, Faltin E, Ryo M, Rillig MC (2022) Tire abrasion particles negatively affect plant growth even at low concentrations and alter soil biogeochemical cycling. Soil Ecol Lett 4(4):409–415

Leistenschneider C, Burkhardt-Holm P, Mani T, Primpke S, Taubner H, Gerdts G (2021) Microplastics in the Weddell Sea (Antarctica): a forensic approach for discrimination between environmental and vessel-induced microplastics. Environ Sci Technol 55(23):15900–15911

Lellis B, Fávaro-Polonio CZ, Pamphile JA, Polonio JC (2019) Effects of textile dyes on health and the environment and bioremediation potential of living organisms. Biotechnol Res Innov 3(2):275–290

Li S, Yang Y, Yang S, Zheng H, Zheng Y, Jun M, Chang J-S (2023) Recent advances in biodegradation of emerging contaminants-microplastics (MPs): feasibility, mechanism, and future prospects. Chemosphere 331:138776

Lian J, Liu W, Meng L, Wu J, Zeb A, Cheng L, Sun H (2021) Effects of microplastics derived from polymer-coated fertilizer on maize growth, rhizosphere, and soil properties. J Clean Prod 318:128571

Liu X, Dai B, Chuai Y, Hu M, Zhang H (2023a) Associations between vitamin D levels and periodontal attachment loss. Clin Oral Investig 27(8):4727–4733

Liu S, Jin R, Zhang J, Zhao Y, Shen M, Wang Y (2023b) Are algae a promising ecofriendly approach to micro/nanoplastic remediation? Sci Total Environ 904:166779

Liu Y, Xu F, Ding L, Zhang G, Bai B, Han Y, Wan S (2023d) Microplastics reduce nitrogen uptake in peanut plants by damaging root cells and impairing soil nitrogen cycling. J Hazard Mater 443:130384

Lozano YM, Aguilar-Trigueros CA, Onandia G, Maaß S, Zhao T, Rillig MC (2021) Effects of microplastics and drought on soil ecosystem functions and multifunctionality. J Appl Ecol 58(5):988–996

Meléndez-Fernández OH, Liu JA, Nelson RJ (2023) Circadian rhythms disrupted by light at night and mistimed food intake alter hormonal rhythms and metabolism. Int J Mol Sci 24(4):3392

Moharir RV, Kumar S (2019) Challenges associated with plastic waste disposal and allied microbial routes for its effective degradation: a comprehensive review. J Clean Prod 208:65–76

Moralia M-A, Quignon C, Simonneaux M, Simonneaux V (2022) Environmental disruption of reproductive rhythms. Front Neuroendocrinol 66:100990

Moyses DN, Teixeira DA, Waldow VA, Freire DM, Castro AM (2021) Fungal and enzymatic biodepolymerization of waste post-consumer poly (ethylene terephthalate)(PET) bottles using Penicillium species. 3 Biotech 11(10):435

Nanthini Devi K, Raju P, Santhanam P, Dinesh Kumar S, Krishnaveni N, Roopavathy J, Perumal P (2021) Biodegradation of low-density polyethylene and polypropylene by microbes isolated from Vaigai River, Madurai, India. Arch Microbiol 203:6253–6265

Nazir MJ, Li G, Nazir MM, Zulfiqar F, Siddique KH, Iqbal B, Du D (2024) Harnessing soil carbon sequestration to address climate change challenges in agriculture. Soil Tillage Res 237:105959

Niu H, Liu S, Jiang Y, Hu Y, Li Y, He L, Chen Z (2023) Are microplastics toxic? A review from eco-toxicity to effects on the gut microbiota. Meta 13(6):739

Nohara NML, Filho HJI, Siqueira AF, Aguiar LGD, Oliveira GCK, Nohara EL, Napoleão DADS (2023) Study of the effectiveness of a ZnO-TiO2 formulation in the degradation of humic substances in mature leachate by solar photocatalysis Brazil. Rev Amb Água 18:e2932

Nohara NML, Ariza-Tarazona MC, Triboni ER, Nohara EL, Villarreal-Chiu JF, Cedillo-González EI (2024) Are you drowned in microplastic pollution? A brief insight on the current knowl-

edge for early career researchers developing novel remediation strategies. Sci Total Environ 918:170382
Oliveira A, Neto AD, Moura M, Dantas TC (2023) Use of surfactant-modified adsorbents in the removal of microplastics from wastewater. J Environ Chem Eng 11(5):110827
Osman AI, Hosny M, Eltaweil AS, Omar S, Elgarahy AM, Farghali M, Batumalaie K (2023) Microplastic sources, formation, toxicity and remediation: a review. Environ Chem Lett 21(4):2129–2169
Palaniappan S, Sadacharan CM, Rostama B (2022) Polystyrene and polyethylene microplastics decrease cell viability and dysregulate inflammatory and oxidative stress markers of MDCK and L929 cells in vitro. Expo Health 14(1):75–85
Palansooriya KN, Sang MK, Igalavithana AD, Zhang M, Hou D, Oleszczuk P, Ok YS (2022) Biochar alters chemical and microbial properties of microplastic-contaminated soil. Environ Res 209:112807
Park SY, Kim CG (2019) Biodegradation of micro-polyethylene particles by bacterial colonization of a mixed microbial consortium isolated from a landfill site. Chemosphere 222:527–533
Peller J, Nevers MB, Byappanahalli M, Nelson C, Babu BG, Evans MA, Shidler S (2021) Sequestration of microfibers and other microplastics by green algae, Cladophora, in the US Great Lakes. Environ Pollut 276:116695
Peng G, Xiang M, Wang W, Su Z, Liu H, Mao Y, Zhang P (2022) Engineering 3D graphene-like carbon-assembled layered double oxide for efficient microplastic removal in a wide pH range. J Hazard Mater 433:128672
Piazza V, Uheida A, Gambardella C, Garaventa F, Faimali M, Dutta J (2021) Ecosafety screening of photo-Fenton process for the degradation of microplastics in water. Front Mar Sci 8:2126
Pinlova B, Nowack B (2023) Characterization of fiber fragments released from polyester textiles during UV weathering. Environ Pollut 322:121012
Qiao R, Sheng C, Lu Y, Zhang Y, Ren H, Lemos B (2019) Microplastics induce intestinal inflammation, oxidative stress, and disorders of metabolome and microbiome in zebrafish. Sci Total Environ 662:246–253
Ran C, Bai X, Tan Q, Luo G, Cao Y, Wu L, Liu M (2023a) Threat of soil formation rate to health of karst ecosystem. Sci Total Environ 887:163911
Ran T, Li J, Liao H, Zhao Y, Yang G, Long J (2023b) Effects of biochar amendment on bacterial communities and their function predictions in a microplastic-contaminated Capsicum annuum L. soil. Environ Technol Innov 31:103174
Ren X, Tang J, Wang L, Sun H (2021) Combined effects of microplastics and biochar on the removal of polycyclic aromatic hydrocarbons and phthalate esters and its potential microbial ecological mechanism. Front Microbiol 12:647766
Rolsky C, Kelkar V, Driver E, Halden RU (2020) Municipal sewage sludge as a source of microplastics in the environment. Curr Opin Environ Sci Health 14:16–22
Roursgaard M, Hezareh Rothmann M, Schulte J, Karadimou I, Marinelli E, Møller P (2022) Genotoxicity of particles from grinded plastic items in Caco-2 and HepG2 cells. Front Public Health 10:906430
Roy T, Dey TK, Jamal M (2023) Microplastic/nanoplastic toxicity in plants: an imminent concern. Environ Monit Assess 195(1):27
Senathirajah K, Kandaiah R, Panneerselvan L, Sathish C, Palanisami T (2023) Fate and transformation of microplastics due to electrocoagulation treatment: impacts of polymer type and shape. Environ Pollut 334:122159
Shen M, Zhang Y, Almatrafi E, Hu T, Zhou C, Song B, Zeng G (2022) Efficient removal of microplastics from wastewater by an electrocoagulation process. Chem Eng J 428:131161
Shi R, Liu W, Lian Y, Wang X, Men S, Zeb A, Zheng Z (2023) Toxicity mechanisms of nanoplastics on crop growth, interference of phyllosphere microbes, and evidence for foliar penetration and translocation. Environ Sci Technol 58(2):1010–1021
Sinha JK, Ghosh S, Swain U, Giridharan NV, Raghunath M (2014) Increased macromolecular damage due to oxidative stress in the neocortex and hippocampus of WNIN/Ob, a novel rat model of premature aging. Neuroscience 269:256–264

Sinha JK, Vashisth K, Ghosh S (2022) The importance of sleep studies in improving the health indices of a nation. Sleep Med X 4:100049

Sridharan S, Kumar M, Bolan NS, Singh L, Kumar S, Kumar R, You S (2021) Are microplastics destabilizing the global network of terrestrial and aquatic ecosystem services? Environ Res 198:111243

Su L, Xiong X, Zhang Y, Wu C, Xu X, Sun C, Shi H (2022) Global transportation of plastics and microplastics: a critical review of pathways and influences. Sci Total Environ 831:154884

Swart B, Pihlajamäki A, Chew YJ, Wenk J (2022) Microbubble-microplastic interactions in batch air flotation. Chem Eng J 449:137866

Tariq M, Iqbal B, Khan I, Khan AR, Jho EH, Salam A, Du D (2024) Microplastic contamination in the agricultural soil—mitigation strategies, heavy metals contamination, and impact on human health: a review. Plant Cell Rep 43(3):65

Thacharodi A, Hassan S, Hegde TA, Thacharodi DD, Brindhadevi K, Pugazhendhi A (2023) Water a major source of endocrine-disrupting chemicals: an overview on the occurrence, implications on human health and bioremediation strategies. Environ Res 231:116097

Thompson RC, Moore CJ, Vom Saal FS, Swan SH (2009) Plastics, the environment and human health: current consensus and future trends. Philos Trans R Soc B Biol Sci 364(1526):2153–2166

Tiwari N, Santhiya D, Sharma JG (2020) Microbial remediation of micro-nano plastics: current knowledge and future trends. Environ Pollut 265:115044

Tomczyk A, Sokołowska Z, Boguta P (2020) Biochar physicochemical properties: pyrolysis temperature and feedstock kind effects. Rev Environ Sci Biotechnol 19(1):191–215

Tong M, He L, Rong H, Li M, Kim H (2020) Transport behaviors of plastic particles in saturated quartz sand without and with biochar/Fe3O4-biochar amendment. Water Res 169:115284

Tursi A, Baratta M, Easton T, Chatzisymeon E, Chidichimo F, De Biase M, De Filpo G (2022) Microplastics in aquatic systems, a comprehensive review: origination, accumulation, impact, and removal technologies. RSC Adv 12(44):28318–28340

Vanapalli KR, Sharma HB, Ranjan VP, Samal B, Bhattacharya J, Dubey BK, Goel S (2021) Challenges and strategies for effective plastic waste management during and post COVID-19 pandemic. Sci Total Environ 750:141514

Veropalumbo R, Oreto C, Viscione N, Pirozzi F, Pontoni L, Trancone G, Russo F (2023) Exploring the effect on the environment of encapsulated micro-and nano-plastics into asphalt mastics for road pavement. Environ Res 216:114466

Walker TR, Fequet L (2023) Current trends of unsustainable plastic production and micro (nano) plastic pollution. TrAC Trends Anal Chem 160:116984

Wan Y, Wu C, Xue Q, Hui X (2019) Effects of plastic contamination on water evaporation and desiccation cracking in soil. Sci Total Environ 654:576–582

Wan Y, Chen X, Liu Q, Hu H, Wu C, Xue Q (2022) Informal landfill contributes to the pollution of microplastics in the surrounding environment. Environ Pollut 293:118586

Wan L, Cheng H, Liu Y, Shen Y, Liu G, Su X (2023) Global meta-analysis reveals differential effects of microplastics on soil ecosystem. Sci Total Environ 867:161403

Wang Y, Baynes A, Renner KO, Zhang M, Scrimshaw MD, Routledge EJ (2022a) Uptake, elimination and effects of cosmetic microbeads on the freshwater gastropod Biomphalaria glabrata. Toxics 10(2):87

Wang X, Dan Y, Diao Y, Liu F, Wang H, Sang W, Zhang Y (2022b) Transport characteristics of polystyrene microplastics in saturated porous media with biochar/Fe3O4-biochar under various chemical conditions. Sci Total Environ 847:157576

Wu J, Yang C, Zhao H, Shi J, Liu Z, Li C, Song F (2023) Efficient removal of microplastics from aqueous solution by a novel magnetic biochar: performance, mechanism, and reusability. Environ Sci Pollut Res 30(10):26914–26928

Xia Y, Xiang X-M, Dong K-Y, Gong Y-Y, Li Z-J (2020) Surfactant stealth effect of microplastics in traditional coagulation process observed via 3-D fluorescence imaging. Sci Total Environ 729:138783

Yang J, Li L, Li R, Xu L, Shen Y, Li S, Luo Y (2021) Microplastics in an agricultural soil following repeated application of three types of sewage sludge: a field study. Environ Pollut 289:117943

Yao J, Peng Z, Chen W, Lin Q, Cheng M, Li H, Yang HY (2023) Surface characteristics of polystyrene microplastics mainly determine their coagulation performances. Mar Pollut Bull 186:114347

Ye S, Cheng M, Zeng G, Tan X, Wu H, Liang J, Yang H (2020) Insights into catalytic removal and separation of attached metals from natural-aged microplastics by magnetic biochar activating oxidation process. Water Res 179:115876

Yu H, Fan P, Hou J, Dang Q, Cui D, Xi B, Tan W (2020) Inhibitory effect of microplastics on soil extracellular enzymatic activities by changing soil properties and direct adsorption: an investigation at the aggregate-fraction level. Environ Pollut 267:115544

Yuan J, Cao J, Yu F, Ma J (2022) Microbial degradation of polystyrene microplastics by a novel isolated bacterium in aquatic ecosystem. Sustain Chem Pharm 30:100873

Zhang G, Zhang F, Li X (2019) Effects of polyester microfibers on soil physical properties: perception from a field and a pot experiment. Sci Total Environ 670:1–7

Zhang Q, He Y, Cheng R, Li Q, Qian Z, Lin X (2022a) Recent advances in toxicological research and potential health impact of microplastics and nanoplastics in vivo. Environ Sci Pollut Res 29(27):40415–40448

Zhao W, Li J, Liu M, Wang R, Zhang B, Meng X-Z, Zhang S (2024) Seasonal variations of microplastics in surface water and sediment in an inland river drinking water source in southern China. Sci Total Environ 908:168241

Zheng B, Li B, Wan H, Lin X, Cai Y (2022) Coral-inspired environmental durability aerogels for micron-size plastic particles removal in the aquatic environment. J Hazard Mater 431:128611

Zhou Y, Wang J, Zou M, Jia Z, Zhou S, Li Y (2020) Microplastics in soils: a review of methods, occurrence, fate, transport, ecological and environmental risks. Sci Total Environ 748:141368

Zhu F, Zhu C, Wang C, Gu C (2019) Occurrence and ecological impacts of microplastics in soil systems: a review. Bull Environ Contam Toxicol 102:741–749

Ziani K, Ioniță-Mîndrican C-B, Mititelu M, Neacşu SM, Negrei C, Moroşan E, Preda O-T (2023) Microplastics: a real global threat for environment and food safety: a state of the art review. Nutrients 15(3):617

Socioeconomic Impacts of Microplastics Pollution in the Marine Environment

7

Ramzi H. Amran, Fotoon Sayegh,
and Sathianeson Satheesh

Abstract

Plastics is contributing a major portion of the litter found in marine environments throughout the world. Microplastics (plastic particles less than 5 mm in size) are causing disquietude because of their capability to harm ecosystems, organisms, and human health. Many studies have reported that microplastics build up in marine organisms and carry through food webs. Through the food chain, microplastics get into the body of humans and cause human wellness problems. Accumulation of microplastics in marine fish is a socioeconomic issue affecting the economy and livelihoods of the fishing community. While the impacts of microplastics on aquatic organisms are studied extensively, the socioeconomic impacts of microplastic pollution in marine ecosystems received little attention. This chapter provides an overview of microplastic pollution in marine ecosystems and its socioeconomic impacts.

Keywords

Plastic pollution · Marine litter · Marine ecosystems · Seafood · Marine fishes

R. H. Amran · S. Satheesh (✉)
Faculty of Marine Sciences, Department of Marine Biology, King Abdulaziz University, Jeddah, Saudi Arabia
e-mail: ssathianeson@kau.edu.sa

F. Sayegh
Faculty of Sciences, Department of Biological Sciences, King Abdulaziz University, Jeddah, Saudi Arabia

Marine Natural Products Research Unit, King Fahad Medical Research Centre, King Abdulaziz University, Jeddah, Saudi Arabia

Faculty of Marine Science and Environments, Department of Marine Biology and Fisheries, Hodeidah University, Hodeidah, Yemen

M. Oves et al. (eds.), *Marine Microplastics and the Quest for Remediation*,
https://doi.org/10.1007/978-981-96-8979-8_7

7.1 Introduction

Plastic is currently the third most produced material in the world, after concrete and steel (Watts 2019). With its many diverse applications around the world, plastic production is likely to continue into the future (Kumar et al. 2021). The commercial manufacture of plastics began in 1950 when yearly output attained two million tons (Mt) (Geyer et al. 2017). By 2015, worldwide plastic production had risen to 380 (mt/year) between 1950 and 2015 (Hossain and Tuha 2020; Cverenkárová et al. 2021), and more than 7800 Mt. of plastic were manufactured. From 1950 and 2015, this increasing growth of the plastics industry produced around 6300 Mt. of waste which is usually thrown into landfills or the environment, about 9% of which was recycled (Cverenkárová et al. 2021; Hossain et al. 2021; Nyam and Ayeleru 2024). China is the leading producer of plastics globally, representing approximately 25% of worldwide production. Over the past 5 years, China's plastic production has overridden 77 Mt., whereas Europe's production has exceeded 60 Mt. (Rhodes 2018). Plastic waste is a global sustainability challenge. Overall, it is estimated that approximately 60% of plastics manufactured over time end up in landfills or are freed into the environment without proper treatment, causing their impact to persist for centuries as these pollutants degrade very slowly (Chamas et al. 2020; Lozano and García-Verdugo 2023; Sen and Yadav 2024). This large amount of long-lasting plastic waste leads to a diversity of critical environmental and health problems (Peng et al. 2018; Khoaele et al. 2023; Nguyen et al. 2023; Wang and Wu 2024). Poorly managed plastic waste can seep into the oceans, where it continues to degrade into microplastics (less than 5 mm) (Andrady 2011). Given the enormous amounts of plastic debris entering aquatic ecosystems, it is no surprise that microplastics are now common in these environments.

Microplastic (MPs) has recently become the focus of much research interest. Plastic litter was first identified in the shape of tiny beads in 1972, when these particles were discovered in the Sargasso Sea (Carpenter and Smith Jr 1972). In 2009, the U.S. National Oceanic and Atmospheric Administration (NOAA) identified particles up to 5 mm in size as MPs (Arthur et al. 2009; Collignon et al. 2014). Nonetheless, there is still discussion regarding the minimum size for MPs, and there is no consensus. Plastic waste infiltrates the aquatic environment either directly or indirectly through landfills, from where it is spread by wind to water bodies (sea and river) (Crawford and Quinn 2017). It is approximated that about 8 Mt. of plastic waste enters the world's seas and oceans annually (MacArthur 2017). MPs can be classified into two principal categories based on their origin, primary and secondary MPs, with the main distinction between the two categories being how they are introduced into the ecosystem (Cverenkárová et al. 2021). Primary MPs are emitted into the environment in their ultimate form, whereas secondary MPs arise due to the weathering and the degradation of larger plastics into smaller particles within the environment (Boucher 2017). The existence of plastic in the ecosystem is a critical worry, as it is persistent material (Shah et al. 2008; MacLeod et al. 2021), often associated with potentially toxic chemicals, and is consumed by aquatic organisms (Gallo et al. 2020; Rafa et al. 2023; Shi et al. 2024).

7.2 Sources of MPs

Plastic is spread all over the world, which may result in MPs and nanoplates that spread in the air, water, and soil where they constitute a key source of plastic pollution, which results in many risks to human health and other living organisms (Saha and Saha 2024). MPs pollution originates from industrial products that are often intentionally or unintentionally dumped or discharged into aquatic environments, these particles can likewise be conveyed into these environments via runoff, sewage, drainages, effluent discharge, waves, and wind and are often referred to as marine debris or litter (Galgani et al. 2013; García-Rivera et al. 2017; Onyena et al. 2021; Jiang et al. 2024; Salgaonkar et al. 2024). MPs in the aquatic environment can arise from both primary and secondary sources (Li et al. 2016). Primary sources involve the intentional or unintentional discharge of micro-pellets, microbeads, and microfibers into the aquatic ecosystem (Gasperi et al. 2018). Plastic pellets are extensively utilized in daily human life, including clothing, automobile manufacturing, agriculture, electrical industry, medical equipment, chemical industry, telecommunications, packaging, household appliances, transportation, and building materials industries (An et al. 2020; Dey et al. 2024; Oladele et al. 2024). Synthetic materials contribute to the release of MPs particles and fibers into the aquatic environment through clothing and textiles, which can leak into an aquatic ecosystem and the air (De Falco et al. 2019; Weis 2021; Kaur and Dandautiya 2024; Surana et al. 2024). Microbeads, after being processed into fine particles, are widely used as an alternative to synthetic dyes that can be added to personal care products such as washing, peeling, whitening, and cleansing products (Zhou et al. 2023; Giustra et al. 2024). Cosmetics and personal care products also contribute to the aquatic ecosystem with MPs particles, as they are among the factors involved in environmental pollution with MPs (Bashir et al. 2021; Losetty et al. 2024; Ranjan et al. 2024). Many cosmetics and personal care products also contain MPs beads, including hair dyes, toothpaste, shower gel, soaps, shampoos, dishwashing detergents, make-up, and sunscreens (McCallum 2018; Anagnosti et al. 2021; Nawalage and Bellanthudawa 2022; Chaudhari et al. 2024). According to Boucher (2017) plastic microbeads have been estimated in many countries and regions of the world, where they are estimated in South Asia (18.3%), > North America (17.2%), > Europe and Central Asia (15.9%), > China (15.8%), > East Asia and Oceania (15.0%), > South America and Africa (9.1%), > Middle East (8.7%). Another source of environmental MPs is paint, which is one of the sources that provide the environment with these particles. House painting processes and their falling into the environment, as well as automotive coatings, signs, and marine coatings (coating ships and boats), are among the sources of these pollutants (Wang et al. 2019a, b; Tamburri et al. 2022). Among other sources, laundry wastewater (home laundry and washing plant wastewater) is one of the most prominent sources of MPs entering the aquatic environment, releasing large quantities of plastic particles and fibres derived from textiles (Galvão et al. 2020; Le et al. 2022; Šaravanja et al. 2022), it is forecasted that nearly 1900 microfibers are released in each wash cycle from a solitary piece of clothing to wastewater management plants (Cheung and Fok 2017).

MPs cannot be removed from wastewater management plants and are, therefore, released into the environment through the discharging of effluent or sludge. A few investigations suggest that MPs fibres found in oceans, rivers, and soil constitute the bulk of MPs in the aquatic environment (Horton et al. 2017). Over 30 types of MPs have been discovered in the effluents of wastewater treatment plants, including polyethylene (PE), polystyrene (PS), and polypropylene (PP) (Li et al. 2019a, b). According to Leslie et al. (2017) they found that the average concentration of MPs in the effluent from sewage at seven studied sites in the Netherlands spanned from 51–81 to 68–910 n/L. In addition, the erosion of tires releases MPs into the environment, which is not taken care of, but may exacerbate the pollution problem (Kole et al. 2017; Giechaskiel et al. 2024).

Secondary sources of MPs are important sources that supply the aquatic environment with these pollutants, as they are fragments derivative from larger plastic materials that have not been effectively discarded (Pettipas et al. 2016). The introduction of MPs into the aquatic environment occurs from secondary sources and is caused by biophysiochemical forces such as heat, biodegradation, mechanical forces, UV irradiation, and oxidation processes (Rillig et al. 2017; Sathyamohan et al. 2023; Sutkar et al. 2023). Plastic bags are one of the largest environmental MPs problems and are an important element that is used constantly in daily life (Yurtsever and Yurtsever 2018). Due to their wide use, ubiquity, light weight, low cost, and easy storage, approximately 5 trillion plastic bags are utilized globally, with roughly 100 million of these bags utilized in the United States each year (EPI 2014), and less than 10% of them are recycled (UNEP 2018). Plastic bags are disposed of in the environment and can be conveyed to the surrounding aquatic ecosystem, so the utilization of plastic bags has been gradually prohibited in many countries due to their long decomposition cycle (Jalil et al. 2013; Ahsan et al. 2020), although there is a different danger of plastic bags, their presence in the environment may be a source of concern and a source of environmental MPs pollutants (An et al. 2020). Similarly, plastic packaging and bags for many consumer goods contribute to MPs contamination during the unpacking and disposal phase (Sobhani et al. 2020; Sharma 2024). Other sources that supply the aquatic environment with MPs are plastic fishing wastes such as fishing nets, buoys, farming cages, floating plastic boxes, fishing lines, ropes, cables, etc. (Niaounakis 2017; Krüger et al. 2020; Sharma et al. 2024). The amount of commercial fishing gear discarded worldwide annually is estimated to range from 0.13 and 135,000 tons (Merrell Jr 1980). Agricultural activities contribute significantly to MPs pollution, mainly through the use of organic fertilizers and soil additives (Rana et al. 2023). Studies show that various types of fertilizers, such as biowaste compost and sewage ash, contain significant amounts of MPs, which can accumulate in agricultural soils and negatively impact soil health and crop production (Zhang et al. 2020, 2022; Guo et al. 2023). Organic fertilizers such as biowaste compost and sewage sludge used in agricultural fertilizers are principal sources of MPs in agricultural soils, which can reach up to 4050 particles per kg (Medyńska-Juraszek and Szczepańska 2023). Most of the above-mentioned sources of MPs enter the aquatic ecosystem, which can cause

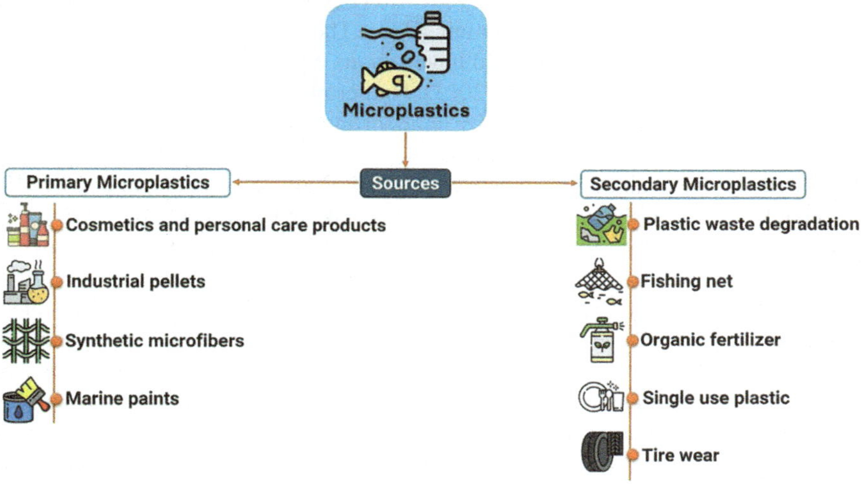

Fig. 7.1 Schematic diagram of some sources of MPs in the aquatic environment

pollution and the resulting effects on humans and other living organisms. Figure 7.1 shows the schematic diagram of some sources of MPs in the aquatic environment.

Modern society's widespread reliance on plastic has led to the proliferation of MPs in the environment, creating significant social and economic challenges. MPs, identified as plastic particles less than 5 mm, have been found to accumulate across a variety of ecosystems, including aquatic, terrestrial, and agricultural systems (Arias-Andres et al. 2018; Sarker et al. 2020). The stability and widespread distribution of MPs have elevated growing apprehensions about their prospective consequences on both public health and the ecosystem (Wang et al. 2021a, b; Sharma and Vidyarthi 2024). The scale of MPs pollution is alarming, with studies showing it is present in large quantities even in remote, sparsely populated areas (Arias-Andres et al. 2018). This widespread pollution stems from the persistent nature of MPs, which takes long periods of time to decompose, as well as its ability to be ingested, inhaled or absorbed through the food chain (Coffin and Weisberg 2022; Eze et al. 2024). The existence of MPs in the environment has far-reaching consequences, particularly for human health and the sustainability of our food systems. MPs can be swallowed by organisms as food, resulting in reduced reproduction, growth, and survival (Coffin and Weisberg 2022). Furthermore, MPs can act as a vehicle for the conveyance and accumulation of various pollutants, such as pesticides, heavy metals, and antibiotics (Arias-Andres et al. 2018; Imran et al. 2019; Rafa et al. 2023).

The socioeconomic impact of MPs pollution is complex and far-reaching. One major concern is the contamination of agricultural soils with MPs, which can directly influence crop yields and the overall productivity of agricultural systems (Zang et al. 2020; Pérez-Reverón et al. 2022). Several studies have shown that MPs can negatively affect the physical, chemical, and biological characteristics of soil, which can undermine plant health and growth (Zang et al. 2020; Gao et al. 2022; Hanif et al. 2024; Hasan and Tarannum 2024). The presence of MPs in the food chain also presents a threat that could have serious consequences for human health (Pironti et al. 2021; Al Mamun et al. 2023). Numerous studies have documented the ingestion and accumulation of MPs by an assortment of marine species, including those consumed by humans (Carbery et al. 2018; Egbeocha et al. 2018; Ribeiro et al. 2019; Miller et al. 2020; Jitkaew et al. 2023; Radhakrishnan et al. 2024). MPs are broadly spread in the aquatic environment, making them difficult to remove or treat, which in turn leads to higher costs for waste management and environmental clean-up (Hettiarachchi and Meegoda 2023). Local authorities and governments may face greater financial pressures in efforts to reduce pollution and address environmental damage. The annual cost of the externalities associated with plastic pollution is estimated at between USD 300-460 billion, covering loss of marine ecosystem services, healthcare costs, and ocean clean-up (Kim 2024). Marine MPs pollution could reduce gross domestic product (GDP(by up to US $7 billion in 2018, due to reduced revenues from tourism, aquaculture, and fishing (Gikandi 2021; Mittempergher et al. 2022). MPs pollution of oceans and coastal areas presents a substantial threat to the aquatic environment and economic well-being of many communities around the world, especially countries that rely on tourism as a primary source of income (Díaz-Mendoza et al. 2020; Ronda et al. 2023; Haldar et al. 2025). The presence of MPs in marine environments can degrade the aesthetic appeal and recreational value of tourist destinations, leading to significant declines in visitor numbers and significant economic losses to the local economy (Karbalaei et al. 2018; Arabi and Nahman 2020; Pilapitiya and Ratnayake 2024). Therefore, the effects of MPs pollution are multiple, and efforts must be made to combat such pollution and reduce its damage to the aquatic environment.

7.3 Environmental Consequences of MPs

MPs pose a number of significant environmental threats to aquatic ecosystems due to their widespread distribution resulting from the degradation of larger plastics and their direct release into the aquatic environment, the most important of which is the impact on marine life (Ogunola et al. 2018; Sridharan et al. 2021). Many aquatic organisms can metabolize persistent organic pollutants (POPs) on MPs, among these compounds are polybrominated diphenyl ethers (PBDEs), which can accumulate in the tissues of marine amphipods, fish, crabs, and marine amphipods (Chua et al. 2014; Auta et al. 2017; Tanaka et al. 2023; Ayoola et al. 2024), also, sharp MPs can cause damage to the digestive system and gill tissues of aquatic organisms (Yuan et al. 2020; Xie et al. 2021; Ammar et al. 2022). MPs and other pollutants can pass into the food chain via living organisms (Eze et al. 2024; Liaqat et al. 2024). Many aquatic invertebrates such as clams, mollusks, shrimp, mussels, and vertebrates such as birds, fish, crustaceans, whales, turtles, and zooplankton can consume MPs without realizing it and recognize them as food (Auta et al. 2017; Lehner et al. 2018; Devi and Kumar 2024), which can reach MPs through the feeding relationships between predator and prey or through direct consumption (Farrell and Nelson 2013; Hollman et al. 2013; Zhu et al. 2018). Yu et al. (2020) reported that MPs ingestion by MPs-consuming organisms may occur at higher concentrations than other chemical pollutants such as POPs in aquatic systems compared to unpolluted environments. Concerns about MPs pollution are increasing due to the nature of these pollutants, which increases the speed of their spread in the aquatic environment and their integration into the food web, which may cause health risks to humans as a result of eating contaminated fish, mollusks and crustaceans, as well as other environmental consequences on the environment and living organisms (Kershaw and Rochman 2015; Phillips and Bonner 2015; Egbeocha et al. 2018; Karbalaei et al. 2018, 2020; Pironti et al. 2021; Yuan et al. 2022; Doshi et al. 2024).

Over the past years, many researchers have documented the existence and accumulation of MPs in the digestive systems of various species of fish (Abbasi et al. 2018, 2023; Hastuti et al. 2019; Klangnurak and Chunniyom 2020; Heshmati et al. 2021; Seetapan and Prommi 2023; Amponsah et al. 2024; Hassine et al. 2024; Tepe et al. 2024). The existence of MPs in the digestive system of fish is usually not a cause for concern because they are not consumed by humans, but what is interesting is the presence of these pollutants in crustaceans, which are consumed by the digestive system (Rainieri and Barranco 2019). In a study conducted by Bessa et al. (2018) they found approximately 157 MPs fibers (96%) and fragments (4%) out of a total of 120 fish, and they were identified as polyester, polypropylene, polyacrylonitrile, polyethylene, nylon and rayon in Mondego estuary, Portugal. In addition, Pellini et al. (2017) documented the dominance of polypropylene (PP), PVC, and polyethylene (PE) in the digestive tract of 95% of benthic flatfish in the Adriatic Sea. Liu et al. (2011) confirmed the uptake of MPs additives in fish from seas surrounding China, resulting in increased concentrations of PBDEs in affected fish. Constant et al. (2022) reported MPs concentrations in 18 native fish species in the Northwestern Mediterranean Sea, with concentrations spanning from 0.00 to 5.15 items per fish, the results revealed the dominance of fibers made of polypropylene (PP), polyester (PES), acrylic (A), polyamide (PA), and polypropylene (PP) at 91% of the particles. Baalkhuyur et al. (2018) documented the existence of 26 MPs fragments (61.5% films and 38.5% fishing thread) in 178 fish species throughout the Red Sea coastline in Saudi Arabia, where polymer analysis results revealed the dominance of polyethylene and polypropylene. Fish subjected to polyethylene and various chemical contaminants are affected by liver disease and toxicity, which causes bioaccumulation of these pollutants (Hu et al. 2022).

Many plankton ingest MPs from the water because they consider them food, which can affect the physiology of the organism because they contain toxic chemicals (Deudero et al. 2014; Cole et al. 2019). These impacts can include respiratory function, feeding habits, reproduction, dysfunctions, and molecular pathways, and these pollutants can move from lower to higher trophic levels through the food web (Zheng et al. 2020; Onyena et al. 2021). Numerous studies have documented the consumption of MPs from the surrounding environment by zooplankton (Desforges et al. 2015; Sun et al. 2017; Kosore et al. 2018; Amin et al. 2020; Botterell et al. 2022; Goswami et al. 2023; Park et al. 2024). Sun et al. (2018) found that there were three types of MPs (fragments, pellets, and fibres), with fibres dominating the bulk, while 19 polymers were found in zooplankton (fish fry, crustaceans, shrimp, and jellyfish) in the northern South China Sea. The bioavailability of MPs in zooplankton is usually determined by the species of plankton and their feeding habits, and thus these particles move to different levels in the food chains and ultimately reach humans, which may endanger their health and lives.

Mussels are widely consumed bivalve mollusks in many countries around the world and are filter-feeding and can ingest MPs and other pollutants from the water (Avio et al. 2015; Kibria et al. 2022). These organisms are considered good bioindicators of environmental pollution by MPs and other pollutants due to their ability to bioaccumulate these pollutants (Viarengo and Canesi 1991; Cunha et al. 2017; Li et al. 2019a, b). Hariharan et al. (2021) documented reduced feeding rates and accumulation of weathered Polyethylene (wPE) in the intestines of *Perna viridis* mussels when exposed to it for 30 days without causing mortality, in contrast, the accumulation of these pollutants affected internal organs, especially the intestine, in addition to reproduction and growth.

There are many biological effects that MPs can cause, including long-term toxicity and interference with the ability of living organisms to grow and reproduce (Yin et al. 2021; Hong et al. 2023). Espinosa et al. (2017) documented the effect of polyvinyl chloride (PVC) plasticizers on the disturbance of biological enzyme activity, growth, oviposition, and oxidative damage in seabream (*Sparus aurata*). Xia et al. (2020) reported the toxic effect of PVC on freshwater *Cyprinus carpio* var. larvae, which was represented by suppression of weight gain and alterations in the activities of antioxidant enzymes in the liver with a change in gene expression, in addition, during histological analysis, damage and cytoplasmic vacuolation were shown in the liver when exposed to concentrations of 20% and 30% of MPs. Plastic debris affects many marine organisms, with impact rates reported in the literature reaching 86% of fish, crustaceans, and sea turtles, 43% of mammals, and 44% of seabirds (Thushari and Senevirathna 2020). The biological effects of MPs on aquatic ecosystems can be recapped as ingestion, decreased reproduction, stress, impaired motility, entanglement, wounds, and ulcers, and eventually death, in addition to the loss of food and habitat (Schnurr et al. 2018; Bhuyan et al. 2021). Because of the significant between MPs and phytoplankton in terms of their white or blue and transparent colour, which in most cases mixes with the desired food of many cetaceans and fish (Boerger et al. 2010), as a result, cetaceans may become vulnerable to a variety of threats such as frailty and disease and thus become more susceptible to predation (Sigler 2014). Filter feeding rates in blue mussels are reduced due to the ingestion of polystyrene (PS) nano-plastics, which also affects regulatory and cellular processes (Rossi et al. 2014). Sussarellu et al. (2016) found that polystyrene reduced the size, number of eggs and larvae, and sperm motility of oysters. In contrast, Thomas et al. (2020) noted a decline in health indicators of juvenile oysters *Crassostrea gigas* with stable weight when exposed to three different concentrations of PS, as high exposure to MPs led to instability of the lysosomal, which plays an important role in defense and deterioration of constituents, and an increase in the mortality rate of oysters. Furthermore, MPs can cause changes in temperature-related biological processes that have led to changes in sex ratios during incubation periods in turtles (Hu et al. 2019). Various studies have investigated the effects of MPs on invertebrates and zooplankton, which have caused developmental abnormalities in embryos (Martínez-Gómez et al. 2017). These are time (Sussarellu et al. 2016), size (La Pietra et al. 2024), and dose- (Balbi et al. 2017) dependent for adults and larvae of many marine invertebrates, dry weight loss in lugworms (Besseling et al. 2013, 2017), and malformations or stillbirths (Rendell-Bhatti 2019), and severe growth retardation of both larvae and juveniles in ascidians and sea urchins (Nobre et al. 2015; Martínez-Gómez et al. 2017; Messinetti et al. 2018).

MPs in the aquatic ecosystem pose many risks and risks to human health, as there is a lot of evidence of human exposure to them through diet (Yuan et al. 2022; Al Mamun et al. 2023; Kadac-Czapska et al. 2024). In the literature, human exposure to MPs has been documented through the ingestion process, adhesion, contact, and inhalation (Thompson et al. 2009; Domenech and Marcos 2021). MPs cause several effects on human health, including carcinogenic effects (Dzierżyński et al. 2024), neurotoxic (Prüst et al. 2020), organ dysfunction (Li et al. 2023), ulcers and skin irritation (Wu et al. 2022), which are caused by the chemicals that enter into the composition of these materials and are released into the aquatic environment as a result of decomposition processes, including phthalates, styrene and bisphenol-A (Benjamin et al. 2017; Barboza et al. 2020; Hu et al. 2021). Due to the danger of MPs and their various effects on humans and the aquatic environment, many countries have established legislation and regulations that limit the use of plastic and work to eliminate it in order to reduce and eliminate this pollution.

Additionally, the aquaculture industry is among the most significant industries that supply animal protein to humans, which leads to enhancing the economic system and combating hunger and poverty in many parts of the world (Naylor et al. 2000; Zhou et al. 2021). MPs are found in many aquatic organisms, with plastic particles identified in more than 100 species of fish (commercial and non-commercial), raising concerns about food safety for humans (Lusher et al. 2017). Numerous literatures have documented the existence of MPs in many environments such as water, soil, and air (Wang et al. 2021a, b; Akbari and Jaafari 2024; Dagwar et al. 2024; Mangla et al. 2025). Huang et al. (2024) documented the presence of MPs in aquaculture ponds, which recorded mean abundances of 36.25 ± 6.79 items/L and 271.65 ± 164.83 items/kg in both water and sediment, respectively. Higher levels of MPs were noted in shrimp and crab ponds compared to fishponds, where fibers dominated the MPs in the water. Due to the increase in aquaculture activities, it has been observed that the concentration of plastic materials is higher in aquaculture environments compared to the surrounding environment (Priscilla and Patria 2020; Miao et al. 2023). MPs contamination of aquaculture occurs through a number of sources, including medicines, feed, fishing nets, and plastic equipment (Lv et al. 2020). Weather and geography can also affect the distribution and accumulation of these pollutants in aquaculture systems (Wang et al. 2019a, b). Figure 7.2 shows the schematic diagram of MPs' environmental impacts and consequences in the aquatic environment.

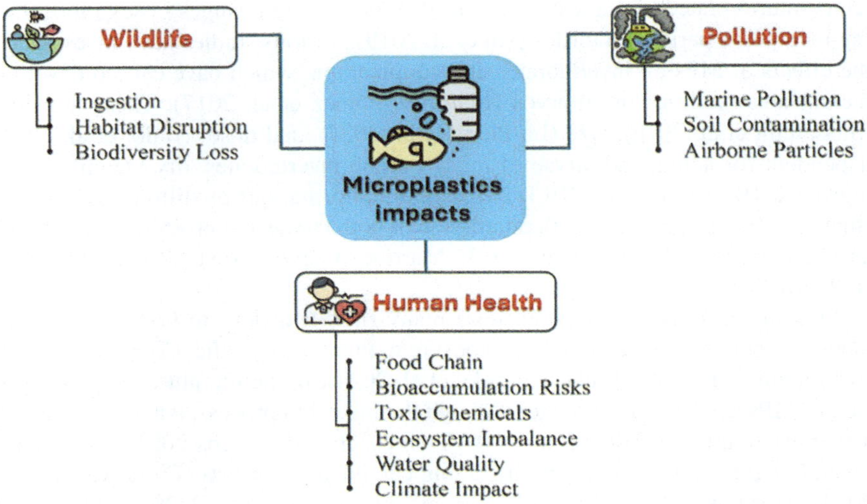

Fig. 7.2 Environmental impacts and consequences of MPs in the aquatic environment

7.4 Economic Impacts

The dissemination of MPs particles in the aquatic environment and their various effects pose many different economic challenges that affect many sectors and local communities, all the way to the global community (Ogunola and Palanisami 2016; Ajibade et al. 2021; Sarma et al. 2022). MPs accumulation in aquatic environments occurs worldwide, causing far-reaching environmental repercussions for the surrounding environment, living organisms, human health, and also causing many economic impacts (Tursi et al. 2022; Pal et al. 2024). Mitigation of high MPs concentrations ranging from ~3–23 μg L^{-1} to ~4500 μg L^{-1} in seawater is highly polluted, requiring a lot of resources, materials and financial resources, which can be a burden on coastal communities and governments (Collignon et al. 2012; Reisser et al. 2015; Green et al. 2017). These governments and coastal communities are often responsible for various environmental clean-up operations, such as organizing beach clean-ups, developing infrastructure, and upgrading wastewater processing facilities to prevent more MPs from entering the aquatic environment (Rizal et al. 2022; Purba et al. 2023). These efforts may impose many pressures on these communities and local budgets, in addition to the degradation of these habitats and environments, adding more costs associated with remediation, mitigation, cleanup, and conservation operations (Burt et al. 2020; Onyena et al. 2021). Furthermore, MPs impact the entertainment and tourism industries in many coastal communities and countries that depend on a clean and healthy coastal environment, as the existence of these particles and waste distorts the environment and deters tourists and visitors (Krelling et al. 2017; Gounden 2021; Issac and Kandasubramanian 2021). In addition, the process of bioaccumulation of MPs particles in seafood causes many concerns, which in turn reduces the enjoyment of these activities and thus reduces the attraction towards these environments (De-la-Torre 2020; Walkinshaw et al. 2020). The existence of MPs in the aquatic environment leads to the deterioration of these environments, causing a decrease in the quality and aesthetics of these environments, which in turn affects the profitability and sustainability of these industries, tourism, and entertainment that depend on these natural environments (Agamuthu et al. 2019; McIlgorm and Xie 2023; Ben-Haddad et al. 2024). The decline in tourism and recreational activities in these locations also causes severe economic consequences, as these industries greatly support local communities (Ekayani et al. 2019).

In the seafood and fishing industry, MPs pose a significant threat to these industries due to their accumulation in many commercial species of fish, mollusks, and other seafood (Wu et al. 2020; Kibria 2023; Santonicola et al. 2023). The existence of MPs in seafood products can cause concern and fear among consumers of these products, which leads to a decrease in demand for these industries, which in turn leads to a decrease in the economics of this industry and its consumption by humans (Mercogliano et al. 2020). The decrease in demand for fish and seafood due to the decrease in consumer confidence in these industries causes severe economic consequences that affect the industries related to these organisms and affect local fishermen and seafood factories (Jennings et al. 2016), as a result, it leads to the loss of many jobs, a decrease in economic activity and limited income for these communities that depend heavily on these industries (Agamuthu et al. 2019; Song et al. 2020). Therefore, the economic impacts of MPs are far-reaching and multifaceted in aquatic environments, as addressing these problems and challenges imposed by this pollution from the high financial costs and efforts extended by management and mitigation to the negative harmful effects on the entertainment, tourism, seafood and fish industries, which impose strenuous efforts to manage this problem and to mitigate these effects on local people, companies and governments (Mihai et al. 2021; Nyadjro et al. 2023; Sinha 2024). Figure 7.3 illustrates the impacts and financial burdens of MPs pollution in the aquatic ecosystem.

7.5 Social Implications

The spread of MPs is a problem that is spreading around the world, affecting human health and the aquatic environment, as a result of its spread throughout the food chain, especially in seafood and various types of commercial fish, which has caused the emergence of serious health problems for humans (Campanale et al. 2020). Food contaminated with MPs is one of the most important points of entry of these pollutants into the human diet (Silva-Cavalcanti et al. 2017; Garrido Gamarro and Constanzo 2022). Cox et al. (2019) documented the existence of different levels of MPs particles in daily foods such as salt, sugar, and bottled water, recording the presence of 0.11, 0.44, and 0.09 MPs/g, respectively. MPs uptake in humans is generally reasonable after exposure, with particles smaller than 150 μm being observed to be able to cross the gastrointestinal epithelium in mammals, causing systemic

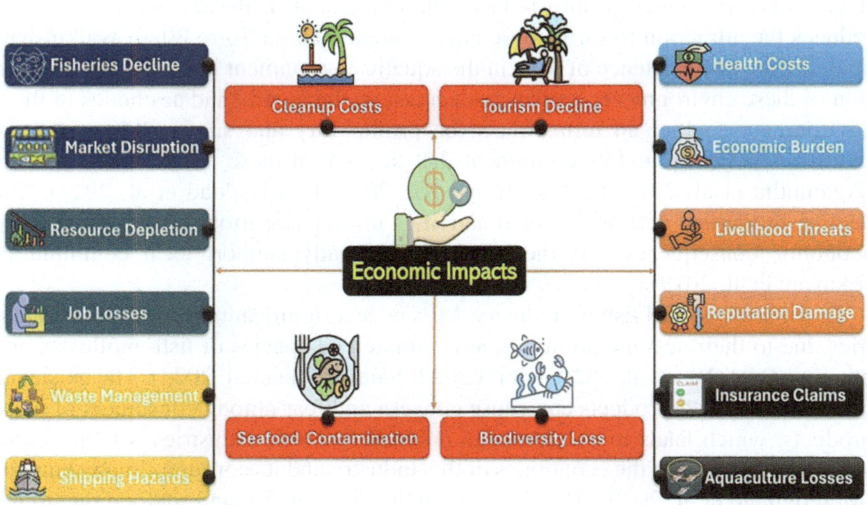

Fig. 7.3 Economic impacts of MPs pollution in the aquatic ecosystem

exposure (Campanale et al. 2020). In contrast, scientists expect that about 0.3% of MPs are absorbed, while about 0.1% (10 μm) is transported to cellular membranes via the blood–brain barrier and placenta and other organs (Barboza et al. 2018). MPs smaller than 2.5 μm enter the gastrointestinal tract upon ingestion, through the gastrointestinal tract by M cells, which transport these particles from the intestinal lumen to mucosal lymphoid tissues through cellular uptake (Campanale et al. 2020). Public awareness and understanding of the social impacts of MPs are important to address this issue, as many studies have shown that the majority of people are concerned about the consequences of this problem, along with the resulting effects on humans and the surrounding environment (Tang et al. 2022). The level of awareness and knowledge may vary between different population groups and regions, so in order to enhance community engagement, it is essential to promote educational initiatives that educate the public about the sources of entry of these pollutants, their effects, and solutions to mitigate and treat MPs pollution (Mehinto et al. 2022).

Education and awareness-raising efforts are important to address such issues related to environmental pollution, which is represented in targeting the community, schools and the general public to raise awareness of the dangers of MPs particles through many awareness and educational programs that will work to raise awareness in society and bring about positive changes in behavior. These programs should aim not only to enhance awareness about the prevalence and impacts of MPs but also to empower the community to take concrete actions, such as reducing daily plastic consumption and adopting appropriate waste management practices (Thushari and Senevirathna 2020). In addition, governments and decision-makers can adopt stricter regulations and laws and promote the development and adoption of sustainable substitutes for plastic products (Watkins et al. 2019; Thushari and Senevirathna 2020).

7.6 Conclusion

MPs pollution in the marine environment is one of the significant socio-economic issues that affects various sectors. From an economic point of view, government and corporate sectors need to invest more funds in waste management and water treatment facilities. The fisheries and tourism sectors face several challenges due to plastic pollution. The safety concerns on kinds of seafood will reduce the income of fishermen. While creating awareness for avoidance of single-use plastic products may decrease the discharge of plastic waste into the seawater, proper safety guidelines are necessary for monitoring the MPs accumulation in seafood and other marine biota. A multifaceted approach that involves scientific research, awareness, and regulations is essential for addressing socio-economic issues of MPs pollution in the marine environment.

References

Abbasi S, Soltani N, Keshavarzi B, Moore F, Turner A, Hassanaghaei M (2018) Microplastics in different tissues of fish and prawn from the Musa estuary, Persian gulf. Chemosphere 205:80–87

Abbasi A, Sadeghi P, Abadi ZTR (2023) Characterization of microplastics in digestive tract of commercial fish species from the Oman Sea. Mar Pollut Bull 197:115769

Agamuthu P, Mehran S, Norkhairah A, Norkhairiyah A (2019) Marine debris: a review of impacts and global initiatives. Waste Manag Res 37(10):987–1002

Ahsan MU, Nasir M, Abbas J (2020) Examining the causes of plastic bags usages and public perception about its effects on the natural environment. Int J Acad Res Business Soc Sci 10(10):80–96

Ajibade FO, Adelodun B, Lasisi KH, Fadare OO, Ajibade TF, Nwogwu NA, Sulaymon ID, Ugya AY, Wang HC, Wang A (2021) Environmental pollution and their socioeconomic impacts. In: Microbe mediated remediation of environmental contaminants. Elsevier, Amsterdam, pp 321–354

Akbari A, Jaafari J (2024) Entry pathway and potential impacts of microplastics in air, water, soil and human health: a review. Caspian J Health Res 9(4):181–198

Al Mamun A, Prasetya TAE, Dewi IR, Ahmad M (2023) Microplastics in human food chains: food becoming a threat to health safety. Sci Total Environ 858:159834

Amin RM, Sohaimi ES, Anuar ST, Bachok Z (2020) Microplastic ingestion by zooplankton in Terengganu coastal waters, southern South China Sea. Mar Pollut Bull 150:110616

Ammar E, Mohamed MS, Sayed AE-DH (2022) Polyethylene microplastics increases the tissue damage caused by 4-nonylphenol in the common carp (Cyprinus carpio) juvenile. Front Mar Sci 9:1041003

Amponsah AK, Afrifa EA, Essandoh PK, Enyoh CE (2024) Evidence of microplastics accumulation in the gills and gastrointestinal tract of fishes from an estuarine system in Ghana. Heliyon 10(3):e25608

An L, Liu Q, Deng Y, Wu W, Gao Y, Ling W (2020) Sources of microplastic in the environment. In: Microplastics in terrestrial environments: emerging contaminants and major challenges. Springer, Cham, pp 143–159

Anagnosti L, Varvaresou A, Pavlou P, Protopapa E, Carayanni V (2021) Worldwide actions against plastic pollution from microbeads and microplastics in cosmetics focusing on European policies. Has the issue been handled effectively? Mar Pollut Bull 162:111883

Andrady AL (2011) Microplastics in the marine environment. Mar Pollut Bull 62(8):1596–1605

Arabi S, Nahman A (2020) Impacts of marine plastic on ecosystem services and economy: state of south African research. S Afr J Sci 116(5-6):1–7

Arias-Andres M, Klümper U, Rojas-Jimenez K, Grossart H-P (2018) Microplastic pollution increases gene exchange in aquatic ecosystems. Environ Pollut 237:253–261

Arthur C, Baker JE, Bamford HA (2009) Effects, and fate of microplastic marine debris. In: Proceedings of the international research workshop on the occurrence. University of Washington Tacoma, Tacoma

Auta HS, Emenike CU, Fauziah SH (2017) Distribution and importance of microplastics in the marine environment: a review of the sources, fate, effects, and potential solutions. Environ Int 102:165–176

Avio CG, Gorbi S, Milan M, Benedetti M, Fattorini D, d'Errico G, Pauletto M, Bargelloni L, Regoli F (2015) Pollutants bioavailability and toxicological risk from microplastics to marine mussels. Environ Pollut 198:211–222

Ayoola TO, Bamidele A, Excellence A, Junaid SO, Ovioma K, Onoriode G, Otubanjo ZA, Uwayo M, Omoregie P (2024) Comparative bioaccumulation of polybrominated diphenyl ethers (Pbdes) in synodontis clarias and tilapia zilli in Lekki lagoon, Lagos, Nigeria. Egypt Acad J Biol Sci B Zool 16(2):159–172

Baalkhuyur FM, Dohaish E-JAB, Elhalwagy ME, Alikunhi NM, AlSuwailem AM, Røstad A, Coker DJ, Berumen ML, Duarte CM (2018) Microplastic in the gastrointestinal tract of fishes along the Saudi Arabian Red Sea coast. Mar Pollut Bull 131:407–415

Balbi T, Camisassi G, Montagna M, Fabbri R, Franzellitti S, Carbone C, Dawson K, Canesi L (2017) Impact of cationic polystyrene nanoparticles (PS-NH2) on early embryo development of Mytilus galloprovincialis: effects on shell formation. Chemosphere 186:1–9

Barboza LGA, Vethaak AD, Lavorante BR, Lundebye A-K, Guilhermino L (2018) Marine microplastic debris: an emerging issue for food security, food safety and human health. Mar Pollut Bull 133:336–348

Barboza LGA, Cunha SC, Monteiro C, Fernandes JO, Guilhermino L (2020) Bisphenol A and its analogs in muscle and liver of fish from the north East Atlantic Ocean in relation to microplastic contamination. Exposure and risk to human consumers. J Hazard Mater 393:122419

Bashir SM, Kimiko S, Mak C-W, Fang JK-H, Gonçalves D (2021) Personal care and cosmetic products as a potential source of environmental contamination by microplastics in a densely populated Asian city. Front Mar Sci 8:683482

Ben-Haddad M, Abelouah MR, Hajji S, Abou Oualid J, Alla AA, Rangel-Buitrago N (2024) Scenic degradation and visual pollution along the Agadir coastline (Morocco): analysis and management. Mar Pollut Bull 205:116629

Benjamin S, Masai E, Kamimura N, Takahashi K, Anderson RC, Faisal PA (2017) Phthalates impact human health: epidemiological evidences and plausible mechanism of action. J Hazard Mater 340:360–383

Bessa F, Barría P, Neto JM, Frias JP, Otero V, Sobral P, Marques JC (2018) Microplastics in juvenile commercial fish from an estuarine environment. In: Proceedings of the international conference on microplastic pollution in the mediterranean sea. Springer, Cham

Besseling E, Wegner A, Foekema EM, Van Den Heuvel-Greve MJ, Koelmans AA (2013) Effects of microplastic on fitness and PCB bioaccumulation by the lugworm Arenicola marina (L.). Environ Sci Technol 47(1):593–600

Besseling E, Foekema EM, van den Heuvel-Greve MJ, Koelmans AA (2017) The effect of microplastic on the uptake of chemicals by the lugworm Arenicola marina (L.) under environmentally relevant exposure conditions. Environ Sci Technol 51(15):8795–8804

Bhuyan MS, Nabi MR-U, Alam MW, Islam MN, Cáceres-Farías L, Bat L, Musthafa MS, Chung SY, Senapathi V, Alfaro-Núñez A (2021) Impacts of microplastics on marine organisms and in human health. Res Sq 1:84

Boerger CM, Lattin GL, Moore SL, Moore CJ (2010) Plastic ingestion by planktivorous fishes in the North Pacific central gyre. Mar Pollut Bull 60(12):2275–2278

Botterell ZL, Bergmann M, Hildebrandt N, Krumpen T, Steinke M, Thompson RC, Lindeque PK (2022) Microplastic ingestion in zooplankton from the Fram Strait in the Arctic. Sci Total Environ 831:154886

Boucher J (2017) Primary microplastics in the oceans: a global evaluation of sources. IUCN, Gland

Burt AJ, Raguain J, Sanchez C, Brice J, Fleischer-Dogley F, Goldberg R, Talma S, Syposz M, Mahony J, Letori J (2020) The costs of removing the unsanctioned import of marine plastic litter to small Island states. Sci Rep 10(1):14458

Campanale C, Massarelli C, Savino I, Locaputo V, Uricchio VF (2020) A detailed review study on potential effects of microplastics and additives of concern on human health. Int J Environ Res Public Health 17(4):1212

Carbery M, O'Connor W, Palanisami T (2018) Trophic transfer of microplastics and mixed contaminants in the marine food web and implications for human health. Environ Int 115:400–409

Carpenter EJ, Smith K Jr (1972) Plastics on the Sargasso Sea surface. Science 175(4027):1240–1241

Chamas A, Moon H, Zheng J, Qiu Y, Tabassum T, Jang JH, Abu-Omar M, Scott SL, Suh S (2020) Degradation rates of plastics in the environment. ACS Sustain Chem Eng 8(9):3494–3511

Chaudhari S, Parmar H, Samnani P (2024) Microplastics in personal care products and cosmetics. In: Microplastics pollution and its remediation. Springer, Cham, pp 171–188

Cheung PK, Fok L (2017) Characterisation of plastic microbeads in facial scrubs and their estimated emissions in mainland China. Water Res 122:53–61

Chua EM, Shimeta J, Nugegoda D, Morrison PD, Clarke BO (2014) Assimilation of polybrominated diphenyl ethers from microplastics by the marine amphipod, Allorchestes compressa. Environ Sci Technol 48(14):8127–8134

Coffin S, Weisberg SB (2022) Understanding health effects pathways and thresholds: filling a critical need to support microplastics management. Microplast Nanoplast 2(1):11

Cole M, Coppock R, Lindeque PK, Altin D, Reed S, Pond DW, Sørensen L, Galloway TS, Booth AM (2019) Effects of nylon microplastic on feeding, lipid accumulation, and moulting in a Coldwater copepod. Environ Sci Technol 53(12):7075–7082

Collignon A, Hecq J-H, Glagani F, Voisin P, Collard F, Goffart A (2012) Neustonic microplastic and zooplankton in the North Western Mediterranean Sea. Mar Pollut Bull 64(4):861–864

Collignon A, Hecq J-H, Galgani F, Collard F, Goffart A (2014) Annual variation in neustonic micro-and meso-plastic particles and zooplankton in the bay of Calvi (Mediterranean–Corsica). Mar Pollut Bull 79(1-2):293–298

Constant M, Reynaud M, Weiss L, Ludwig W, Kerhervé P (2022) Ingested microplastics in 18 local fish species from the northwestern Mediterranean Sea. Microplastics 1(1):186–197

Cox KD, Covernton GA, Davies HL, Dower JF, Juanes F, Dudas SE (2019) Human consumption of microplastics. Environ Sci Technol 53(12):7068–7074

Crawford CB, Quinn B (2017) Plastic production, waste and legislation. Microplast Pollut 30:39–56

Cunha S, Pena A, Fernandes J (2017) Mussels as bioindicators of diclofenac contamination in coastal environments. Environ Pollut 225:354–360

Cverenkárová K, Valachovičová M, Mackuľak T, Žemlička L, Bírošová L (2021) Microplastics in the food chain. Life 11(12):1349

Dagwar PP, Saole P, Jeevanasai A, Seetha Rama Raju M, Sharma H, Bahukhandi KD (2024) Sources, types, and occurrences of microplastics in soil, water, and air. In: Microplastics: environmental pollution and degradation process. Springer, Cham, pp 23–56

De Falco F, Di Pace E, Cocca M, Avella M (2019) The contribution of washing processes of synthetic clothes to microplastic pollution. Sci Rep 9(1):6633

De-la-Torre GE (2020) Microplastics: an emerging threat to food security and human health. J Food Sci Technol 57(5):1601–1608

Desforges J-PW, Galbraith M, Ross PS (2015) Ingestion of microplastics by zooplankton in the Northeast Pacific Ocean. Arch Environ Contam Toxicol 69:320–330
Deudero S, Nadal MA, Estarellas F, Alomar C (2014) Microplastic exposure in pelagic fishes and holothurians: a Mediterranean case study. Centro Oceanográfico de Baleares, Barcelona
Devi SS, Kumar AB (2024) Impacts of mesoplastics and microplastics on marine biodiversity. In: Meso-and microplastic risk assessment in marine environments. Elsevier, Amsterdam, pp 309–340
Dey S, Veerendra G, Babu PA, Manoj AP, Nagarjuna K (2024) Degradation of plastics waste and its effects on biological ecosystems: a scientific analysis and comprehensive review. Biomed Mater Devices 2(1):70–112
Díaz-Mendoza C, Mouthon-Bello J, Pérez-Herrera NL, Escobar-Díaz SM (2020) Plastics and microplastics, effects on marine coastal areas: a review. Environ Sci Pollut Res 27(32):39913–39922
Domenech J, Marcos R (2021) Pathways of human exposure to microplastics, and estimation of the total burden. Curr Opin Food Sci 39:144–151
Doshi M, Rabari V, Patel A, Yadav VK, Sahoo DK, Trivedi J (2024) A systematic review on microplastic contamination in marine Crustacea and Mollusca of Asia: current scenario, concentration, characterization, polymeric risk assessment, and future prospectives. Water Environ Res 96(5):e11029
Dzierżyński E, Gawlik PJ, Puźniak D, Flieger W, Jóźwik K, Teresiński G, Forma A, Wdowiak P, Baj J, Flieger J (2024) Microplastics in the human body: exposure, detection, and risk of carcinogenesis: a state-of-the-art review. Cancer 16(21):3703
Egbeocha CO, Malek S, Emenike CU, Milow P (2018) Feasting on microplastics: ingestion by and effects on marine organisms. Aquat Biol 27:93–106
Ekayani M, Pramudita D, Istiqomah A, Tampubolon B (2019) Economic benefits of natural tourism: driving community's participation for conservation in Halimun Salak National Park, Indonesia. IOP Conf Ser Earth Environ Sci 285:012003
EPI (2014) Plastic bags fact sheet. http://www.earth-policy.org/images/uploads/press_room/Plastic_Bags.pdf
Espinosa C, Cuesta A, Esteban MÁ (2017) Effects of dietary polyvinylchloride microparticles on general health, immune status and expression of several genes related to stress in gilthead seabream (Sparus aurata L.). Fish Shellfish Immunol 68:251–259
Eze CG, Nwankwo CE, Dey S, Sundaramurthy S, Okeke ES (2024) Food chain microplastics contamination and impact on human health: a review. Environ Chem Lett 22(4):1889–1927
Farrell P, Nelson K (2013) Trophic level transfer of microplastic: Mytilus edulis (L.) to Carcinus maenas (L.). Environ Pollut 177:1–3
Galgani F, Hanke G, Werner S, Oosterbaan L, Nilsson P, Fleet D, Kinsey S, Thompson R, Van-Franeker J, Vlachogianni T (2013) Monitoring guidance for marine litter in European seas
Gallo F, Fossi C, Weber R, Santillo D, Sousa J, Ingram I, Nadal A, Romano D (2020) Marine litter plastics and microplastics and their toxic chemicals components: the need for urgent preventive measures. In: Analysis of Nanoplastics and microplastics in food. CRC Press, Boca Raton, pp 159–179
Galvão A, Aleixo M, De Pablo H, Lopes C, Raimundo J (2020) Microplastics in wastewater: microfiber emissions from common household laundry. Environ Sci Pollut Res 27:26643–26649
Gao H, Liu Q, Yan C, Mancl K, Gong D, He J, Mei X (2022) Macro-and/or microplastics as an emerging threat effect crop growth and soil health. Resour Conserv Recycl 186:106549
García-Rivera S, Lizaso JLS, Millán JMB (2017) Composition, spatial distribution and sources of macro-marine litter on the Gulf of Alicante seafloor (Spanish Mediterranean). Mar Pollut Bull 121(1-2):249–259
Garrido Gamarro E, Constanzo V (2022) Microplastics in food commodities: a food safety review on human exposure through dietary sources. FAO, Rome
Gasperi J, Wright SL, Dris R, Collard F, Mandin C, Guerrouache M, Langlois V, Kelly FJ, Tassin B (2018) Microplastics in air: are we breathing it in? Curr Opin Environ Sci Health 1:1–5
Geyer R, Jambeck JR, Law KL (2017) Production, use, and fate of all plastics ever made. Sci Adv 3(7):e1700782

Giechaskiel B, Grigoratos T, Mathissen M, Quik J, Tromp P, Gustafsson M, Franco V, Dilara P (2024) Contribution of road vehicle Tyre Wear to microplastics and ambient air pollution. Sustain For 16(2):522

Gikandi L (2021) The hidden cost of plastic. https://updates.panda.org/the-hidden-cost-of-plastic

Giustra M, Sinesi G, Spena F, De Santes B, Morelli L, Barbieri L, Garbujo S, Galli P, Prosperi D, Colombo M (2024) Microplastics in cosmetics: open questions and sustainable opportunities. ChemSusChem 17:e202401065

Goswami P, Selvakumar N, Verma P, Saha M, Suneel V, Vinithkumar NV, Dharani G, Rathore C, Nayak J (2023) Microplastic intrusion into the zooplankton, the base of the marine food chain: evidence from the Arabian Sea, Indian Ocean. Sci Total Environ 864:160876

Gounden D (2021) An assessment of visitor profiles, consumption patterns and perceptions as well as the state of coastal and marine tourism (specifically beach) sites in KwaZulu-Natal Province, South Africa

Green DS, Boots B, O'Connor NE, Thompson R (2017) Microplastics affect the ecological functioning of an important biogenic habitat. Environ Sci Technol 51(1):68–77

Guo S, Zhang J, Liu J, Guo N, Zhang L, Wang S, Wang X, Zhao M, Zhang B, Chen Y (2023) Organic fertilizer and irrigation water are the primary sources of microplastics in the facility soil. Beijing Sci Total Environ 895:165005

Haldar S, Chanda V, Muralidaran Y, Mishra P (2025) Impact of microplastics on economic condition in underdeveloped nations. In: Microplastics. Elsevier, Amsterdam, pp 481–508

Hanif M, Aijaz N, Azam K, Akhtar M, Laftah W, Babur M, Abbood N, Benitez I (2024) Impact of microplastics on soil (physical and chemical) properties, soil biological properties/soil biota, and response of plants to it: a review. Int J Environ Sci Technol. https://doi.org/10.1007/s13762-024-05656-y

Hariharan G, Purvaja R, Anandavelu I, Robin R, Ramesh R (2021) Accumulation and ecotoxicological risk of weathered polyethylene (wPE) microplastics on green mussel (Perna viridis). Ecotoxicol Environ Saf 208:111765

Hasan MM, Tarannum MN (2024) Adverse impacts of microplastics on soil physicochemical properties and crop health in agricultural systems. J Hazard Mater Adv 17:100528

Hassine IB, Abidli S, Lahbib Y, El Menif NT (2024) First characterization and spatial distribution of microplastics in Sardina pilchardus fish gathered along the northern Tunisian coast. Ecohydrol Hydrobiol 25:414–421

Hastuti AR, Lumbanbatu DT, Wardiatno Y (2019) The presence of microplastics in the digestive tract of commercial fishes off Pantai Indah Kapuk coast, Jakarta, Indonesia. Biodivers J Biol Divers 20(5):1233–1242

Heshmati S, Makhdoumi P, Pirsaheb M, Hossini H, Ahmadi S, Fattahi H (2021) Occurrence and characterization of microplastic content in the digestive system of riverine fishes. J Environ Manag 299:113620

Hettiarachchi H, Meegoda JN (2023) Microplastic pollution prevention: the need for robust policy interventions to close the loopholes in current waste management practices. Int J Environ Res Public Health 20(14):6434

Hollman PC, Bouwmeester H, Peters RJB (2013) Microplastics in aquatic food chain: sources, measurement, occurrence and potential health risks

Hong Y, Wu S, Wei G (2023) Adverse effects of microplastics and nanoplastics on the reproductive system: a comprehensive review of fertility and potential harmful interactions. Sci Total Environ 903:166258

Horton AA, Walton A, Spurgeon DJ, Lahive E, Svendsen C (2017) Microplastics in freshwater and terrestrial environments: evaluating the current understanding to identify the knowledge gaps and future research priorities. Sci Total Environ 586:127–141

Hossain MI, Tuha MASM (2020) Biodegradable plastic production from daily household waste materials and comparison the decomposing time with synthetic polyethylene plastic. Int J Adv Life Sci Res 3:16–19

Hossain MU, Ng ST, Dong Y, Amor B (2021) Strategies for mitigating plastic wastes management problem: a lifecycle assessment study in Hong Kong. Waste Manag 131:412–422

Hu D, Shen M, Zhang Y, Li H, Zeng G (2019) Microplastics and nanoplastics: would they affect global biodiversity change? Environ Sci Pollut Res 26:19997–20002

Hu X, Biswas A, Sharma A, Sarkodie H, Tran I, Pal I, De S (2021) Mutational signatures associated with exposure to carcinogenic microplastic compounds bisphenol A and styrene oxide. NAR Cancer 3(1):zcab004

Hu J, Zuo J, Li J, Zhang Y, Ai X, Zhang J, Gong D, Sun D (2022) Effects of secondary polyethylene microplastic exposure on crucian (Carassius carassius) growth, liver damage, and gut microbiome composition. Sci Total Environ 802:149736

Huang J-N, Xu L, Wen B, Gao J-Z, Chen Z-Z (2024) Characteristics and risks of microplastic contamination in aquaculture ponds near the Yangtze estuary, China. Environ Pollut 343:123288

Imran M, Das KR, Naik MM (2019) Co-selection of multi-antibiotic resistance in bacterial pathogens in metal and microplastic contaminated environments: an emerging health threat. Chemosphere 215:846–857

Issac MN, Kandasubramanian B (2021) Effect of microplastics in water and aquatic systems. Environ Sci Pollut Res 28:19544–19562

Jalil MA, Mian MN, Rahman MK (2013) Using plastic bags and its damaging impact on environment and agriculture: an alternative proposal. Int J Learn Dev 3(4):1–14

Jennings S, Stentiford GD, Leocadio AM, Jeffery KR, Metcalfe JD, Katsiadaki I, Auchterlonie NA, Mangi SC, Pinnegar JK, Ellis T (2016) Aquatic food security: insights into challenges and solutions from an analysis of interactions between fisheries, aquaculture, food safety, human health, fish and human welfare, economy and environment. Fish Fish 17(4):893–938

Jiang J, He L, Zheng S, Liu J, Gong L (2024) A review of microplastic transport in coastal zones. Mar Environ Res 196:106397

Jitkaew P, Pradit S, Noppradit P, Sengloyluan K, Yucharoen M, Suwanno S, Tanrattanakul V, Sornplang K, Nitiratsuwan T, Geindre M (2023). Accumulation of microplastics in stomach, intestine, and tissue of two shrimp species (Metapenaeus moyebi and Macrobrachium rosenbergii) at the Khlong U-Taphao, southern Thailand

Kadac-Czapska K, Knez E, Grembecka M (2024) Food and human safety: the impact of microplastics. Crit Rev Food Sci Nutr 64(11):3502–3521

Karbalaei S, Hanachi P, Walker TR, Cole M (2018) Occurrence, sources, human health impacts and mitigation of microplastic pollution. Environ Sci Pollut Res 25:36046–36063

Karbalaei S, Golieskardi A, Watt DU, Boiret M, Hanachi P, Walker TR, Karami A (2020) Analysis and inorganic composition of microplastics in commercial Malaysian fish meals. Mar Pollut Bull 150:110687

Kaur G, Dandautiya R (2024) Microplastics in synthetic textile industries: origin, occurrence, routes and their fates on environment and human. E3S Web of Conf 509:01010

Kershaw PJ, Rochman CM (2015) Sources, fate and effects of microplastics in the marine environment: part 2 of a global assessment. Reports and studies-IMO/FAO/Unesco-IOC/WMO/IAEA/UN/UNEP joint group of experts on the scientific aspects of marine environmental protection (GESAMP) Eng No 93

Khoaele KK, Gbadeyan OJ, Chunilall V, Sithole B (2023) The devastation of waste plastic on the environment and remediation processes: a critical review. Sustain For 15(6):5233

Kibria G (2023) Impacts of microplastic on fisheries and seafood security—global analysis and synthesis. Sci Total Environ 904:166652

Kibria G, Nugegoda D, Haroon AY (2022) Microplastic pollution and contamination of seafood (including fish, sharks, mussels, oysters, shrimps and seaweeds): a global overview. In: Microplastic pollution: environmental occurrence and treatment technologies. Springer, Cham, pp 277–322

Kim L (2024) Unpacking the true cost of plastics. UBS Global

Klangnurak W, Chunniyom S (2020) Screening for microplastics in marine fish of Thailand: the accumulation of microplastics in the gastrointestinal tract of different foraging preferences. Environ Sci Pollut Res 27(21):27161–27168

Kole PJ, Löhr AJ, Van Belleghem FG, Ragas AM (2017) Wear and tear of tyres: a stealthy source of microplastics in the environment. Int J Environ Res Public Health 14(10):1265

Kosore C, Ojwang L, Maghanga J, Kamau J, Kimeli A, Omukoto J, Ngisiage N, Mwaluma J, Ongada H, Magori C (2018) Occurrence and ingestion of microplastics by zooplankton in Kenya's marine environment: first documented evidence. Afr J Mar Sci 40(3):225–234

Krelling AP, Williams AT, Turra A (2017) Differences in perception and reaction of tourist groups to beach marine debris that can influence a loss of tourism revenue in coastal areas. Mar Policy 85:87–99

Krüger L, Casado-Coy N, Valle C, Ramos M, Sánchez-Jerez P, Gago J, Carretero O, Beltran-Sanahuja A, Sanz-Lazaro C (2020) Plastic debris accumulation in the seabed derived from coastal fish farming. Environ Pollut 257:113336

Kumar R, Verma A, Shome A, Sinha R, Sinha S, Jha PK, Kumar R, Kumar P, Shubham, and Das, S. (2021) Impacts of plastic pollution on ecosystem services, sustainable development goals, and need to focus on circular economy and policy interventions. Sustain For 13(17):9963

La Pietra A, Fasciolo G, Lucariello D, Motta CM, Venditti P, Ferrandino I (2024) Polystyrene microplastics effects on zebrafish embryological development: comparison of two different sizes. Environ Toxicol Pharmacol 106:104371

Le L-T, Nguyen K-QN, Nguyen P-T, Duong HC, Bui X-T, Hoang NB, Nghiem LD (2022) Microfibers in laundry wastewater: problem and solution. Sci Total Environ 852:158412

Lehner R, Petri-Fink A, Rothen-Rutishauser B (2018) Nanoplastic impact on human health—a 3D intestinal model to study the interaction with nanoplastic particles. In: Proceedings of the international conference on microplastic pollution in the mediterranean sea. Springer, Cham

Leslie H, Brandsma S, Van Velzen M, Vethaak A (2017) Microplastics en route: field measurements in the Dutch river delta and Amsterdam canals, wastewater treatment plants, North Sea sediments and biota. Environ Int 101:133–142

Li WC, Tse HF, Fok L (2016) Plastic waste in the marine environment: a review of sources, occurrence and effects. Sci Total Environ 566:333–349

Li J, Lusher AL, Rotchell JM, Deudero S, Turra A, Bråte ILN, Sun C, Hossain MS, Li Q, Kolandhasamy P (2019a) Using mussel as a global bioindicator of coastal microplastic pollution. Environ Pollut 244:522–533

Li X, Ji Y, Mei Q, Chen B, Zhang X, Dong B, Dai X (2019b) Review of microplastics in wastewater and sludge of wastewater treatment plant. Water Purif Technol 38(7):13–22

Li Y, Tao L, Wang Q, Wang F, Li G, Song M (2023) Potential health impact of microplastics: a review of environmental distribution, human exposure, and toxic effects. Environ Health 1(4):249–257

Liaqat S, Hussain M, Riaz J (2024) Entry of the microplastics in food chain and food web. In: Microplastic pollution. Springer, Cham, pp 289–306

Liu Y-P, Li J-G, Zhao Y-F, Wen S, Huang F-F, Wu Y-N (2011) Polybrominated diphenyl ethers (PBDEs) and indicator polychlorinated biphenyls (PCBs) in marine fish from four areas of China. Chemosphere 83(2):168–174

Losetty V, Reddy GN, Yadav CH, Sivakumar K (2024) Study on microplastic content in cosmetic products and their detrimental effect on human health. In: Toxic effects of micro-and nanoplastics: Environment, food and human health. Wiley, Hoboken, pp 253–264

Lozano P, García-Verdugo E (2023) From green to circular chemistry paved by biocatalysis. Green Chem 25(18):7041–7057

Lusher A, Hollman P, Mendoza-Hill J (2017) Microplastics in fisheries and aquaculture: status of knowledge on their occurrence and implications for aquatic organisms and food safety. FAO, Rome

Lv W, Yuan Q, He D, Lv W, Zhou W (2020) Microplastic contamination caused by different rearing modes of Asian swamp eel (Monopterus albus). Aquac Res 51(12):5084–5095

MacArthur E (2017) The new plastics economy: rethinking the future of plastics & catalysing action. Ellen MacArthur Foundation, Cowes, p 68

MacLeod M, Arp HPH, Tekman MB, Jahnke A (2021) The global threat from plastic pollution. Science 373(6550):61–65

Mangla S, Pathak G, Bhan V (2025) Microplastics contamination in soil, water, air, and environmental health risk assessment. In: Microplastics. Elsevier, Amsterdam, pp 3–30

Martínez-Gómez C, León VM, Calles S, Gomáriz-Olcina M, Vethaak AD (2017) The adverse effects of virgin microplastics on the fertilization and larval development of sea urchins. Mar Environ Res 130:69–76

McCallum W (2018) How to give up plastic: a conscious guide to changing the world, one plastic bottle at a time. Penguin, London

McIlgorm A, Xie J (2023) The costs of environmental degradation from plastic pollution in selected coastal areas in the United Republic of Tanzania

Medyńska-Juraszek A, Szczepańska A (2023) Microplastic pollution in EU farmland soils: preliminary findings from agricultural soils (southwestern Poland). Agriculture 13(9):1733

Mehinto AC, Coffin S, Koelmans AA, Brander SM, Wagner M, Thornton Hampton LM, Burton AG Jr, Miller E, Gouin T, Weisberg SB (2022) Risk-based management framework for microplastics in aquatic ecosystems. Microplast Nanoplast 2(1):17

Mercogliano R, Avio CG, Regoli F, Anastasio A, Colavita G, Santonicola S (2020) Occurrence of microplastics in commercial seafood under the perspective of the human food chain. A review. J Agric Food Chem 68(19):5296–5301

Merrell TR Jr (1980) Accumulation of plastic litter on beaches of Amchitka Island. Alaska Mar Environ Res 3(3):171–184

Messinetti S, Mercurio S, Parolini M, Sugni M, Pennati R (2018) Effects of polystyrene microplastics on early stages of two marine invertebrates with different feeding strategies. Environ Pollut 237:1080–1087

Miao C, Zhang J, Jin R, Li T, Zhao Y, Shen M (2023) Microplastics in aquaculture systems: occurrence, ecological threats and control strategies. Chemosphere 340:139924

Mihai F-C, Gündoğdu S, Markley LA, Olivelli A, Khan FR, Gwinnett C, Gutberlet J, Reyna-Bensusan N, Llanquileo-Melgarejo P, Meidiana C (2021) Plastic pollution, waste management issues, and circular economy opportunities in rural communities. Sustain For 14(1):20

Miller ME, Hamann M, Kroon FJ (2020) Bioaccumulation and biomagnification of microplastics in marine organisms: a review and meta-analysis of current data. PLoS One 15(10):e0240792

Mittempergher D, Raes L, Jain A, Mittempergher D, Raes L, Jain A (2022) The economic impact of marine plastic pollution in Antigua and Barbuda: impacts on the fisheries and tourism sectors, and the benefits of reducing mismanaged waste, vol 31. IUCN, Gland, pp 880–893

Nawalage NSK, Bellanthudawa BKA (2022) Synthetic polymers in personal care and cosmetics products (PCCPs) as a source of microplastic (MP) pollution. Mar Pollut Bull 182:113927

Naylor RL, Goldburg RJ, Primavera JH, Kautsky N, Beveridge MC, Clay J, Folke C, Lubchenco J, Mooney H, Troell M (2000) Effect of aquaculture on world fish supplies. Nature 405(6790):1017–1024

Nguyen M-K, Lin C, Nguyen H-L, Le V-R, Priya K, Singh J, Chang SW, Um M-J, Nguyen DD (2023) Emergence of microplastics in the aquatic ecosystem and their potential effects on health risks: the insights into Vietnam. J Environ Manag 344:118499

Niaounakis M (2017) Management of marine plastic debris. William Andrew, Norwich

Nobre C, Santana M, Maluf A, Cortez F, Cesar A, Pereira C, Turra A (2015) Assessment of microplastic toxicity to embryonic development of the sea urchin Lytechinus variegatus (Echinodermata: Echinoidea). Mar Pollut Bull 92(1-2):99–104

Nyadjro ES, Webster JA, Boyer TP, Cebrian J, Collazo L, Kaltenberger G, Larsen K, Lau YH, Mickle P, Toft T (2023) The NOAA NCEI marine microplastics database. Sci Data 10(1):726

Nyam TT, Ayeleru OO (2024) Conversion of waste plastics into value-added materials: a global perspective. University of Johannesburg, Doornfontein

Ogunola O, Palanisami T (2016) Microplastics in the marine environment: current status, assessment methodologies, impacts and solutions. J Pollut Effects Control 4(161):2

Ogunola OS, Onada OA, Falaye AE (2018) Mitigation measures to avert the impacts of plastics and microplastics in the marine environment (a review). Environ Sci Pollut Res 25:9293–9310

Oladele IO, Okoro CJ, Adelani SO, Agbeboh NI, Betiku OT (2024) Current application of recycled waste plastics as a sustainable materials: a review on availability, processing and application. J Thermoplast Compos Mater 38(1):277–301

Onyena A, Aniche D, Ogbolu B, Rakib M, Uddin J, Walker T (2021) Governance strategies for mitigating microplastic pollution in the marine environment: a review. Microplastics 1:15–46
Pal D, Prabhakar R, Barua VB, Zekker I, Burlakovs J, Krauklis A, Hogland W, Vincevica-Gaile Z (2024) Microplastics in aquatic systems: a comprehensive review of its distribution, environmental interactions, and health risks. Environ Sci Pollut Res 32:56–88
Park J-S, Yoo J-W, Lee Y-H, Park C, Lee Y-M (2024) Size-and shape-dependent ingestion and acute toxicity of fragmented and spherical microplastics in the absence and presence of prey on two marine zooplankton. Mar Pollut Bull 206:116768
Pellini G, Gomiero A, Fortibuoni T, Fabi G, Grati F, Tassetti AN, Polidori P, Vega CF, Scarcella G (2017) Plastic soles: microplastic litter in the gastrointestinal tract of Solea solea from the Adriatic Sea. In: Proceedings of the international conference on microplastic pollution in the mediterranean sea. Springer, Cham
Peng X, Chen M, Chen S, Dasgupta S, Xu H, Ta K, Du M, Li J, Guo Z, Bai S (2018) Microplastics contaminate the deepest part of the world's ocean. Geochem Perspect Lett 9(1):1–5
Pérez-Reverón R, Álvarez-Méndez SJ, Kropp RM, Perdomo-González A, Hernández-Borges J, Díaz-Peña FJ (2022) Microplastics in agricultural systems: analytical methodologies and effects on soil quality and crop yield. Agriculture 12(8):1162
Pettipas S, Bernier M, Walker TR (2016) A Canadian policy framework to mitigate plastic marine pollution. Mar Policy 68:117–122
Phillips MB, Bonner TH (2015) Occurrence and amount of microplastic ingested by fishes in watersheds of the Gulf of Mexico. Mar Pollut Bull 100(1):264–269
Pilapitiya PNT, Ratnayake AS (2024) The world of plastic waste: a review. Clean Mater 11:100220
Pironti C, Ricciardi M, Motta O, Miele Y, Proto A, Montano L (2021) Microplastics in the environment: intake through the food web, human exposure and toxicological effects. Toxics 9(9):224
Priscilla V, Patria M (2020) Comparison of microplastic abundance in aquaculture ponds of milkfish Chanos chanos (Forsskål, 1775) at Muara Kamal and Marunda, Jakarta Bay. IOP Conf Ser Earth Environ Sci 404:012027
Prüst M, Meijer J, Westerink RH (2020) The plastic brain: neurotoxicity of micro-and nanoplastics. Part Fibre Toxicol 17:1–16
Purba NP, Pasaribu B, Faizal I, Martasuganda MK, Ilmi MH, Febriani C, Alfarez RR (2023) Coastal clean-up in Southeast Asia: lessons learned, challenges, and future strategies. Front Mar Sci 10:1250736
Radhakrishnan NKK, Sangeetha J, Alabhai JM, Jayasree P (2024) Accumulation of microplastics in bivalves within the Chandragiri River in South-Western India. Anthr Coasts 7(1):5
Rafa N, Ahmed B, Zohora F, Bakya J, Ahmed S, Ahmed SF, Mofijur M, Chowdhury AA, Almomani F (2023) Microplastics as carriers of toxic pollutants: source, transport, and toxicological effects. Environ Pollut 343:123190
Rainieri S, Barranco A (2019) Microplastics, a food safety issue? Trends Food Sci Technol 84:55–57
Rana MM, Haque MR, Tasnim SS, Rahman MM (2023) The potential contribution of microplastic pollution by organic fertilizers in agricultural soils of Bangladesh: quantification, characterization, and risk appraisals. Front Environ Sci 11:1205603
Ranjan VP, Joseph A, Srivastava S, Sharma HB, Biswas B, Goel S, Kumar S (2024) From cosmetics to contamination: microplastics in personal care products as vectors for chromium in aquatic environments. Waste Manage Bull 2(3):229–240
Reisser J, Slat B, Noble K, Du Plessis K, Epp M, Proietti M, de Sonneville J, Becker T, Pattiaratchi C (2015) The vertical distribution of buoyant plastics at sea: an observational study in the North Atlantic gyre. Biogeosciences 12(4):1249–1256
Rendell-Bhatti FA (2019) Developmental toxicity of microplastic leachates on marine larvae. University of Exeter, Exeter
Rhodes CJ (2018) Plastic pollution and potential solutions. Sci Prog 101(3):207–260
Ribeiro F, O'Brien JW, Galloway T, Thomas KV (2019) Accumulation and fate of nano-and microplastics and associated contaminants in organisms. TrAC Trends Anal Chem 111:139–147

Rillig MC, Ingraffia R, de Souza Machado AA (2017) Microplastic incorporation into soil in agroecosystems. Front Plant Sci 8:1805
Rizal NIM, Kamaruddin SA, Nazri R (2022) Microplastic pollution in marine waters: a malaysian perspective. JTHEM 7(28):76–88
Ronda AC, Menéndez MC, Tombesi N, Álvarez M, Tomba JP, Silva LI, Arias AH (2023) Microplastic levels on sandy beaches: are the effects of tourism and coastal recreation really important? Chemosphere 316:137842
Rossi G, Barnoud J, Monticelli L (2014) Polystyrene nanoparticles perturb lipid membranes. J Phys Chem Lett 5(1):241–246
Saha G, Saha SC (2024) Tiny particles, big problems: the threat of microplastics to marine life and human health. PRO 12(7):1401
Salgaonkar S, Botle A, Barabde G, Herlekar M (2024) Micro-Nanoplastics from Stormwater runoffs to water bodies: an in-depth investigation. In: Management of micro and nano-plastics in soil and biosolids: fate, occurrence, monitoring, and remedies. Springer, Cham, pp 95–117
Santonicola S, Volgare M, Cocca M, Dorigato G, Giaccone V, Colavita G (2023) Impact of fibrous microplastic pollution on commercial seafood and consumer health: a review. Animals 13(11):1736
Šaravanja A, Pušić T, Dekanić T (2022) Microplastics in wastewater by washing polyester fabrics. Materials 15(7):2683
Sarker A, Deepo DM, Nandi R, Rana J, Islam S, Rahman S, Hossain MN, Islam MS, Baroi A, Kim J-E (2020) A review of microplastics pollution in the soil and terrestrial ecosystems: a global and Bangladesh perspective. Sci Total Environ 733:139296
Sarma H, Hazarika RP, Kumar V, Roy A, Pandit S, Prasad R (2022) Microplastics in marine and aquatic habitats: sources, impact, and sustainable remediation approaches. Environ Sustain 5(1):39–49
Sathyamohan G, Sewwandi M, Ambade B, Vithanage M (2023) Sources and circulation of microplastics in the aerosphere–atmospheric transport of microplastics. In: Microplastics in the ecosphere. Elsevier, Amsterdam, pp 125–146
Schnurr RE, Alboiu V, Chaudhary M, Corbett RA, Quanz ME, Sankar K, Srain HS, Thavarajah V, Xanthos D, Walker TR (2018) Reducing marine pollution from single-use plastics (SUPs): a review. Mar Pollut Bull 137:157–171
Seetapan K, Prommi TO (2023) Microplastics in commercial fish digestive tracts from freshwater habitats in northern Thailand. Ecol Montenegrina 68:48–65
Sen SB, Yadav V (2024) Social and economic impacts of plastics in the environment. In: Waste-to-wealth. CRC Press, Boca Raton, pp 251–271
Shah AA, Hasan F, Hameed A, Ahmed S (2008) Biological degradation of plastics: a comprehensive review. Biotechnol Adv 26(3):246–265
Sharma P (2024) Microplastic contamination in food processing: role of packaging materials. Food Sci Eng 5:271–287
Sharma P, Vidyarthi VK (2024) Impact of microplastic intake via poultry products: environmental toxicity and human health. J Hazard Mater Adv 14:100426
Sharma D, Dhanker R, Bhawna T, Raza S, Sharma A (2024) Fishing gears and nets as a source of microplastic. In: Microplastic pollution. Springer, Cham, pp 127–140
Shi C, Liu Z, Yu B, Zhang Y, Yang H, Han Y, Wang B, Liu Z, Zhang H (2024) Emergence of nanoplastics in the aquatic environment and possible impacts on aquatic organisms. Sci Total Environ 906:167404
Sigler M (2014) The effects of plastic pollution on aquatic wildlife: current situations and future solutions. Water Air Soil Pollut 225:1–9
Silva-Cavalcanti JS, Silva JDB, de França EJ, de Araújo MCB, Gusmão F (2017) Microplastics ingestion by a common tropical freshwater fishing resource. Environ Pollut 221:218–226
Sinha S (2024) Food security challenges and microplastics: a comprehensive review. In: Microplastics in African and Asian environments: the influencers, challenges, and solutions. Springer, Cham, pp 361–371

Sobhani Z, Lei Y, Tang Y, Wu L, Zhang X, Naidu R, Megharaj M, Fang C (2020) Microplastics generated when opening plastic packaging. Sci Rep 10(1):4841

Song JA, Choi CY, Park H-S (2020) Exposure of bay scallop Argopecten irradians to micro-polystyrene: bioaccumulation and toxicity. Compar Biochem Physiol Part C Toxicol Pharmacol 236:108801

Sridharan S, Kumar M, Bolan NS, Singh L, Kumar S, Kumar R, You S (2021) Are microplastics destabilizing the global network of terrestrial and aquatic ecosystem services? Environ Res 198:111243

Sun X, Li Q, Zhu M, Liang J, Zheng S, Zhao Y (2017) Ingestion of microplastics by natural zooplankton groups in the northern South China Sea. Mar Pollut Bull 115(1-2):217–224

Sun X, Liu T, Zhu M, Liang J, Zhao Y, Zhang B (2018) Retention and characteristics of microplastics in natural zooplankton taxa from the East China Sea. Sci Total Environ 640:232–242

Surana D, Patel P, Ghosh P, Sharma S, Kumar V, Kumar S (2024) Microplastic fibers in different environmental matrices from synthetic textiles: ecotoxicological risk, mitigation strategies, and policy perspective. J Environ Chem Eng 12:112333

Sussarellu R, Suquet M, Thomas Y, Lambert C, Fabioux C, Pernet MEJ, Le Goïc N, Quillien V, Mingant C, Epelboin Y (2016) Oyster reproduction is affected by exposure to polystyrene microplastics. Proc Natl Acad Sci 113(9):2430–2435

Sutkar PR, Gadewar RD, Dhulap VP (2023) Recent trends in degradation of microplastics in the environment: a state-of-the-art review. J Hazard Mater Adv 11:100343

Tamburri MN, Soon ZY, Scianni C, Øpstad CL, Oxtoby NS, Doran S, Drake LA (2022) Understanding the potential release of microplastics from coatings used on commercial ships. Front Mar Sci 9:1074654

Tanaka N, Takada N, Takahashi M, Yeo BG, Oya Y, Watanabe I, Fujita Y, Takada H, Mizukawa K (2023) Bioaccumulation and metabolism of polybrominated diphenyl ethers (PBDEs) in coenobitid hermit crabs from marine litter-polluted beaches in remote islands. Mar Pollut Bull 190:114812

Tang X-W, Yu R, Fan M-H, Yang Z, Liao Z, Yang Z-X, Xie C-Y, Xuan Y-K, Wang J-X, Zhang X-L (2022) Physiological and transcriptome analysis of Mytilus coruscus in response to Prorocentrum lima and microplastics. Front Mar Sci 9:1087667

Tepe Y, Aydın H, Ustaoğlu F, Kodat M (2024) Occurrence of microplastics in the gastrointestinal tracts of four most consumed fish species in Giresun, the southeastern Black Sea. Environ Sci Pollut Res 31(43):55336–55345

Thomas M, Jon B, Craig S, Edward R, Ruth H, John B, Dick VA, Heather LA, Matthew S (2020) The world is your oyster: low-dose, long-term microplastic exposure of juvenile oysters. Heliyon 6(1):e03103

Thompson RC, Moore CJ, Vom Saal FS, Swan SH (2009) Plastics, the environment and human health: current consensus and future trends. Philos Trans R Soc B Biol Sci 364(1526):2153–2166

Thushari GGN, Senevirathna JDM (2020) Plastic pollution in the marine environment. Heliyon 6(8):e04709

Tursi A, Baratta M, Easton T, Chatzisymeon E, Chidichimo F, De Biase M, De Filpo G (2022) Microplastics in aquatic systems, a comprehensive review: origination, accumulation, impact, and removal technologies. RSC Adv 12(44):28318–28340

UNEP (2018) Banning single-use plastic: lessons and experiences from countries. https://www.unep.org/interactives/beat-plastic-pollution/

Viarengo A, Canesi L (1991) Mussels as biological indicators of pollution. Aquaculture 94(2-3):225–243

Walkinshaw C, Lindeque PK, Thompson R, Tolhurst T, Cole M (2020) Microplastics and seafood: lower trophic organisms at highest risk of contamination. Ecotoxicol Environ Saf 190:110066

Wang X, Wu L (2024) Intergenerational differences in the environmental concerns of plastic waste business owners: environmental knowledge, environmental risk exposure, and community connection as mediators. Hum Soc Sci Commun 11(1):1–13

Wang J, Lu L, Wang M, Jiang T, Liu X, Ru S (2019a) Typhoons increase the abundance of microplastics in the marine environment and cultured organisms: a case study in Sanggou Bay, China. Sci Total Environ 667:1–8
Wang T, Li B, Zou X, Wang Y, Li Y, Xu Y, Mao L, Zhang C, Yu W (2019b) Emission of primary microplastics in mainland China: invisible but not negligible. Water Res 162:214–224
Wang C, Zhao J, Xing B (2021a) Environmental source, fate, and toxicity of microplastics. J Hazard Mater 407:124357
Wang Y, Wang X, Li Y, Li J, Liu Y, Xia S, Zhao J (2021b) Effects of exposure of polyethylene microplastics to air, water and soil on their adsorption behaviors for copper and tetracycline. Chem Eng J 404:126412
Watkins E, Schweitzer J-P, Leinala E, Börkey P (2019) Policy approaches to incentivise sustainable plastic design. OECD, Paris
Watts J (2019) Concrete: the most destructive material on earth. Guardian 25:1–9
Weis JS (2021) Most microplastics come from clothes. Bioscience 71(4):321–321
Wu F, Wang Y, Leung JY, Huang W, Zeng J, Tang Y, Chen J, Shi A, Yu X, Xu X (2020) Accumulation of microplastics in typical commercial aquatic species: a case study at a productive aquaculture site in China. Sci Total Environ 708:135432
Wu P, Lin S, Cao G, Wu J, Jin H, Wang C, Wong MH, Yang Z, Cai Z (2022) Absorption, distribution, metabolism, excretion and toxicity of microplastics in the human body and health implications. J Hazard Mater 437:129361
Xia X, Sun M, Zhou M, Chang Z, Li L (2020) Polyvinyl chloride microplastics induce growth inhibition and oxidative stress in Cyprinus carpio var. larvae. Sci Total Environ 716:136479
Xie M, Lin L, Xu P, Zhou W, Zhao C, Ding D, Suo A (2021) Effects of microplastic fibers on Lates calcarifer juveniles: accumulation, oxidative stress, intestine microbiome dysbiosis and histological damage. Ecol Indic 133:108370
Yin K, Wang Y, Zhao H, Wang D, Guo M, Mu M, Liu Y, Nie X, Li B, Li J (2021) A comparative review of microplastics and nanoplastics: toxicity hazards on digestive, reproductive and nervous system. Sci Total Environ 774:145758
Yu Q, Hu X, Yang B, Zhang G, Wang J, Ling W (2020) Distribution, abundance and risks of microplastics in the environment. Chemosphere 249:126059
Yuan W, Zhou Y, Chen Y, Liu X, Wang J (2020) Toxicological effects of microplastics and heavy metals on the Daphnia magna. Sci Total Environ 746:141254
Yuan Z, Nag R, Cummins E (2022) Human health concerns regarding microplastics in the aquatic environment-from marine to food systems. Sci Total Environ 823:153730
Yurtsever M, Yurtsever U (2018) Commonly used disposable plastic bags as a source of microplastic in environment. In: Proceedings of the international conference on microplastic pollution in the mediterranean sea. Springer, Cham
Zang H, Zhou J, Marshall MR, Chadwick DR, Wen Y, Jones DL (2020) Microplastics in the agroecosystem: are they an emerging threat to the plant-soil system? Soil Biol Biochem 148:107926
Zhang L, Xie Y, Liu J, Zhong S, Qian Y, Gao P (2020) An overlooked entry pathway of microplastics into agricultural soils from application of sludge-based fertilizers. Environ Sci Technol 54(7):4248–4255
Zhang S, Li Y, Chen X, Jiang X, Li J, Yang L, Yin X, Zhang X (2022) Occurrence and distribution of microplastics in organic fertilizers in China. Sci Total Environ 844:157061
Zheng S, Zhao Y, Liangwei W, Liang J, Liu T, Zhu M, Li Q, Sun X (2020) Characteristics of microplastics ingested by zooplankton from the Bohai Sea, China. Sci Total Environ 713:136357
Zhou A, Zhang Y, Xie S, Chen Y, Li X, Wang J, Zou J (2021) Microplastics and their potential effects on the aquaculture systems: a critical review. Rev Aquac 13(1):719–733
Zhou Y, Ashokkumar V, Amobonye A, Bhattacharjee G, Sirohi R, Singh V, Flora G, Kumar V, Pillai S, Zhang Z (2023) Current research trends on cosmetic microplastic pollution and its impacts on the ecosystem: a review. Environ Pollut 320:121106
Zhu D, Bi Q-F, Xiang Q, Chen Q-L, Christie P, Ke X, Wu L-H, Zhu Y-G (2018) Trophic predator-prey relationships promote transport of microplastics compared with the single Hypoaspis aculeifer and Folsomia candida. Environ Pollut 235:150–154

Towards Blue Futures: A Call for Collective Remediation

8

Beatriz Pérez-Aragón, Jocelyn Tapia-Fuentes, Juan Carlos Alvarez-Zeferino, Alethia Vázquez-Morillas, Maribel Velasco-Pérez, and Arely Areanely Cruz-Salas

Abstract

This chapter addresses the need for a blue future and the urgency of remedying the problem of microplastics in our oceans. In addition, it presents an optimistic and proactive vision to address this challenge, highlighting the importance of collective collaboration and technological innovation in the search for solutions. Different strategies and actions being taken worldwide at different scales, for example: at industrial, governmental, and non-governmental organizations level, for the remediation and prevention of microplastic pollution in the sea are discussed. Ultimately, a call to action is issued, urging individuals, communities, and governments to unite in the fight for a cleaner and more sustainable future for our oceans, since this is not an option, it is a necessity that requires concrete actions both locally and internationally in which all the actors involved participate.

Keywords

Environmental education · Industry · Recycling · Regulations · Strategies

B. Pérez-Aragón · A. Vázquez-Morillas · M. Velasco-Pérez · A. A. Cruz-Salas (✉)
Universidad Autónoma Metropolitana, Alcaldía Azcapotzalco Ciudad de México, Mexico City, Mexico
e-mail: aacs@azc.uam.mx

J. Tapia-Fuentes
Universidad Autónoma de Baja California, Campus Mexicali, Mexicali, Mexico

J. C. Alvarez-Zeferino
Universidad Autónoma Metropolitana, Alcaldía Azcapotzalco Ciudad de México, Mexico City, Mexico

Secretaría de la Defensa Nacional, Universidad del Ejército y Fuerza Aérea. Escuela Militar de Ingeniería, Mexico City, Mexico

M. Oves et al. (eds.), *Marine Microplastics and the Quest for Remediation*,
https://doi.org/10.1007/978-981-96-8979-8_8

8.1 Introduction

The marine system as we know it has been changed. The delicate balance of marine life is being threatened by an almost invisible enemy: microplastics. These tiny fragments of plastic, infiltrating every corner of the marine environment, might represent one of the biggest environmental crises we have ever seen, with consequences we are only beginning to understand. While the magnitude of the problem is overwhelming, growing awareness of the impacts of microplastics has begun to inspire a global movement towards remediation.

The path toward a blue future (one in which our oceans are free of plastic pollution) requires an unprecedented collective effort. This chapter explores the strategies, collaborations, and innovations needed to address this crisis. From cutting-edge scientific research to community action, there is an urgent call for governments, organizations, and citizens around the world to join forces in a common mission: to restore the health and balance of our oceans. This effort is not only crucial for the survival of countless marine species and the ones that rely on oceans too, but also for the long-term well-being of humanity. By addressing the challenge of microplastics, we have the opportunity to reframe our relationship with the environment and build a sustainable legacy for future generations. This chapter is an invitation to participate in this monumental task, to transform consciousness into action, and to turn the threat of microplastics into an opportunity for regeneration and hope.

8.2 Understanding the Microplastic Crisis

Oceans are fundamental for the well-being of humanity, they are a source of food and energy, regulate the climate of the planet, and provide space for economic, recreational, sporting, and spiritual activities. Oceans are a key element of sustainable development, and their annual economic value is estimated at USD 2.5 trillion (UNESCO-UNEP 1976). However, they are highly affected by the triple planetary crisis (climate change, pollution, and biodiversity loss) caused by anthropogenic activities. In 2016 the Organization for Economic Cooperation and Development (OCDE) in the report "The Ocean Economy in 2030" identified the following pressures and drivers for the ocean economy (OECD 2016):

- Changes in sea temperature and sea levels.
- Acidification.
- Decreasing concentration of oxygen in oceans.
- Alterations of water currents and circulation patterns.
- Changes in the hydrological cycle.
- Unsustainable fishing.
- Pollution.

The types of pollution include nitrogen and phosphorous from agriculture runoff, plastic pollution, chemical pollution, nuclear radiation, noise (from military activities, seabed mapping, and shipping), oil pollution, seabed mining, and tourism

(OECD 2016). Plastic pollution has been recognized as an emerging problem (UNEP 2021a, b). The extensive use of short-time plastic products and mismanagement of plastic waste are some of the most significant drivers of plastic pollution. The amount of plastic produced and plastic waste generated from 1950 to 2017 was estimated at 9.2 and 7 billion tons, respectively (Geyer et al. 2017). Globally, 10% of plastic waste was recycled, 14% was incinerated, and the remaining 76% is in landfills, dumping sites, or the environment (Geyer et al. 2017).

Leakage of plastic waste to the environment has been estimated in several studies (Borrelle et al. 2020; Sonke 2025; Ryberg et al. 2019). For instance, Ryberg et al. (2019) found total plastic leakage in 2015 at 9.2 million tons. Borrelle et al. (2020) calculated that in 2016 between 19 and 23 million tons of plastic waste reached the aquatic environment. The OCDE calculated plastic leakage in 2019 at 22 million tons, with a projection of 44 million tons by 2060 (Sonke 2025).

Microplastics are of special interest due to their ubiquity in the environment. They are plastic particles with a size less than 5 mm. Microplastics can be classified as primary and secondary. Primary microplastics are manufactured to be used or added to other products, examples of this type of microplastic are glitter, pellets, and plastic particles added to cosmetics. Secondary microplastics are produced by the fragmentation or degradation of plastics of macroplastics (plastics with a size greater than 5 mm), such as fibers from textiles, called microfibers, and particles generated by tire abrasion (UNEP 2021a). Most microplastics in the oceans are secondary.

The OECD estimated that microplastics released to the oceans will double from 2019 to 2060, to 5.8 million tons (Sonke 2025). Microplastics leakage to the oceans has land- and sea-based sources (UNEP 2021a). Land-based sources include the fragmentation of plastics in landfills, dumps, or the environment, the abrasion of synthetic polymers (road markings, tires, artificial turf, and textiles), the release of plastic particles in cosmetics and personal hygiene products to wastewater, loss of pellets during transportation or manufacturing, plastics used in agriculture, and degradation of macroplastics on beaches. Sea-based sources of microplastics include lost or abandoned fishing gear, degradation and fragmentation of such equipment, waste from the maritime industry (such as paint and garbage from ships), and the abrasion of aquaculture gear. However, most microplastics found in the oceans originate from land-based sources (UNEP 2021a).

8.2.1 Environmental, Social, and Economic Effects of Microplastics

Pollution of the oceans with microplastics has environmental, health, social, and economic negative impacts (UNEP 2021b). Assessment of these impacts is difficult to carry out due to the complexity of the marine socioecological systems. Ecological impacts can be assessed by studying the toxicity of specific types of microplastics on different species of marine biota in laboratory settings and estimating the reduction of ecological services. There is a growing body of literature on toxicity studies, however, results show inconsistencies; some of these can be attributed to the type, dose, and size of the microplastics used in the experiments, the studied species, and the duration of the tests (UNEP 2021b). Studies in crustaceans, mollusks, and polychaetes have found no detectable changes, changes in protein expression, feeding behavior, larval development, inflammation, and reduction in growth, reproductive success, and survival rates (UNEP 2021b). The presence of additives and sorbed pollutants such as persistent organic pollutants and heavy metals can also affect microplastic's toxicity. It is important to acknowledge that there are important gaps in the literature, Ajith et al. (2020) reviewed 220 articles published from 2010 to 2019 and found that only 44 out of 192 countries have conducted research on microplastics, and 38% of the organisms studied were fish. Regarding the assessment of ecological services, a landmark study was published by Halpern et al. (2008), which found that all areas of the marine environment were affected by human influence, and 41% are strongly affected by multiple drivers. More recently Beaumont et al. (2019) conducted a literature review of 1191 datapoints, finding that there is global evidence of impact with medium to high frequency on eight ecological subjects studied, with a medium to high degree of irreversibility. Humans are exposed to microplastics through the ingestion of seafood, water, marine salt, and other foods. Marine organisms can ingest microplastics directly from the water or from other organisms that have ingested microplastics. The toxicity of microplastics has been extensively studied in-vitro and animal models, some studies have found effects on cell viability, stress response, cancer, reproductive system, and genomic systems (Cruz Salas et al. 2024). Social effects of ocean pollution with plastics include higher exposure to microplastics in communities that depend on seafood for protein intake (UNEP 2021b); reduction in productivity, viability, profitability, and safety of the fishing and aquaculture industry; reduction in human well-being from biodiversity loss; and reduction of time spent in recreational activities in marine environments (Beaumont et al. 2019). There is a growing concern that social impacts are unevenly distributed in different sectors of society, affecting disproportionally the most vulnerable sectors (Bennett et al. 2021). Economic costs related to microplastic pollution are due to leakage prevention, welfare related to negative effects on human health from microplastic exposure, loss of ecosystem services, and negative effects on the fishing and aquaculture industries. Several studies estimate the cost of plastic pollution in the marine environment, but to the best of our knowledge, there is none specific to the cost of microplastics. Currently, there is not enough information to provide a robust estimate of plastic pollution globally (UNEP 2021b). In the face of the many challenges that face the oceans, Blue Future is an initiative that is focused on fostering the sustainability of the marine ecosystem. Research on the many aspects related to microplastic pollution could promote sustainable practices such as biodiversity conservation; management of fisheries, and fisheries plastic waste; and implementation of policies that protect the most vulnerable sectors, among others.

8.2.2 Environmental Education as a Starting Point

Environmental education has its origins in the 1950s and 1960s, mainly driven by the actions of NGOs in different countries. However, it was recognized at the international level from the Stockholm Conference in 1972 and from the International Workshop on Environmental Education (EE), held in Belgrade in October 1975, when a frame of reference, objectives, and principles were built from which the most accepted definition of this field of science was proposed (Carter 2010). This definition states that EE is a process aimed at the world population to create awareness and concern about the environment in general and its associated problems, through tools related to knowledge, attitude changes, commitments, motivations, and development of individual and collective skills to solve current problems and prevent new situations (UNESCO-UNEP 1976). This laid the groundwork for the definitive approach to be established in 1977 in the document known as The Tbilisi Declaration, which added the importance of the interdependence of economic, social, political, and ecological aspects in rural and urban areas within EE strategies (Carter 2010).

Particularly, among the principles published in the Stockholm Declaration (1972), the seventh stated that States should take measures to prevent the pollution of the seas by substances that could endanger human health and marine life, but plastic was not mentioned as one of the pollutants (United Nations, 1972). In addition, Principle 19 stresses the importance of environmental education to broaden well-informed public opinion on environmental issues and to motivate changes in the behavior of individuals, companies, and communities. It also stressed the importance of the role of the media in the dissemination of educational information (United Nations Organization (ONU) 1992).

From 1980 onwards, there was an increase in scientific publications on the pollution of seas and oceans by plastic debris, which highlighted the increase in this problem. Multiple studies documented the accumulation of plastics in different oceans, with evidence of effects on marine mammals, fish and birds, both by entanglement in plastic debris and by ingestion (Browne et al. 2015; Geyer et al. 2017; Jambeck et al. 2015).

Despite this increasing scientific evidence, there were only a few local efforts, mainly motivated by NGOs, on environmental education related to marine and ocean pollution. The International Convention for the Prevention of Pollution from Ships (MARPOL) was one of the first international protocols to integrate in its Annex V a total ban on the dumping at sea of any type of ship-generated plastics. Although this convention was adopted as early as 1973, it entered into force in 1983, but it was not until 1988 that Annex V became official (MARPOL 1973). Despite being one of the first advances in controlling plastic waste dumped at sea, this protocol was only focused on ships, omitting other sources of pollution, mainly those located in cities and industrial areas.

During the 1980s, local beach cleanup initiatives began, particularly starting on US beaches in 1986 with the International Coastal Cleanup Day, founded by Linda Maraniss and Kathy O'Hara, and motivated by the Marine Conservation Center (now Ocean Conservancy). By 1989, the initiative was already being implemented on multiple beaches in Canada and Mexico, permeating internationally with the official recognition of the International Coastal Cleanup Day (ICC) (Ocean Conservancy n.d.). This environmental education initiative raised awareness in various sectors of society about the increase in marine pollution and collected data that would serve as evidence to motivate the reduction of waste on beaches and the reduction of single-use plastics. Although its international recognition triggered awareness and implementation in several countries, historically it was NGOs that started EE strategies. However, this meant that the momentum of EE depended on national political agendas, leaving EE strategies lagging for several years in different countries, with NGOs and academia being the main drivers of these initiatives (Maraniss and O'Hara 1986).

The United Nations Conference on Environment and Development, held in Rio de Janeiro in 1992, promoted multiple changes at the international level with the publication of Agenda 21, which emerged as a proposal for a global, national, and local action plan to promote sustainable development, in which it was proposed to address multiple environmental problems from an integral perspective with society, the economic system and governmental decisions (United Nations Organization (ONU) 1992). Although this program was not legally binding, many countries adopted sustainability strategies based on the conceptual framework defined in the chapters of Agenda 21, and recognized the active role of NGOs in the implementation of these strategies.

Particularly, chapter 17 of Agenda 21 points out the importance of protecting the oceans and seas, highlighting actions in marine and coastal territory management, protection, sustainable use, and strengthening international cooperation. It also emphasizes that land-based sources account for 70% of marine pollution, including solid waste, wastewater, sediments, plastic materials, and metals, among others, as the main pollutants. In addition, in Chapter 36, the importance of promoting education, training, and awareness of sustainability and environmental issues was again emphasized (United Nations Organization (ONU) 1992).

Despite international advances in this area, environmental education efforts were still focused on beach cleanup until 1997, when oceanographer Charles Moore discovered the large garbage patch located in the North Pacific Gyre (North Pacific Ocean Current System). This discovery was the watershed for increased research and global public awareness about plastic pollution in the oceans, due to the impact of the first estimates of the size of the slick, which was estimated to be around 80 thousand tons of plastic in volume and a variable size but comparable to the size of the state of Texas (Moore 1999).

The media coverage of this discovery highlighted the impact of human and plastic waste in the oceans and led to the discovery of other garbage islands in different ocean gyres around the world. This also influenced the political will to generate international agreements and local policies to reduce plastic waste, as well as the incorporation of the topic in environmental education programs to illustrate the consequences of plastic waste (Moore 1999).

The increase in scientific publications on plastic pollution since 1980 has evidenced the magnitude of the problem, documenting the devastating impact on marine wildlife and ocean ecosystems (Browne et al. 2015; Geyer et al. 2017; Jambeck et al. 2015). In this context, environmental education stands as an essential starting point to address the plastics crisis. By providing people with the necessary knowledge about reduction of plastic use, proper waste management, and participation in clean-up initiatives, it fosters a change in attitude and behavior that can help to significantly reduce plastic pollution in the seas. Environmental education not only increases awareness of the plastics crisis but also empowers individuals and communities to take effective action, thus contributing to the protection and restoration of marine ecosystems. It is a fundamental component for the development of comprehensive strategies to mitigate and eventually solve plastic pollution in the oceans.

8.3 Efficiency Micro-Solutions: The Impact of Local Solutions

To reduce microplastics' entry into the environment, individual and collective actions must be implemented, including consumer responsibility, business practices, and government policies. Some of the most critical actions are described below.

Reducing the Use of Single-Use Plastics These proposals seek to implement bans on single-use plastics such as bags, straws, and plastic utensils, as well as promote the use of reusable and biodegradable alternatives (UNESCO-UNEP 1976). The above is directly related to microplastics because when most mismanaged conventional plastics degrade, they become secondary microplastics (0.5–5 mm), which are more abundant than primary ones in marine environments (Alvarez-Zeferino et al. 2020), hence, these measures seek to establish a preventive principle.

Improving Urban Solid Waste Management Strengthening recycling systems and ensuring that plastics are properly recycled or disposed of (European Union 2023) could reduce plastic pollution in the oceans. This is because one of the sources of microplastics is solid waste disposed of in landfills, which can fragment when exposed to the outdoors and subsequently be transported by wind or leachate and can also combine with sludge from treatment plants (Golwala et al. 2021).

Establishing Laws and Regulations Ban the production and sale of products that contain microplastics, especially in cosmetics and personal hygiene products, and promote regulations that encourage companies to reduce the use of plastics and develop more sustainable alternatives (FDA 2022).

Promotion of Environmental Education The awareness plays a significant role in the fight against microplastic pollution. Public education campaigns that highlight the detrimental effects of microplastics on the environment and health are instru-

Table 8.1 Examples of citizen science programs related to plastic pollution

Project, *organization*	Locality/country	Objective	References
Microplastics community science pilot project, *Ocean Diagnostics*	Vancouver Island, British Columbia, Canada	Contribute to generating information to understand the presence of microplastics in coastal ecosystems, involving citizens	Flores-Ruiz et al. (2022)
Nurdle patrol, *Nurdle Patrol*	It began on the American coasts of the Gulf of Mexico and has subsequently spread to Mexico, Canada, and Belize	Generate information related to pellets on beaches, to later be geolocated	Patrol-Nurdle (2021)
Clean seas odyssey, *Clean Seas Odyssey*	South-western coast of England, UK	Generate plastic pollution information to generate possible solutions	Nel et al. (2020)
Citizen science, *Microplásticos en Ambientes Marinos*	Mexican coasts, Mexico	Obtain information on microplastic concentrations to generate diagnoses of the country's situation	Alvarez-Zeferino et al. (2023)

mental in encouraging responsible behavior among consumers, thereby reducing the demand for plastic products (EPA 2020). A prime example is the 2012 campaign against microbeads, which successfully raised awareness about the environmental risks they posed. By 2024, the campaign had convinced 448 brands from 119 manufacturers to remove microbeads from their products (Plastic Soup Foundation 2024b).

Citizen Science Programs have emerged as a powerful tool in the fight against microplastics. It presents a great opportunity to cover larger areas, but it is crucial to provide volunteers with affordable and reliable techniques for their collections (Nel et al. 2020). This ensures the generation of solid data that can be utilized by the scientific community, underlining the importance of accessibility in scientific research (Table 8.1).

Investigation and Development Fund research to better understand the impact of microplastics and develop new technologies for their elimination (European Union 2023). Several research have been carried out on the removal of microplastics, especially in water (Rodríguez-Villa et al. 2024). Within the treatments, coagulation with different salts has been studied; the results have shown to be promising (Moraczewska-majkut and Popenda 2024), reverse osmosis process (Lee et al. 2024) and through oxidation by Yao et al. (2023) has also been studied.

Cleaning of Natural Environments Organize and participate in cleaning activities on beaches, rivers, and other natural environments to remove existing plastics, in addition to implementing cleaning technologies in bodies of water, such as containment barriers and automated collection systems (NOAA - National Oceanic and Atmospheric Administration 2018). This is a global effort, with one of the most widespread programs being that of the Ocean Conservancy, which organizes the International Coastal Cleanup annually with the support of volunteers from 97 countries (Ocean Conservancy n.d.).

Corporate Responsibility It is crucial that we encourage companies to adopt sustainable practices, such as reducing plastic packaging and adopting circular economy models. We need to promote transparency in the supply chain and require that products are free of microplastics (Comisión Europea 2023).

Monitoring and Evaluation Establish monitoring systems to evaluate the presence of microplastics in the environment and the effects of implemented policies, in addition to adapting and improving policies based on the results of monitoring and continuous research (UNESCO-UNEP 1976). Some efforts have focused on creating international microplastics monitoring networks and laboratories. One of the objectives of these networks is to provide infrastructure, sampling methodologies, and sample processing to compare the information obtained (Bakir et al. 2024).

8.4 International Collaboration and Government Participation

National governments have implemented various measures to prevent and control plastic pollution, including bans, requirements for specific material characteristics, and taxes (see Table 8.2; UNESCO-UNEP 1976). As of 2020, plastic bag bans were the most prevalent policy at both national and subnational levels. In 2018, an estimated 3.7 billion people lived in areas with some form of plastic bag restriction (Karasik et al. 2020).

Despite the varying effects of these policies, which are highly dependent on the national context, plastic pollution is recognized as a global problem that transcends geopolitical boundaries. A concerted global effort is necessary to address this issue (UNESCO-UNEP 1976). In fact, fragmented regulatory responses to this global challenge have been identified as a contributing factor to the increase of plastics in the environment (Maes et al. 2023).

Despite the varying effects of these policies, which depend heavily on the national context, plastic pollution is recognized as a global problem that transcends geopolitical boundaries. Therefore, a concerted global effort is necessary to address this issue (UNEP 2021). In fact, fragmented regulatory responses to this global challenge have been identified as a contributing factor to the increase of plastics in the environment (Maes et al. 2023).

Although not specifically directed at plastic pollution, different multilateral environmental agreements (MEAs) developed in the twentieth century can be applied to plastics: the Convention on the Prevention of Marine Pollution by Dumping Wastes and Other Matter (London Convention or LC 72), the International Convention for the Prevention of Pollution from Ships (MARPOL 1973), the Convention on the Conservation of Migratory Species of Wild Animals (1979), the United Nations

Table 8.2 Policies applied to prevent and control plastic pollution. Adapted from Karasik et al. (2020)

Type of instrument	Category	Examples
Regulatory	Affirmative: rules that describe what shall be done	• Capture of plastic post-leakage • Responsible for handling plastic waste • Extended producer responsibility • Improvement of processes or products
	Prohibitive: rules disallowing specific actions	• Limit on production of flow in a specific stage of the life cycle • Ban • Prohibition of specific management practices
Economic	Incentives	• Subsidies of advantageous processes or products • Cash for return • Tax break
	Disincentives	• Fee • Tax • Levy
Informative		• Research and monitoring • Education • Ecolabeling

Convention on the Law of the Sea (UNCLOS), the Convention on Biological Diversity (1992), Agenda 21 (1992), the FAO Code of Conduct for Responsible Fisheries (1995), and the UN Global Programme of Action for the Protection of the Marine Environment from Land-Based Activities (1995) (Karasik et al. 2020).

The twenty-first century has witnessed an evolution from those instruments, which focused on other specific environmental problems, to a growing body of policies that specifically address plastics at the global level. The resolutions from the United Nations Environment Assembly (Table 8.3) are particularly noteworthy for their relevance and their recommendations for states to take urgent action regarding plastic pollution.

In March 2022, a landmark resolution was achieved at UNEA 5.2, mandating the development of a legally binding Global Plastics Treaty. To facilitate this process, an Intergovernmental Negotiating Committee (INC) was established to guide and conduct the negotiations. The Treaty must create synergies with existing MEAs, such as the Basel and Stockholm Conventions (Maes et al. 2023). Currently, the Basel Convention covers the movement and disposal of hazardous plastic waste, while the Stockholm Convention addresses certain toxic and persistent additives and chemicals in plastics. In this regard, the new Treaty should avoid duplicating regulations on these issues, focusing instead on aspects that are not already covered.

The development of the Treaty faces significant challenges, as Member States hold differing positions on specific objectives and views, particularly regarding the inclusion of the entire plastics life cycle, the extended producer responsibility

Table 8.3 UNEA resolutions regarding plastic pollution

Year	UNEA resolution
2014	1/6 "marine plastic debris and microplastics"
2016	2/11 "marine plastic litter and microplastics"
2018	3/7 "marine litter and microplastics"
2019	4/6 "marine plastic litter and microplastics"
2019	4/9 "addressing single-use plastic products pollution"
2022	5/14 "end plastic pollution: Towards an international legally binding instrument"

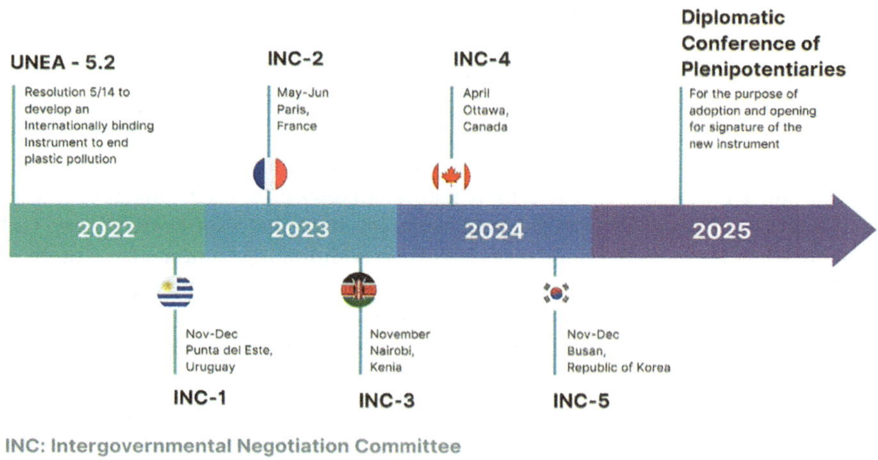

Fig. 8.1 Calendar of negotiation for the global treaty on plastics

principle, and the potential for capping plastics production (Wang 2023). A calendar has been established to discuss the content of the treaty (Fig. 8.1) with five INC sessions planned. The Treaty is expected to be signed and adopted by mid-2025 (UNESCO-UNEP 1976). This new MEA would establish a novel global approach to tackling plastic pollution in a more coordinated and effective manner, applying lessons learned from previous global agreements.

8.5 A Call for Industries to Be Part of the Change

Faced with the problem of microplastic pollution, companies have also developed different strategies with the aim of minimizing the entry of MP into different ecosystems. Strategies can serve different purposes, for example, the manufacture of plastic-free personal care products, the recycling of certain plastic waste into new products, and the implementation of action plans to prevent plastic leakage into the environment. Some examples are described below.

8.5.1 Personal Care Products without Added Microplastics

In February 2011, the Plastic Soup Foundation (PSF) was established, based in Amsterdam, but with an international focus (Plastic Soup Foundation 2024b). PSF seeks to combat plastic waste (PW) from its origin, to this end it has set three objectives: (1) prevent PW from reaching the environment, (2) disseminate findings on the health risks derived from the use of plastics and their additives and (3) reduce the production and use of plastic, all of this is based on scientific research and also emphasizing the precautionary principle (Plastic Soup Foundation 2024a).

One of the campaigns that PSF has is Beat the Microbead, which began in 2012 with the aim of raising public awareness about the problem and providing them with tools when purchasing their personal care products so that they are informed about which of them contain MP and thus decide consciously (Plastic Soup Foundation 2024e). The above is achieved through an application, launched in October 2013. This app scans the ingredients on the packaging and after a few seconds allows us to find out if it contains MP and their names. Likewise, through this application, consumers can browse to know the brands and products that are free of MP and that contain MP (Plastic Soup Foundation 2024f). In addition, PSF initiated a petition in 2019, through change.org, to the European Commission (EC) to prohibit industries throughout the European Union from intentionally adding MP to cosmetics. This petition had a total of 60,203 signatures, which enabled the EC to take the said petition into consideration (Change.org 2019).

Derived from the above and from initiatives by other NGOs, in the European Union as of October 17, 2023, Regulation (EU) 2023/2055 came into force, gradually restricting MPs from intentionally adding to cosmetics, care products personal care, and detergents in order to reduce emissions of MP from everyday products (Comisión Europea 2023; European Union 2023). In this sense, several companies, focused on personal care products, have migrated to manufacturing these without MP content. PSF has a "Zero Plastic Inside" list, which in turn is a logo, to which one refers to products without MP certified by the Soup Foundation. To date a total of 158 MP-free brands have been reported (Plastic Soup Foundation 2024c) (Table 8.4).

All the brands mentioned in the previous table also cover different types of products, those reported in the Beat the Microbead campaign are a total of 4635 (Plastic Soup Foundation 2024d) and are distributed as follows (Fig. 8.2).

8.5.2 Zero Pellet Loss Program

Operation Clean Sweep (OCS) is an international program led by the plastics industry. It was founded in the United States in 1991 by the Plastics Industry Association and since 2004 it has been jointly administered by said association and the American Chemistry Council (ACC)'s Plastics Division, the latter representing the main manufacturers of plastic resins. The OCS aims to support companies in various countries to prevent the loss of pellets in their different operations (PLASTICS and

Table 8.4 Personal care product brands without microplastics ordered by alphabet

	No. of brands	Brand(s)
A	15	Aleph Beauty, Amazing, Adama, Aminu, Absolution, Abloom Skincare, Aimée de Mars, Ananné, Avirida, Auromère, Aurelia London, Africa organics, Arman prana, AnnemariE Börlind, Aphyta Natural Skincare
B	14	Beauty Kitchen, Baya Baya, Bareluxe Skincare, Beyond Men's Care, Bare i Natural, Booming_Bob, Bitener, Beauty Kubes, Bybi, By Selina, Beauty Cleanse, Botanique, Bee Honest Cosmetics de Traay, Bare Nature Japan
C	5	Choose, Catteco, Chi Natural Life, Cîme Honestly Beautiful, Conscious Chemist
D	4	De Kleine Badkamer, Dee's Organics, Dr. Lipp, Dado Sens
E	6	Eau Thermale Jonzac, Eco.Kid, Earth Line, Envy u Skincare, Esse Probiotic Skincare, ever kind
F	5	Fill refill Co., Fat Forest, Fresh Therapies, Food for Skin, Flow cosmetics
G	3	Gaia & Vie, Goodness, Green People,
H	6	Happy Soaps, Happy Earth, Happy habit, Hemp Collective, HelemaalShea, Harmonie
I	4	Ivy & Lou Lou, Into the Beard, Inika, Iconic Elements
J	3	Jungle Culture, Juni Cosmetics, Jimseido
K	6	Koseli, Kind2, Korento, Kinvara, Kaerel, Karma mama
L	9	La Pink, Living Nature, Loveli, Loïs Lee, Lady Green, Lemon Pharma, Linden Leaves, Logona, Loverock
M	10	Marcel's Green Soap, Muze, Maryve Natural Skincare, Mary's Jane, Mane Methods, Mgc Derma, Miracles by Stella, Marley's Amsterdam, Magical Tree, Mas Newen, Mokosh
N	12	Namza, Nature Bar, Noé Skincare, Nono Lab, Neal's Yard Remedies, Nuud, Naïf, Niyok, Novexpert, Naissance, Naya, Needcrystals
O	1	Odylique
P	5	Plume, Pipper, Puuur, Petit & Jolie, Plantae
R	4	Ringana, Relevant, R10labs, Roos Cosmetics
S	22	Seepje, Superzero, Seventy One Percent, Scrubbee, Sol de Ibiza, So'bio étic, Skirr Skin, Sooth Care, SkinR Natural Care, Smpl Skin, Shrub, Sofia Latif, Smyle, Skin w/o, Sante, Sanatio Naturalis, Self, Silk Therapeutics, Soap7, Sea Line, Speick, Sunbutter
T	7	Toot!, The Ohm Collection, Tanorganic, Trashy Soaps, Tribe Skincare, Trilogy, The jojoba
U	5	Urb'n Nature, UpCircle Beauty, Uoga Uoga, Urban Botanicals, Urtekram
V	1	Valley Mist
W	8	Weleda, Witlof Skincare, We Love the Planet, Wild Sage + Co, White Rabbit Skincare, Whitfords, Werfzeep, Wol(k)
Y	1	Yepoda
Z	2	Zero Waste Path, Zindeo

ACC - Plastics Industry Association and American Chemistry Council 2024b), through the implementation of a program specific to each country, where the company agrees to carry out a series of activities to achieve zero pellet loss (PLASTICS and ACC - Plastics Industry Association and American Chemistry Council 2024a).

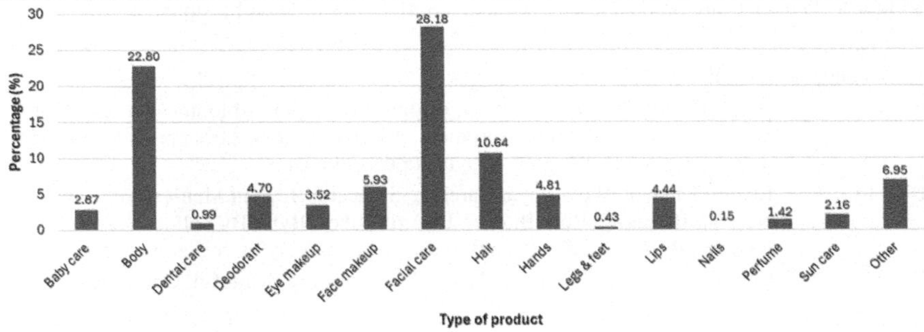

Fig. 8.2 Types of personal care products without microplastics

OCS's international expansion began in 2010 and currently has more than 22 active programs in five continents, more than 65 countries, and around 4000 companies. By 2024 it is planned to continue expanding with the inclusion of programs in India, China, and Ghana (PLASTICS and ACC - Plastics Industry Association and American Chemistry Council 2024a). Table 8.8 presents a summary of the different countries/regions that already adhered to the OCS, as well as information on the license holder and main achievements.

8.5.3 Manufacture of Products from Marine Plastic Waste

This latest initiative focuses on the recovery of plastic waste from some natural ecosystems for its subsequent transformation into new products. It is important to highlight that in this type of initiative, the collaboration of different actors such as NGOs, government, companies, and the public is necessary. Here are two examples.

NetPlus®

NetPlus® is the raw material obtained from the recycling of worn and broken fishing nets through an alliance between Bureo (private sector), fishermen, and different companies focused on the manufacturing of clothing, shoes, sunglasses, and water sports accessories (Bureo 2024d). The process to obtain the raw material consists of the following stages (Bureo 2024b):

Stage 1: Collection. Through collection programs in fishing communities in nine countries (Chile, Peru, Panama, Argentina, Ecuador, Mexico, United States, Seychelles, and Japan), fishermen hand over their fishing nets, which have reached the end of their useful life, in specific centers and they receive a payment based on the number of networks they deliver (Bureo 2024c).

Stage 2: Pretreatment. The nets in turn go through a sequential process of classification, cleaning, crushing, and packaging at Bureo facilities. They are then sent to the recycling partners with whom the Bureo works.

Stage 3: Pelletization. Through the depolymerization process, the networks are transformed into pellets of nylon or HDPE, which are called NetPlus.

Stage 4: Manufacturing of new products. Nylon pellets go through an extrusion process to produce high-quality threads which are used to manufacture high-performance fabrics. For its part, HDPE pellets are used for the manufacture of injection molded products, such as sunglasses, surf fins, board games, and frisbees. Bureo currently has more than 25 brand partners, including Patagonia, Costa, Yeti, futures, carver, Rei Co-op, Rivian, Dakine, Orvis, Outerknown, Quiksilver, allbirds, Jenga, Trek, Eddie Bauer, Mountain Hardwear, L.L. Bean, Smartwool, prAna, and Bass Pro Shops. In addition, to date, more than 3.9 million products manufactured with NetPlus have entered the market, including outerwear, activewear, hats, caps, sunglasses, shorts, and fins (Bureo 2024a), so it is common for the aforementioned brands to promote these products on their websites (Costa del Mar 2024; Futures 2022; Patagonia 2024; YETI COOLERS 2024).

Adidas X Parley

Parley for the Oceans is an environmental organization and network whose goal is to raise awareness about the beauty and fragility of the oceans and develop projects to end their destruction. This organization works together with communities, companies, governments, and environmentalists based on creativity, collaboration, and eco-innovation (Parley for the Oceans 2024b). One of the initiatives with companies is "Adidas & Parley", announced on Earth Day in 2015 with the presentation to the United Nations of the "Ultra Boost" footwear prototype whose upper part was made of threads and filaments recovered from ocean waste and illegal deepwater gillnets (Parley for the Oceans 2024a). From the above, Parley created Ocean Plastic® which consists of a range of high-quality materials for the sports, fashion, design, and luxury industries manufactured from intercepted and recycled marine plastic waste (Parley for the Oceans 2024d). To be able to manufacture it, there are a series of previous steps that include: collecting plastic waste through cleanups on beaches, remote islands, rivers, mangroves, wetlands, and coastal communities; classification carried out by partners in each country which consists of grouping by type and color; recycling carried out by certified partners who clean, flake and pelletize the plastics in various colors; obtaining intermediate products which consist of threads and fabrics with a variety of colors, masses, and finishes; and manufacturing of final products such as accessories, apparel, eyewear, footwear, furniture, among others (Parley for the Oceans 2024c). As a result, Adidas no longer distributes plastic bags in its retail stores, no longer sells microbead body care products, and has suspended the use of plastic bottles at all meetings at its headquarters.

Additionally, in 2017 adidas manufactured one million pairs of shoes with Parley Ocean Plastic®, subsequently increasing to five million in 2018, 11 million in 2019, and around 15 million in 2020 (Parley for the Oceans 2024a). It is worth noting that several other brands and artists have also begun to manufacture their products using Parley Ocean Plastic® materials. Among these are Dior, American Express, Byborre, the Australian FinTech bank WLTH, as well as Niccolo Casas and Iris van Herpen (Parley for the Oceans 2024d).

8.6 The Role of Non-Governmental Organizations

Plastic pollution in the seas and oceans has reached alarming levels and is expected to increase in the future (Jambeck et al. 2015). The problem is extremely complex, ranging from inadequate waste disposal, lack of regulation in the production, and use of plastics, to low environmental awareness (Jambeck et al. 2015). This crisis represents a serious threat to marine and terrestrial ecosystems (Geyer et al. 2017). However, there are individuals and organizations that have come together with a common vision: plastic-free oceans, through the implementation of various actions that seek to understand and solve the problem. Although the idea of plastic-free oceans may seem utopian, these organizations maintain their conviction to fight and work for a sustainable future for these ecosystems. In this section, we will delve into the participation and impact of NGOs in the fight against plastic pollution in the seas and oceans.

In the fight against plastic pollution, we find diverse actors such as government, industry, non-governmental civil society organizations (NGOs), and academia. These actors face one of the greatest challenges in recent environmental history: making the seas plastic-free. Each actor focuses on different strategies and scales throughout the entire plastic waste management cycle.

The government contributes by implementing public policies aimed at reducing consumption, improving waste disposal, and mitigating pollution. Citizen participation, through environmental NGOs, is responsible for covering the aspects that the government leaves unattended. Academia, on the other hand, is mainly dedicated to describing the problem and developing technologies for cleaning up the seas. In general, all these actors carry out various actions to remedy the problem with a sense of urgency.

In this context, NGOs have a key function in driving collaborative cleanup efforts, gathering resources and expertise, and tackling environmental issues (Vince and Hardesty 2018). NGOs, with their ability to operate independently of governments and their flexibility to adapt to different contexts, are particularly effective in implementing environmental projects (Haward 2018). Through awareness campaigns, their political advocacy, and the implementation of practical projects, these organizations act as catalysts for change, encouraging greater citizen participation and multi-sector collaborations (Vince and Hardesty 2018).

Throughout their history, NGOs have triggered significant social changes, such as the fall of the Pinochet dictatorship in Chile and the struggle against apartheid in South Africa (Harvard Law School n.d.). NGOs, as their name suggests, are organizations independent from government, generally focused on solving social, cultural, or environmental issues (Agencia de la ONU para los Refugiados (ACNUR) España 2017; Habitats for Humanity n.d.). They may arise from civil society initiatives with the aim of generating change at the community, regional, or international level (Agencia de la ONU para los Refugiados (ACNUR) España 2017).

NGOs and the government are not separate entities; NGOs often influence decision-making, act as consultants, and sometimes collaborate with the government to meet their common goals. One of the most notable characteristics of NGOs is that they do not seek to make a profit; all benefits, including monetary ones, must go to the target community to which they direct their work (Habitats for Humanity n.d.). In addition, they encourage the effort and participation of communities as drivers of development to boost their independence and self-sustainability (Agencia de la ONU para los Refugiados (ACNUR) España 2017).

The inaction of some governments in the face of environmental deterioration and mismanagement of natural resources has led NGOs to take actions of greater impact and transcendence in the climate fight. With their unique role, NGOs serve as critical forces in organizing large-scale remediation and spearheading efforts to address environmental threats. Thanks to their work, issues such as ocean pollution and excessive plastic consumption have gained greater visibility in the media.

Today, there are multiple NGOs dedicated to the preservation of the environment, including the ocean. However, in this section we will focus on those non-profit organizations that focus exclusively on plastic pollution in seas and oceans, leaving for-profit organizations out of the scope of this analysis.

Generally, NGOs working on behalf of the ocean do not focus their efforts on a single task but tend to address the problem from multiple approaches. Some are dedicated to making the problem visible through awareness and education campaigns, documentaries, and scientific dissemination; others focus on practical solutions such as cleaning up seas and beaches; smaller group conducts research on the problem; and finally, others promote and develop public policies. In Table 8.5 we listed some examples of different NGOs focused on addressing plastic pollution in oceans.

Within the awareness and education category, we find the NGO Blutopia, an association focused on making documentaries about plastic pollution in the ocean (Blutopia n.d.). We also have Algalita, a pioneer in this field, who was the first to draw attention to the now-called "Great Pacific Garbage Patch" almost 30 years ago. Today, it focuses on research and education for the prevention of marine plastic pollution (Algalita, n.d.). Another NGO that conducts research is Expedition seventh Continent, noted for its work in collecting data and understanding the problem through scientific expeditions.

NGOs focused on cleaning beaches and oceans tend to be highly visible and recognized for the great impact they generate, as many of them have even removed tons of plastic from the sea. Such is the case of The Ocean Cleanup, which concentrates its efforts on cleaning the great Pacific garbage patch from which it has removed 452,269 kg in total with its patented System 03 system. This NGO stands out for its development of innovative technology to collect plastic trash and recycle it once collected (The Ocean Cleanup 2024). Another notable organization in this category is the Ocean Voyages Institute, which in 2020 conducted the largest open ocean cleanup in history, collecting about 170 tons of plastics (Ocean Voyages Institute (OVI) 2024). Another NGO, Take 3 for the Sea has removed 56 million items of rubbish (Take 3 for the Sea 2024).

Finally, we can talk about NGOs that have a closer relationship with the government, promoting public policies; here we find the NGO 5 Gyres, an association that has contributed to the approval of several local and state bills, such as the ban of

Table 8.5 Examples of NGOs that carry out activities aimed at cleaning plastic pollution from the oceans

Activities	NGO	Year	Relevant projects/achievements
Awareness and education	Blutopia	2018	The film "*eating: The power to save the ocean*" and the series "*the new convenience*" in 2020
	Algalita	1994	The film "*the story of plastic*" in 2020
	Expedition seventh continent (Expédition 7eme Continent 2024)	2010	The 2024 educational tour in Occitanie
	Lonely whale	2015	Unwrap the future; NextWave plastics
	Oceanoplastic	NS	NS
	Plastic oceans	NS	Blue communities, trees and seas festival
	Plastic soup	<2017	Beat the microbead and world cleanup day
	Stop plastic pollution	NS	NS
Research and data collection	Expedition seventh continent (Expédition 7eme Continent 2024)	2010	Blue exploration (L'exploration bleue)
	eXXpedition	2014	eXXpedition round the world
	5 gyres	<2004	Investigating additive leaching from bioplastics and "better alternatives 3.0: a case study on bioplastic products and packaging"
	Stop plastic pollution	NS	NS
Practical solutions	Ocean voyages institute	1979	Project Kaisei
	Oceanoplastic	NS	Multiple cleanings in beaches
	Ocean sole	NS	Transform plastic into art
	The ocean cleanup	2013	System 03
	The Seacleaners	2016	Expedition Mobula
	Take 3 for the sea	NS	Take 3 clean up
Public policies	5 gyres	<2004	Led the #BeadFree campaign that helped pass the micro bead-free waters act of 2015
	Oceana	2001	Initiative for United States to eliminate single-use plastics across the Federal Government

expanded polystyrene in Los Angeles and the returnable bottle systems in California, both proposals with a great impact on the generation of plastic waste.

Table 8.5 lists 16 associations according to their activities. As can be seen, there are more associations working in the field of education and awareness of the problem, most of the organizations in this category are dedicated to giving workshops, talks, and dissemination of the problem, which implies a lower monetary investment, and a smaller organization compared to those that carry out research expeditions where it is necessary to pay the expenses of the expedition. These awareness-raising activities help to understand and make the problem visible, in order to prevent pollution from consumption by society. On the other hand, for research activities and practical solutions, where associations focus mainly on the removal of plastics from beaches and oceans, the category of public policies is where we found fewer associations, possibly because political management can be complicated and requires a long time to pass a law. Of all the NGOs investigated, only Plastic Soup is fully focused on the fight against microplastics. With its campaign "Beat the microbead" it managed to get microbeads banned in 19 countries; no less important was the congress "The Plastic Health Summit", organized in 2019, which was the first of its kind worldwide, where the effects of plastic on health were discussed (Plastic Oceans International n.d.). Some others, such as The Ocean Cleanup, mention in their speech how macroplastics can become microplastics but do not focus on these. There is no single NGO that has a unique solution to this problem alone can achieve a plastic-free ocean, however, through their diverse initiatives and visions, such as the promotion of environmental awareness, the implementation of cleanup and removal projects, and scientific research, NGOs have proven to be key agents in the fight against marine plastic pollution. The positive impact of NGOs in the remediation of marine plastic pollution is evident in the successful projects they have led, and their ability to unite communities, governments, and private actors around a common cause. However, to maximize results, these organizations must receive the necessary support, both financial and political, and meet operational challenges with innovation and perseverance.

8.7 Blue Futures: A Necessity, Not an Option

In a current context in which sustainability has acquired greater importance and in which urgent measures are required for the care of ecosystems, the concept of "Blue Future" emerges as a crucial measure to ensure the health of the oceans and the prosperity of coastal areas and communities through harmony between economic and social development, and the preservation and restoration of marine ecosystems (Hart et al. 2018; Zhao and Li 2023). The ocean is essential not only for the great global biodiversity it contains but also for the subsistence of millions of people worldwide. It is currently threatened by various problems such as climate change, overexploitation of marine resources, pollution by plastic waste and microplastics, and the destruction of coastal habitats, among others. If any immediate action measure is implemented, marine ecosystems will continue to deteriorate until they reach a point of no return in which the consequences for biodiversity and coastal communities will be devastating (Hart et al. 2018; Zhao and Li 2023). The development and implementation of a sustainable future is essential not only to stop the deterioration of coastal ecosystems, preserve food security, and contribute to economic growth but also to ensure the well-being of millions of people around the world whose livelihood is linked to the ocean (Queirós et al. 2024; Yu et al. 2024). A Blue Future must be carried out through the adoption of practices and policies that promote the conservation and sustainable use of marine resources such as the protection of key ecosystems such as coral reefs and mangroves (fundamental for the development of biodiversity and protection from natural phenomena), strict regulations against overfishing, promotion of sustainable fishing and aquaculture practices (that allow maintaining ecological balance), designation of marine protected areas (MPA), collaboration in joint research projects between different nations and the development of alliances between civil society, industry, government and academia (Duarte et al. 2020; Hart et al. 2018; Zhao and Li 2023).

Some of the activities that need to be implemented to ensure the success of the Blue Future are described in Table 8.6.

Although the Blue Future only considers certain activities and actions related to the marine environment for now, industries with the greatest potential for pollution and destruction, such as oil and gas extraction, deep-sea mining, and the shipping industry, are being left aside, so measures are required to include these industries within the parameters of the Blue Future (Hart et al. 2018; Pace et al. 2023; Yu et al. 2024). Additionally, it has been suggested to consider all those related to water, such as freshwater management, processing and distribution of drinking water, and wastewater treatment, to mention a few (Hart et al. 2018).

At a time when the future of ecosystems does not look promising, and the damage to the ocean is numerous and serious, it is of special environmental, economic, and social importance to protect and guarantee the health of marine ecosystems

Table 8.6 Activities related to the blue future

Activity	Goal	Benefits	References
Sustainable fishing	Guarantee that fish and shellfish populations are maintained at adequate levels and that there is no overexploitation	Ensures a balance of fish and shellfish populations, as well as a constant supply of food for humans and a source of employment for fishermen	Hart et al. (2018)
Sustainable aquaculture	Breed and cultivate marine organisms while reducing environmental impacts and habitat degradation	Provides an additional source of income and food and reduces pressure on wild fish populations	Zhao and Li (2023)
Restoration of ecosystems	Restore marine ecosystems to protect the habitats of thousands of species and reduce the impacts of climate change	Increases natural barriers against natural phenomena	Duarte et al. (2020), Pace et al. (2023)
Ecotourism	Ensure that tourism practices carried out in marine ecosystems are sustainable and cause the least negative environmental impact on the areas	Provides a source of income for the coastal community and preserve marine ecosystems, the beauty of coastal landscapes, and the culture and traditions of communities	Amon et al. (2022)
Delimitation of marine protected areas (MPA)	Restrict, through specific regulations and standards, certain human activities that may damage the ecosystem of a certain area of the ocean	Preserves the biodiversity of the area, promotes the recovery of species populations, and encourages sustainable fishing	Hart et al. (2018), Queirós et al. (2024)

through the collaboration of different international sectors and local communities (Hart et al. 2018; Queirós et al. 2024; Yu et al. 2024).

The Blue Future will bring with it not only the health of marine ecosystems but also various opportunities such as the creation of new jobs, the economic growth of coastal communities, the development and use of new technological applications such as artificial intelligence and the Internet of things (used to make projections or identify potential risks of marine pollution, to name a few), economic collaborations and the creation of new lines of multidisciplinary research (Pace et al. 2023; Yu et al. 2024).

In some countries, the blue future has already been implemented through different strategies such as the designation of marine protected areas and the application of regulations on fishing and aquaculture practices, among others. Table 8.7 are

Table 8.7 Countries that have implemented the blue future concept

Country	Location	Initiative	Reference
Norway	In northern Europe, in the region known as Scandinavia	Norway has established protected areas and strict regulations for fishing. The country also promotes the blue economy and sustainable aquaculture	Gullestad et al. (2017), Norwegian Ministry of Trade (2019)
Australia	It is in the southern hemisphere, between the Indian and Pacific oceans	Australia has implemented a comprehensive plan that seeks to protect the Greta barrier reef by restricting nearby industrial activities, controlling fishing, and restoring coral reefs, among others	Coggan et al. (2021)
Seychelles	It is in the Indian Ocean, northeast of Madagascar	Seychlees has created an innovative climate change conservation and adaptation trust fund (SeyCCAT) that supports projects such as the expansion of marine areas, sustainable fisheries, biodiversity conservation and protection, and climate change adaptation	The Commonwealth Blue Charter (2018)
Fiji	It is an archipelago composed of more than 300 islands located in the South Pacific Ocean in the Oceania region	Fiji has developed several marine conservation agreements (MCAs) between tourism operators and indigenous resource-owning communities in Fiji, which aim to conserve biodiversity and manage sustainable fisheries. Strategies for mangrove restoration have also been implemented in the country	Devi and Holland (2024), Mangubhai et al. (2020)
Costa Rica	Located in Central America, bordered by the Caribbean Sea to the east and the Pacific Ocean to the west	Between 1960 and 1970, the Tortuguero National Park (PNT) was created in the country to conserve and research sea turtles and promote ecotourism in the town of Tortuguero in Costa Rica	Gutiérrez-Lince et al. (2020)

listed some countries that have succeeded in the blue future practices they have implemented in their territories.

The "Blue Future" represents a pivotal framework for safeguarding the health of marine ecosystems while fostering the sustainable development of coastal communities. As the ocean faces mounting threats from overexploitation, pollution, and habitat destruction, it is imperative that global efforts focus on holistic strategies that balance economic progress with environmental stewardship. The successful implementation of practices like sustainable fishing, ecosystem restoration, and the protection of marine areas demonstrates the potential for meaningful change. However, to fully realize the goals of the Blue Future, it is crucial to expand these efforts to include industries with significant environmental impact, such as oil extraction and deep-sea mining. By embracing a collaborative, interdisciplinary approach, we can not only protect marine biodiversity but also create new economic opportunities, promote innovation, and ensure a prosperous and resilient future for communities worldwide.

Table 8.8 Relevant information from the countries participating in the Operation Clean Sweep program. Elaborated with information from PLASTICS and ACC - Plastics Industry Association and American Chemistry Council (2024a)

Table 8.8 Relevant information from the countries participating in the Operation Clean Sweep program. Elaborated with information from PLASTICS & ACC, 2024a

ID	Country or region	License holder and year established	Achievements
1	Argentina	Cámara Argentina de Industria Plástica (CAIP) – 2019	90 companies
2	Brazil	Associação Brasileira da Indústria do Plástico (ABIPLAST) & Instituto Socioambiental do Plastico (Plastivida) – 2014 and 2019	63 companies, more than 70 of them reported data in 2023
3	Canada	Chemistry Industry Association of Canada (CIAC) – 2020	50 companies. In 2023 an OCS category for transportation partners launched and around 130 data reporting and facility assessments were collected
4	Chile	Association of Plastic Industry Chile (ASIPLA) – 2022	20 companies
5	Colombia	Colombia Acoplasticos – 2016	8 companies, four of them reported successful data in 2023
6	Denmark	Danish Plastics Federation – 2013	132 companies. OCS Europe Certification Scheme documents were translated into Danish for greater ease of implementation by companies
7	Europe	Plastics Europe & European Plastics Converters – 2013	2,900 companies in 48 countries. In 2023 initiated a data reporting program and 40 companies reported OCS Core metrics

(continued)

Table 8.8 (continued)

ID	Country or region	License holder and year established	Achievements
8	Gulf region	Gulf Petrochemical & Chemical Association (GPCA) – 2020	34 companies in 6 countries
9	Malaysia	Malaysian Plastics Manufacturers Association (MPMA) – 2023	16 companies
10	Mexico	Asociación Nacional de Industrias del Plástico A.C. (ANIPAC) – 2011	90 companies. In 2023, six companies had external inspections, 27 companies reported data and four companies got OCS blue membership
11	Netherlands	The Dutch Federation of Rubber and Plastics Industry (NRK) – 2015	110 companies. Participation in World Cleanup Day was encouraged in 2023
12	New Zealand	Plastics New Zealand – 2015	75 companies. In 2023 had a National Clean Up Day with more than 100 participants from different plastics companies
13	Poland	Polish Union of Plastics Converters (PUPC) – 2018	144 companies. In 2023 had a webinar with a participation of more than 80 people
14	Portugal	Associação Portuguesa da Industria de Plasticos (APIP) – 2019	57 companies. In 2023 developed a webinar for 2024
15	Singapore	Singapore Chemical Industry Council – 2019	15 companies
16	South Africa	Plastics South Africa – 2012	17 companies
17	Spain	Asociación Española de Industriales de Plásticos (ANAIP) – 2016	456 companies. In 2023 organized a symposium of good practices for the prevention of pellets releases with more than 100 participants
18	Turkey	Turkish Plastics Industry Foundation (PAGEV) –2016	12 members. In 2023 had an external inspection program and designed a new website
19	United Kingdom	British Plastics Federation – 2010	318 companies of four countries
20	United States	ACC's Plastics Division and Plastics Industry Association – 1991 and blue membership in 2018	250 combined OCS, OCS Blue, and OCS logistics members

8.8 Conclusions

The chapter tries to establish a comprehensive framework to address the crisis of plastic pollution in the oceans, emphasizing the collaboration among multiple stakeholders as a crucial element for change. First, the alarming state of plastic pollution in seas and oceans is highlighted as a pressing issue that demands immediate attention due to its devastating effects on marine and terrestrial ecosystems. The active participation of non-governmental organizations (NGOs) emerges as a vital component in this struggle, operating on various fronts, from public awareness to the implementation of effective environmental policies. NGOs not only mobilize resources and knowledge but also foster greater citizen participation and multi-sector collaborations. Additionally, the concept of "Blue Future" is presented as an urgent necessity to ensure the health of the oceans and the well-being of coastal communities. This approach advocates for the adoption of practices and policies that promote the conservation of marine resources, restoration of ecosystems, and establishment of marine protected areas. Initiatives in sustainable fishing and aquaculture are key examples of how economic development can be balanced with environmental protection. However, we also tried to note the need to extend these strategies to address pollution generated by industries with significant environmental impact, such as oil extraction and deep-sea mining. Throughout the chapter, it is emphasized that there is no single solution to plastic pollution; instead, a combination of awareness initiatives, research, and practical action is required. NGOs have proven their effectiveness by executing concrete cleanup and awareness projects, while collaboration with governments and local communities enhances the impact of their efforts. The successful implementation of sustainable practices in countries such as Norway, Australia, and Costa Rica exemplifies how policies aimed at a "Blue Future" can yield tangible benefits for both the environment and communities. Finally, we want to reiterate that the fight against plastic pollution and the protection of oceans is not just an environmental challenge but also a matter of social and economic justice. By integrating collaborative and multidisciplinary approaches, a future can be built in which the health of our oceans and the well-being of coastal communities are interlinked, ensuring a sustainable legacy for future generations. The unification of efforts and collective commitment are essential to guarantee the prosperity and resilience of our oceans, as well as to promote equitable and sustainable development that benefits all.

References

Amon D, Metaxas A, Stentiford G, Escovar-Fadul X, Walker TR, Diana Z, Karathanasi F, Voyer M, & Hemery L (2022) Blue economy for a sustainable future. One Earth, 5(9):960–963. https://doi.org/10.1016/j.oneear.2022.08.017

Agencia de la ONU para los Refugiados (ACNUR) España (2017) Una ONG: ¿Cuál es su función social? EACNUR. https://eacnur.org/es/blog/una-ong-funcion-social. Accessed 13 June 2024

Ajith, N., Arumugam, S., Parthasarathy, S., Manupoori, S., & Janakiraman, S. (2020). Global distribution of microplastics and its impact on marine environment—a review. Environmental Science and Pollution Research, 27(21):25970–25986. https://doi.org/10.1007/S11356-020-09015-5/METRICS

Algalita (n.d.) Accomplishments. https://algalita.org/accomplishments/. Accessed 13 June 2024

Alvarez-Zeferino, Juan Carlos, Sara Ojeda-Benítez, Arely Areanely Cruz-Salas, Carolina Martínez-Salvador, & Alethia Vázquez Morillas (2020) "Dataset of Quantification and Classification of Microplastics in Mexican Sandy Beaches." Data in Brief 33:106473. https://doi.org/10.1016/j.dib.2020.106473

Alvarez-Zeferino, Juan Carlos, Arely Areanely Cruz-Salas, Hernández-Soriano Ana Isabel, & Alethia VazquezMorillas (2023) "Ciencia Ciudadana. Guía Para La Recolección de Microplásticos En Playas." https://doi.org/10.13140/RG.2.2.15958.46404

Bakir, Adil, Alexandra R Mcgoran, Briony Silburn, Josie Russell, Holly Nel, Amy L Lusher, Ruth Amos, et al (2024) "Creation of an International Laboratory Network towards Global Microplastics Monitoring Harmonisation." Scientific Reports, 1–17. https://doi.org/10.1038/s41598-024-62176-y

Beaumont NJ, Aanesen M, Austen MC, Börger T, Clark JR, Cole M, Hooper T, Lindeque PK, Pascoe C, & Wyles KJ (2019) Global ecological, social and economic impacts of marine plastic. Marine Pollution Bulletin, 142:189–195. https://doi.org/10.1016/J.MARPOLBUL.2019.03.022

Bennett NJ, Blythe J, White CS, & Campero C (2021) Blue growth and blue justice: Ten risks and solutions for the ocean economy. Marine Policy, 125:104387. https://doi.org/10.1016/J.MARPOL.2020.104387

Blutopia (n.d.) Événements. https://blutopia.org/evenements/. Accessed 12 June 2024

Borrelle SB, Ringma J, Lavender Law K, Monnahan CC, Lebreton L, McGivern A, Murphy E, Jambeck J, Leonard GH, Hilleary MA, Eriksen M, Possingham HP, De Frond H, Gerber LR, Polidoro B, Tahir A, Bernard M, Mallos N, Barnes M, & Rochman CM (2020) Predicted growth in plastic waste exceeds efforts to mitigate plastic pollution. Science, 369(6509):1515–1518. https://doi.org/10.1126/SCIENCE.ABA3656/SUPPL_FILE/ABA3656-BORRELLE-SM-DATA-S4.CSV

Browne MA, Crump P, Niven SJ, Teuten E, Tonkin A, Galloway T, Thompson RC (2015) Accumulation of microplastic on shorelines worldwide: sources and sinks. Environ Sci Technol 45(21):9175–9179. https://doi.org/10.1021/es201811s

Bureo (2024a) Brand partners. https://bureo.co/brand-partners

Bureo (2024b) How it works. Processing. https://bureo.co/how-it-works#processing

Bureo (2024c) How it works. Trace https://bureo.co/how-it-works#traceability

Bureo (2024d) NetPlus® material. Fabric. https://bureo.co/material#fabric

Carter RL (2010) The history and philosophy of environmental education. In: The inclusion of environmental education. Science teacher education. Springer, Cham, pp 3–16

Change.org (2019) Petición sign this petition and tell the EU to ban microplastics in cosmetics! https://www.change.org/p/european-commission-sign-this-petition-and-tell-eu-to-ban-microplastics-in-cosmetics

Coggan A, Thorburn P, Fielke S, Hay R, & Smart JCR (2021) Motivators and barriers to adoption of Improved Land Management Practices. A focus on practice change for water quality improvement in Great Barrier Reef catchments. Marine Pollution Bulletin, 170:112628. https://doi.org/10.1016/j.marpolbul.2021.112628

Comisión Europea (2023) Restricción de los microplásticos en la UE a partir del 17 de octubre de 2023. https://trade.ec.europa.eu/access-to-markets/es/news/restriccion-de-los-microplasticos-en-la-ue-partir-del-17-de-octubre-de-2023

Costa del Mar (2024) Untangle our oceans. https://www.costadelmar.com/en-us/inside-costa/protect/untangle-our-oceans

Cruz Salas AA, Velasco Pérez M, Tecorralco Bobadilla A L, Vázquez-Morillas A, & Espinosa-Valdemar RM (2024) Human toxicity of nano and microplastics. In Inamuddin (Ed.), Toxic effects of micro- and nanoplastics: environment, food and human health (1st ed., p. 610). Wiley.

Devi J, & Holland E (2024) Building resilience in a climate crisis: best practices for mangrove restoration along the Coral Coast, Fiji. Journal of Interdisciplinary Research, 8(1).

Duarte CM, Agusti S, Barbier E, Britten GL, Castilla JC, Gattuso JP, Fulweiler RW, Hughes TP, Knowlton N, Lovelock CE, Lotze HK, Predragovic M, Poloczanska E, Roberts C, & Worm B (2020) Rebuilding marine life. Nature, 580(7801):39–51. https://doi.org/10.1038/s41586-020-2146-7

European Union (2023) Commission regulation EU 2023/2055. https://eur-lex.europa.eu/legal-content/EN/TXT/?uri=OJ%3AJOL_2023_238_R_0003&qid=1629107278018

Expédition 7eme Continent (2024) Accueil. https://www.septiemecontinent.com/. Accessed 23 September 2024

Exxpedition (2024) About. https://exxpedition.com/expedition/round-the-world/. Accessed 12 June 2024

FDA (2022) "The Microbead-Free Waters Act: FAQs." https://www.fda.gov/cosmetics/cosmetics-lawsregulations/microbead-free-waters-act-faqs

EPA (2020) "Reducing and Reusing Basics I US EPA." U.S. Environmental Protection Agency. 2020. https://www.epa.gov/recycle/reducing-and-reusing-basics

Flores-Ruiz I, Yang S, Edson W, Wenham D, & Posacka A (2022) "COMMUNITY SCIENTISTS HELP IDENTIFY LOCAL PRIORITIES FOR MICROPLASTICS: A PILOT STUDY ON VANCOUVER ISLAND, BRITISH COLUMBIA, CANADA." VANCOUVER ISLAND, BRITISH COLUMBIA, CANADA. https://24307406.fs1.hubspotusercontentna1.net/hubfs/24307406/Ocean Diagnostics Community Science Pilot Study Scientific Report 2022.pdf.

Futures (2022) Alpha netplus. https://futuresfins.com/pages/alpha-netplus

Geyer R, Jambeck JR, Law KL (2017) Production, use, and fate of all plastics ever made. Sci Adv 3(7):e1700782. https://doi.org/10.1126/sciadv.1700782

Golwala, Harmita, Xueyao Zhang, Syeed Md Iskander, & Adam L Smith (2021) "Solid Waste: An Overlooked Source of Microplastics to the Environment." Science of The Total Environment 769:144581. https://doi.org/10.1016/j.scitotenv.2020.144581

Gullestad P, Abotnes AM, Bakke G, Skern-Mauritzen M, Nedreaas K, & Søvik G (2017) Towards ecosystem-based fisheries management in Norway – Practical tools for keeping track of relevant issues and prioritising management efforts. Marine Policy, 77(December 2016):104–110. https://doi.org/10.1016/j.marpol.2016.11.032

Gutiérrez-Lince J, Dorantes Palacios M, & Valverde RA (2020) Case Study: The Evolution of Tourism and Sea Turtle Conservation at Tortuguero National Park, Costa Rica. Sea Turtle Research and Conservation: Lessons From Working In The Field, 105–111. https://doi.org/10.1016/B978-0-12-821029-1.00011-8

Habitats for Humanity (n.d.) Differences between for-profit and nonprofit organizations. Habitat Broward. https://habitatbroward.org/blog/differences-between-for-profit-and-nonprofit-organizations/. Accessed 23 September 2024

Halpern BS, Walbridge S, Selkoe KA, Kappel CV, Micheli F, D'Agrosa C, Bruno JF, Casey KS, Ebert C, Fox HE, Fujita R, Heinemann D, Lenihan HS, Madin EMP, Perry MT, Selig ER, Spalding M, Steneck R, & Watson R (2008) A global map of human impact on marine ecosystems. Science, 319(5865):948–952. https://doi.org/10.1126/SCIENCE.1149345/SUPPL_FILE/HALPERN_SOM.PDF

Hart M, Conathan M, Johnson B, & Polefka S (2018) Blue Future. Mapping Opportunities for U.S.-China Ocean Cooperation. May, 1–54.

Harvard Law School (n.d.) Nongovernmental organizations (NGOs). Bernard Koteen Office of Public Interest Advising. https://hls.harvard.edu/bernard-koteen-office-of-public-interest-advising/about-opia/what-is-public-interest-law/public-service-practice-settings/international-public-interest-law-practice-setting/nongovernmental-organizations-ngos/. Accessed 13 June 2024

Haward M (2018) Plastic pollution of the world's seas and oceans as a contemporary challenge in ocean governance. Nat Commun 9(1):667. https://doi.org/10.1038/s41467-018-03104-3

Jambeck JR, Geyer R, Wilcox C, Siegler TR, Perryman M, Andrady A, Law KL (2015) Plastic waste inputs from land into the ocean. Science 347(6223):768–771. https://doi.org/10.1126/science.1260352

Karasik R, Vegh T, Diana Z, Bering J, Caldas J, Pickle A, Rittschof D, & Virdin J (2020) 20 Years of Government Responses to the Global Plastic Pollution Problem The Plastics Policy Inventory 20 Years of Government Responses to the Global Plastic Pollution Problem. Duke Nicholas Institute for Environmental Policy Solutions, 1–309.

Lee, Ingyu, Sanoev Khujaniyoz, Heekyong Oh, Hyunook Kim, and Taehyeok Hong. 2024. "The Removal of Microplastics from Reverse Osmosis Wastewater by Coagulation." Journal of Environmental Chemical Engineering 12(4):113198. https://doi.org/10.1016/j.jece.2024.113198

Maes T, Wienrich N, Weiand L, & Cowan E (2023) A little less conversation: How existing governance can strengthen the future global plastics treaty. Cambridge Prisms: Plastics, 1, e22. https://doi.org/10.1017/plc.2023.22
Mangubhai S, Sykes H, Manley M, Vukikomoala K, & Beattie M (2020) Contributions of tourism-based Marine Conservation Agreements to natural resource management in Fiji. Ecological Economics, 171(December 2019), 106607. https://doi.org/10.1016/j.ecolecon.2020.106607
Maraniss L, O'Hara K (1986) International coastal cleanup day. Ocean Conservancy, Washington
MARPOL (1973) International convention for the prevention of pollution from ships. International Maritime Organization, London
Moore CJ (1999) Synthetic polymers in the marine environment: a rapidly increasing ecological threat
Moraczewska-majkut, Katarzyna, & Agnieszka Popenda (2024) "Desalination and Water Treatment." E. Wiśniowska, K. Moraczewska-Majkut, W. Nocoń et Al. Desalination and Water Treatment 319 (April):14–16. https://doi.org/10.1016/j.dwt.2024.100462
Naciones Unidas (1972) Declaración de Estocolmo sobre el Medio Humano. Conferencia de las Naciones Unidas sobre el Medio Humano, Estocolmo
Nel, Holly A, Gregory H Sambrook, Rogan Harmer, Rebecca Sykes, Uwe Schneidewind, Iseult Lynch, Stefan Krause, et al (2020) "Citizen Science Reveals Microplastic Hotspots within Tidal Estuaries and the Remote Scilly Islands, United Kingdom." Marine Pollution Bulletin 161 (June). https://doi.org/10.1016/j.marpolbul.2020.111776
NOAA - National Oceanic and Atmospheric Administration (2018) "Marine Debris. It's a Global Problem." Marine Debris It's a Global Problem. 2018. https://oceanservice.noaa.gov/hazards/marinedebris/
Norwegian Ministry of Trade I, & F (2019) Blue Opportunities.
Ocean Conservancy (n.d.) A brief history of the international coastal cleanup. https://oceanconservancy.org/trash-free-seas/international-coastal-cleanup/
Ocean Voyages Institute (OVI) (2024). https://www.oceanvoyagesinstitute.org/home/. Accessed 12 June 2024
Oceanoplastic (n.d.). https://oceanoplastic.org/qui-sommes-nous/. Accessed 7 July 2024
OECD (2016) The Ocean Economy in 2030. https://doi.org/10.1787/9789264251724-en
Organizacion de las Naciones Unidas (1973) Informe de la conferencia de las Naciones Unidas sobre el medio ambiente. Naciones Unidas, New York
Pace LA, Saritas O, & Deidun A (2023) Exploring future research and innovation directions for a sustainable blue economy. Marine Policy, 148(November 2022), 105433. https://doi.org/10.1016/j.marpol.2022.105433
Parley for the Oceans (2024a) Adidas x Parley. https://parley.tv/initiatives/adidasxparley
Parley for the Oceans (2024b) Mission. https://parley.tv/
Parley for the Oceans (2024c) Ocean plastic. How it works. https://parley.tv/oceanplastic-how-it-works
Parley for the Oceans (2024d) Parley ocean plastic®. https://parley.tv/oceanplastic
Patagonia (2024) NetPlus® - patagonia. https://www.patagonia.com/netplus/
Patrol-Nurdle (2021) "About." Nurdle Patrol. 2021. https://nurdlepatrol.org/app/about
Plastic Oceans International (n.d.) Plastic oceans. https://plasticoceans.org/. Accessed 29 June 2024
Plastic Soup Foundation (2024a) About us. https://www.plasticsoupfoundation.org/en/. Accessed 30 June 2024
Plastic Soup Foundation (2024b) About us - who are we? https://www.plasticsoupfoundation.org/en/about-us/
Plastic Soup Foundation (2024c) Beat the microbead. Look for the zero. https://www.beatthemicrobead.org/zero-products/
Plastic Soup Foundation (2024d) Beat the microbead. Product results. https://www.beatthemicrobead.org/product-results/?c=Zero
Plastic Soup Foundation (2024e) Beat the microbead - about us. https://www.beatthemicrobead.org/about-us/

Plastic Soup Foundation (2024f) Beat the microbead - download the free app. https://www.beatthemicrobead.org/download-btmb-app/
PLASTICS & ACC - Plastics Industry Association & American Chemistry Council (2024a) Operation clean sweep 2023 global impact report. https://opcleansweep.org/about/ocs-annual-report/
PLASTICS & ACC - Plastics Industry Association & American Chemistry Council (2024b) Operation clean sweep. About. https://opcleansweep.org/about/
Queirós AM, Talbot E, Msuya FE, Kuguru B, Jiddawi N, Mahongo S, Shaghude Y, Muhando C, Chundu E, Jacobs Z, Sailley S, Virtanen EA, Viitasalo M, Osuka K, Aswani S, Coupland J, Wilson R, Taylor S, Fernandes-Salvador JA, … Popova E (2024) A sustainable blue economy may not be possible in Tanzania without cutting emissions. Science of the Total Environment, 947(January). https://doi.org/10.1016/j.scitotenv.2024.174623
Ritchie, H., & Roser, M. (2020). Plastic pollution. Our world in data https://ourworldindata.org/plastic-pollution
Rodríguez-Villa AG, Lobato-Rocha SLD, Tapia-Fuentes J, Alvarez-Zeferino JC, & Cruz-Salas AA (2024) "Remediation Strategies for Micro (Nano) Plastics." In Toxic Effects of Micro-and Nanoplastics: Environment, Food and Human Health, 505–39. John Wiley & Sons, Inc.
Ryberg MW, Hauschild MZ, Wang F, Averous-Monnery S, & Laurent A (2019) Global environmental losses of plastics across their value chains. Resources, Conservation and Recycling, 151:104459. https://doi.org/10.1016/J.RESCONREC.2019.104459
Sonke JE, Koenig A, Segur T, Yakovenko N (2025) Global environmental plastic dispersal under OECD policy scenarios toward 2060. Sci Adv 11(16):eadu2396
Take 3 for the Sea. (2024). Take 3—A clean ocean is everybody's business. https://www.take3.org/. Accessed 30 June 2024
The Commonwealth Blue Charter (2018) Sustainable Blue Economy Case Study Innovative Financing-Debt for Conservation Swap, Seychelles' Conservation and Climate Adaptation Trust and the Blue Bonds Plan, Seychelles (on-going).
The Ocean Cleanup. (2024). Oceans. https://theoceancleanup.com/oceans/. Accessed 15 June 2024
UNEP (2021a) Drowning in plastics: Marine litter and plastic waste. Vital graphics. United Nations Environment Programme.
UNEP (2021b) From pollution to solution: A global assessment of marine litter and plastic pollution. United Nations Environment Programme.
UNESCO-UNEP (1976) The Belgrade charter. UNESCO-UNEP Environmental Education Newsletter, Connect
United Nations (1972) Stockholm Declaration on the Human Environment. Conferencia de las Naciones Unidas sobre el Medio Humano, Estocolmo
United Nations Organization (ONU) (1992) Agenda 21. Conference on Environment and Development (UNCED). United Nations, Rio de Janeiro
Vince J, Hardesty BD (2018) Governance solutions to the tragedy of the commons that marine plastics have become. Front Mar Sci 5:214. https://doi.org/10.3389/fmars.2018.00214
Wang S (2023) International law-making process of combating plastic pollution: Status Quo, debates and prospects. Marine Policy, 147:105376. https://doi.org/10.1016/j.marpol.2022.105376
Yao, Jingjing, Bolun Yu, Haipu Li, Ying Yang, and Hui Ying Yang (2023) "Effects of Polypropylene Microplastics on Multiple Sulfonamide Removals by the Fenton Process." Chemical Engineering Science 276:118779. https://doi.org/10.1016/j.ces.2023.118779
YETI COOLERS (2024) YETI built for the wild mid pro retro flat brim hat. https://www.yeti.com/apparel/hats/21023006110.html
Yu H, Zahidi I, Fai CM, Liang D, & Madsen DØ (2024) From waste to wealth: The "Blue Circle" vision for a sustainable ocean future. Science Bulletin, 69(8):993–996. https://doi.org/10.1016/j.scib.2024.02.024
Zhao Y, & Li Y (2023) Blue transition for sustainable marine fisheries: Critical drivers and evidence from China. Journal of Cleaner Production, 421(February), 138535. https://doi.org/10.1016/j.jclepro.2023.138535

Microplastic Control by Personal Choices, Individual Actions Matter

9

Carolina Martínez-Salvador, Alethia Vázquez-Morillas, Alejandro Cueva, Jocelyn Tapia-Fuentes, and María Azahara Mesa-Jurado

Abstract

Microplastics are one of the most relevant environmental problems of our lifetime, being a threat to all kinds of living forms. Their diffuse sources, complex transport, and accumulation processes are still not well understood. Human activities are the sole source of microplastics, and complex processes determine their fate in the environment. Knowing that different human actions generate microplastics, in this chapter, the following questions are explored: What are the most microplastic-generating human activities, particularly those related to personal choices? What does personal choice mean from a microplastic-generation perspective? What are the areas where personal choice and public policies intertwine? This chapter analyses the realm of sources and paths from personal

C. Martínez-Salvador (✉) · M. A. Mesa-Jurado
Laboratorio Transdisciplinario para la Sustentabilidad (LaTSu), Departamento de Ciencias de la Sustentabilidad, El Colegio de la Frontera Sur, Unidad Villahermosa, Villahermosa, Tabasco, Mexico
e-mail: carolina.martinez@posgrado.ecosur.mx

A. Vázquez-Morillas
Área de investigación Tecnologías Sustentables, Departamento de Energía, División de Ciencias Básicas a Ingeniería, Universidad Autónoma Metropolitana Azcapotzalco (UAM-A), Mexico City, Mexico
e-mail: alethia@azc.uam.mx

A. Cueva
Departamento de Ciencias de la Sustentabilidad, El Colegio de la Frontera Sur, Unidad Villahermosa, Villahermosa, Tabasco, Mexico
e-mail: alejandro.cueva@ecosur.mx

J. Tapia-Fuentes
Instituto de Ingeniería, Universidad Autónoma de Baja California (UABC) (Engineering Institute), Mexicali, Baja California, Mexico
e-mail: jocelyn.tapia@uabc.edu.mx

M. Oves et al. (eds.), *Marine Microplastics and the Quest for Remediation*,
https://doi.org/10.1007/978-981-96-8979-8_9

choices, usually linked to consumption, and the interlinks between personal choices and systemic configuration that enhances microplastics' input into the environment. The areas and potential generations and paths are analyzed qualitatively, and some measures that might have an impact on generation, input, and remediation from the individual perspective are proposed. This chapter closes with a general discussion about the need to conceptualize solutions in terms of public policies, a science-driven public discussion, and the acknowledgment that collective effort towards plastics pollution is the way forward.

Keywords

Microplastics · Pollution · Sustainability personal choice · Public policies · Consumption patterns

9.1 Introduction

Microplastics (MPs) are a complex subject with growing interest both from academia and public domains, since they are considered a threat to all kinds of living forms. Their diffuse sources, complex transport, and accumulation processes are still not well understood. Human activities are the sole source of microplastics, and complex processes determine their fate in the environment. Thus, in this chapter, Sects. 9.1 and 9.2 briefly dive into sources, reservoirs, paths, and transport processes, as a way of contextualizing MPs as an environmental problem and identifying the main areas where personal choices have impacts. Section 9.3 addresses questions of what personal choice entails in the context of MPs generation; besides, some ideas on individual prevention measures are discussed; furthermore, Sect. 9.3 describes MPs generation in sectors and activities that might be more prompt to generate MPs and intertwined with personal choices, thus addressing the question on what are the most microplastic-generating human activities, particularly those related to personal choices? Finally, in Sects. 9.4 and 9.5, the question of what areas personal choice and public policies intertwin are briefly addressed. In Sect. 9.4, this is made through a critical lens of freedom in modern markets, while in Sect. 9.5 some institutional measures already on implementation that have, so far, impacted plastic pollution and what is coming.

9.1.1 Brief Capitulation on Microplastics as an Environmental Problem

Microplastics (MPs) are polymeric particles with a size equal to or less than 5 mm that, depending on their sources, can be classified either as primary or secondary (Ghosh et al. 2023; Leslie et al. 2022). Primary MPs are those that are intentionally manufactured and are commonly used as resin pellets or microbeads in hygiene and personal care products (PCP) such as toothpaste, exfoliants, makeup, facial and body soaps, among others (Singh & Mishra 2023; Sun et al. 2020). Collective discussions (The Scientists' Coalition for an Effective Plastics Treaty 2024) suggest that MPs might also be those that enter the environment in small sizes, regardless of the fragmentation process, which would imply classifying tire particles and synthetic fibers from clothes as primary too, although some sources still classify them as secondary MPs (Ryan et al. 2020). Secondary MPs are those that originate through the fragmentation, degradation, and erosion of large plastics by sun exposure, mechanical abrasion, and wave movement, among other physicochemical processes (Anderson et al. 2016; Haque and Fan 2023).

The size of MPs allows them to move and be deposited in different environmental compartments, such as rivers, streams, wetlands, ponds, lakes, seas, soil, and the atmosphere, through surface water runoff, atmospheric deposition, wind, or waves (in the case of marine ecosystems) (Moita Neto and Silva 2023).

The tiny size of MPs, as well as their buoyancy and lightweight, make them vectors for harmful pollutants to the environment, such as polychlorinated biphenyl (PCB) and polycyclic aromatic hydrocarbons (PAH), which can filter into the tissues of organisms (Kida and Koszelnik 2021; Pelamatti et al. 2021). The presence of MPs has not only caused damage to ecosystems but has also had an impact on the economy of certain sectors, like fishing and tourism, as there has been a decrease in visits due to the contamination of the environment, as well as a decrease in the amount of fish and seafood available for human consumption (Rios-Mendoza et al. 2021).

In addition to the effects that MPs cause on ecosystems and their components, they have also caused effects on biota due to their consumption by aquatic species present along the food chain (Okeke et al. 2022). It has been reported that the ingestion of MPs by marine species has caused effects on reproductive systems, tissue damage, alteration of the food chain and the nutrient cycle, and even death (Egbeocha et al. 2018; Wang et al. 2019).

Biota in marine and terrestrial ecosystems have not been the only ones affected by MPs pollution. The presence of these particles has also been reported in the human body, specifically in blood cells (Leslie et al. 2022). MPs enter the body through different routes: contact with the skin through the use and application of hygiene and personal care products (PCP) such as soaps (body and facial), the respiratory tract, and the ingestion through food and drinks (Ghosh et al. 2023; Hongsawat et al. 2024).

Unlike the effects on marine fauna, the effects on human health caused by exposure to MPs have only recently been explored, trying to find their sources, as well as their short- and long-term effects. Some of the reported effects are digestive and respiratory problems, sleep disruption, contribution to obesity, and increased risk of diabetes, cancer, and antibiotic resistance; additionally, MPs can act as endocrine disruptors and alter hormones (Alimba et al. 2021; Dubey et al. 2022; Ebrahimi et al. 2022; Lett et al. 2021; Yuan et al. 2022).

Due to the problems caused by the presence of MPs, it is important to identify the sources of generation, which will allow the establishment of mitigation strategies and reduction of generation and influx. These strategies must consider public participation and their choice of products that potentially contain or become MPs, especially those that do not indicate this on their label.

9.1.2 Microplastics Generation: Source, Paths, and Transport Processes: What Is Known So Far?

In this section, a brief description of what is known of MPs sources, paths, and transport processes is discussed, and it is aimed to provide some context on where and when individual choices might play a role in MPs input into ecosystems. This section also intends to illustrate where more systemic approaches might be better suited to both understand and address MPs' presence in the environment. It can be argued that in a highly connected and interdependent world, local actions might have a global reach, while no individual actions are entirely independent nor outside of the systemic process. In this context, systemic processes refer specifically to practices and public policies that hinder or determine individual actions when it comes to preventing plastic pollution in general. For example, the lack of municipal and differentiated waste collection might persuade individuals to not separate or recycle deeming their efforts wasteful, or even when individuals go beyond municipal practices, the efficiency of valuable and recyclable material might be impacted due to this shortcoming.

9.1.3 Sources and Pathways in the Atmosphere, Marine Environments, Fresh Waters, and Soils

In general, sources or microplastics in any environmental compartment can be direct or indirect, either related to human practices or environmental factors (Ng et al. 2018). The following sections describes briefly some sources, as a way of contextualizing where and when personal choices might play a role in preventing MPs release into the environment. Noteworthy is that the atmosphere, water, and soil are in constant feedback, thus compartmentalizing sources and pathways helps to understand complex processes.

MPs in the atmosphere reach this environmental compartment through a variety of paths and come from a wide range of sources, and airborne MPs are considered a transboundary challenge for international cooperation, given the lack of borders between sources and final reservoirs (Peries et al. 2024). Most airborne MPs come from synthetic textiles (from manufacturing to final usage, and either domestic, industrial, or agricultural) and extended to end-of-life on incineration facilities or landfill areas (Bamigboye et al. 2024). Macroplastics and textiles that end on a landfill are a major source of airborne MPs, generated either when unsupervised facilities burn or due to plastic fragmentation due to weathering (Petrovic et al. 2023).

Marine environments, one of the most studied environmental compartments, might get MPs input either from land-based or water-based sources. In the first case, shortcomings in waste management and littering might play a significant role. At least 80% of plastic pollution that ends up in the marine environment comes from land-based sources (Acarer Arat 2024). Direct discharge from wastewater effluents, either treated or not, is also a significant source of MPs, and it has been found that more MPs are present in wastewater effluents of economically developed areas (Talukdar et al. 2024), which might point out the influence on income, wealth and MPs pollution.

Urban stormwater also contributes to MPs in marine environments (Sewwandi et al. 2024), but also any other water resource, such as rivers, lakes, and increasingly, groundwater. It has been reported that up to 15,500 MPs/L have been found (and as little as 0.02) in stormwater and it is deemed that stormwater plays a crucial role as an input source of plastic pollution towards the aquatic environment. In areas where rainwater reaches the aquifer, MPs can also be deposited.

Groundwater and soils have a close relationship when it comes to traveling MPs. For instance, soils might accumulate MPs through direct or indirect sources (Lwanga et al. 2022). Direct sources are related directly to crop production, such as the usage of plastic greenhouse infrastructure, irrigation pipes, or mulching for soil moisture control. Mulch has been one of the most studied sources of MPs in soil. Some correlations have been found between plastic greenhouses and MPs, and the use of twine to bind fruits and vegetables is also a signal as a source, yet no specific data has been produced (Lwanga et al. 2022). Many other sources can link both soils and groundwater, like sewage systems and septic tanks (Dong et al. 2022), through infiltration and leakage due to poor maintenance, but the main source might be tire wear off in roads, due to heat and friction degradation processes (Brahney et al. 2021). Recently, vegetation has been studied as a temporal sinkhole of MPs in their pathway from the atmosphere to soils (Sunaga et al. 2024) and as a reservoir in water gateways (van Emmerik et al. 2022).

9.1.4 Transport Process and Sinks of Microplastics

Airborne MPs pathways depend on particle size and mostly travel through wind currents. The smaller the particles, the further they can travel and be deposited on the ground, usually by wet or dry deposition (Jones 2023). Dry deposition of smaller particles (including nanoplastics) poses a greater risk of transboundary pollution; it has been estimated these particles can travel thousands of kilometers before being deposited on the ground (Allen et al. 2019), where they can also be re-suspended by several processes, including lift-up and drafting (Rohais et al. 2024).

In water, MPs pathways are mainly a function of size, type of polymer, and interactions with biota, both via biofilm attachment, and increasing ingestion and further expulsion with fecal pellets, that might impact residence times of MPs in water bodies (Koelmans et al. 2017; Kooi et al. 2017). Most plastic pollution, including MPs, reaches marine environments through rivers, which are considered the main pathways, but they also act as reservoirs, although there is no higher concentration than in all kinds of sediments (Schreyers et al. 2024; van Emmerik et al. 2022). The ocean is both a source and reservoir of MPs, in which multiple processes interact, such as UV and thermal degradation, abrasion, physicochemical fragmentation, and uptake and defecation by marine biota (Koelmans et al. 2017; Kooi et al. 2017).

In soil and groundwater, known processes that move MPs include bioturbation (Ng et al. 2018), percolation, and infiltration of both irrigation and pluvial water, which carries pollution downwards (Li et al. 2021). Specifically related to urban design policy, the existence and maintenance of sewage systems and septic tanks (or lack of it), might impact how much MPs are released, particularly towards groundwater (Umeh et al. 2024).

Related to human decisions, in areas where Managed Aquifer Recharge (MAR) or just pluvial injection towards groundwater deposit is actively pursued, MPs might enter the groundwater systems, especially in areas where those resources are greater exposed (Stefan and Ansems 2018). The main mass fluxes seem to be either continental water towards marine environments or from direct deposition of MPs in both water bodies and soils, which implies that great attention should be addressed to sources where personal choice and main output exist.

9.2 Personal Choices and Microplastics

For the purposes of this chapter, personal choice is discussed solely regarding consumers' habits and individual actions on pollution prevention measures. This means that the scope under the topic is analyzing mainly consumption habits, as well as environmental education efforts. This chapter does not intend to discuss the topic exhaustively or rigorously from the economic point of view, but to invite the reader to deepen the consumption theory approach. Given the consensus that systemic actions should be at the core of public policies, individual actions play an important, yet probably, insufficient effort. For instance, it is known that most plastic pollution reaching marine ecosystems comes from land sources (International Union for Conservation of Nature (IUCN) 2024), particularly improper waste management, which falls within the scope of municipal policies, or lack of them (International Union for Conservation of Nature (IUCN) 2024). Moreover, plastics pollution is a transboundary challenge, thus intergovernmental cooperation is needed, which implies efforts that transcendent personal choices.

9.2.1 What Is Personal Choice in the Context of Microplastics Generation, Consumers Habits, and Individual Actions against Pollution?

Consumers' habits are influenced by a plethora of reasons and are not limited just by how much resources are needed to have a suitable life. Instead, they are a function of modern societies and the fallacy of infinite growth and production as keys to wellbeing (Pirgmaier 2020).

What determines consumer's choices? Choices can be influenced both by perception and by knowledge, while perceptions are how everyone interprets the world and its problems through sociodemographic characteristics such as gender, age, political orientation, personality traits, and educational level; knowledge is the information and understanding that is obtained from a topic over time through experience, education, or learning (Davison et al. 2021; Deng et al. 2020; Van Rensburg et al. 2020). Perceptions regarding plastic pollution in marine environments are still scarce and many people remain unaware of the ecological effects of plastic leakage. For instance, Deng et al. (2020) found that only 26% of the surveyed population were somewhat aware of MPs before the survey, while the rest was unfamiliar with it, but were much more aware of marine pollution from more common plastics, such as bags and bottles, particularly related to the Great Pacific Plastic Island (Deng et al. 2020). Dong et al. (2022) also concluded that knowledge is not enough to act or adapt towards sustainable attitudes. Changes regarding consumer choices usually depend on perceptions and values related to environmental problems and risk associations. For example, while MPs present in food or drinking water pose a major concern to participants, plastic particles present in personal care products are not deemed as risky, thus tendencies to regulate and raise awareness are less in these cases (Deng et al. 2020).

The same goes for environmental impact. People care more about the negative impacts of MPs on human health and less so when it comes to effects on marine ecosystems and biota, which can be perceived through funding proxies. People are keener to fund and invest in research aimed at improving MPs impact on human health rather than marine ecosystems (Davison et al. 2021; Deng et al. 2020).

A key variable of perception regarding MPs impact is the proximity to marine environments. People who tend to live or work in marine-related fields might show a greater concern for plastic pollution implications, compared to those people who live in inland cities and who might perceive the issue as distant or even a less urgent matter (Davison et al. 2021; Henderson and Green 2020; Kautish et al. 2021). Gender might also be significant since women (especially those with university degrees) show greater concern about plastic pollution than men (Charitou et al. 2021). This does not mean that local environmental knowledge and awareness do not play a role in caring. The difference might rely on pre-establish social roles, and even with time allocation; people with less leisure of free time to allocate might be more prompt to disregard environmental concerns and, thus care less about their own environmental impacts (Charitou et al. 2021; Davison et al. 2021; Deng et al. 2020).

9.2.2 Personal Choice and Individual Measures to Prevent or Address Pollution per Sector

The Organization for Economic Co-operation and Development (OECD) has concluded that mismanaged plastic is the main driver of plastic pollution (Organisation for Economic Cooperation and Development (OECD) 2022) and, from the estimated 22 million tons of plastic, at least 12% are MPs (2.6 million tons). From this, most leakage is related to transportation activities (tire wear and brake dust)

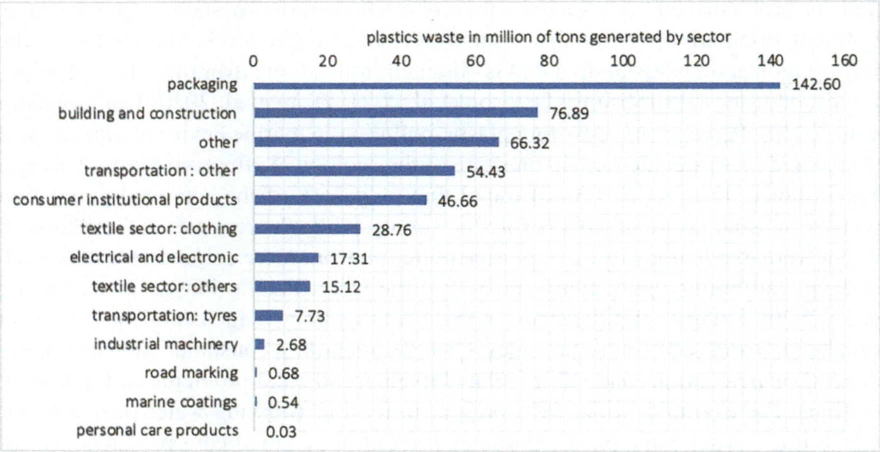

Fig. 9.1 Millions of plastic tons were generated as waste by sector in 2019. Own elaboration based on data from Organisation for Economic Cooperation and Development (OECD) (2022)

(Organisation for Economic Co-operation and Development (OECD) 2022). By 2019, most plastic usage was in the packaging sector (Fig. 9.1), driven largely by e-commerce practices.

Figure 9.1 describes the main drivers of plastic pollution worldwide. There are many sectors that contribute greatly to leakage but for the purposes of where personal choices might have the greater direct impact the following would be briefly described: (1) consumer and institutional products, (2) packaging and single-used plastics, (3) clothing, (4) personal care products, and (5) transportation, which, together, account for more than half of the sources. Some research has suggested that (6) painting might also be one of the main sources of airborne MPs (Paruta et al. 2021), thus a brief section would also be included. Plastic usage varies greatly with region, with the world average reaching 60 kg/year/capita, with higher consumption in the United States and Canada (255.2 and 202.2 kg/capita/year) followed by some OECD European countries (152 kg/capita/year).

9.2.2.1 Consumer and Institutional Plastics

Institutional products are goods used for the maintenance of operation of any economic activity and are an independent category of household goods (Law Insider 2024), while consumer plastics are a wide range of items, such as food containers, trash bags and cans, tumblers, toys, household items and furniture, electronics and even medical applications (Gardner Business Media Inc 2024).

9.2.2.2 Medical Items

Medical uses are a complex subject and would not be addressed in the light of personal choices, mainly because most practices come from an institutional framework. Although some of the applications are needed to ensure sterile packaging and safety procedures, there is a growing push to find nonplastic alternatives, when possible, even in the healthcare sector (Rasheed and Walraven 2023).

Some items include disposable syringes that arguably are a *must have* in health care, while others, such as disposable linens, might be avoidable in the pursuit of sustainability in the sector (Shelton 2023). What can be noted is that, in the same way, that urban waste must be properly handled to avoid leakage and plastic pollution, the same applies to plastic-related biological hazardous waste, a special type of waste with multiple bioenvironmental implications, which might be around 30% of all medical waste produced (Rizan et al. 2020).

9.2.2.3 Trash Bags

There is no specific data found exclusively from trash bags, but it has been estimated that each year, at least 500 billion plastic bags are used globally (United Nations Environment Programme (UNEP) 2024). This might be an example of the intertwines of personal choice, active participation in environmental practices, and systemic hinders on individual efforts. Plastic bags, when no ban has been enacted, are still a common occurrence, used in grocery shopping, packaging, and trash management. And, increasingly so, compostable or eco-friendly alternatives are being presented to consumers. Alternative options might still have an environmental impact and a hard time fulfilling their promises. For instance, compostable trash bags could be a possible solution assuming compostable conditions are always met in waste management municipal facilities, which might not always be the case, especially with countries con already struggling with lacking municipal infrastructure.

Moreover, the throwaway culture that has been mentioned before is not being addressed if the bags are still single-use, no matter their compostable nature, thus a false perception of eco-friendliness might not change the structural problem, which is the single use of materials and the short life span. Food containers, as part of the packaging, will be addressed in the following sections, thus, some data on other consumer and institutional plastics will be discussed.

9.2.2.4 Toys

Data on the toy industry is worth mentioning. It has been estimated that around 6% of the composition of plastic waste in landfills is made up of discarded toys (Karademir 2023), and plastic toys have been directly linked with MPs and nanoplastics emissions, just by their being used (Luo et al. 2024). In fact, toys are, for the most part, designed with no intention of recycling, thus ending up at the end of life, in landfills, or being incinerated (Luo et al. 2024). Moreover, toys are rapidly discarded, with the exceptions of collectible toys and toys with emotional bonds attached to them, which last longer in the hands of consumers. Nevertheless, these types of toys might only account for less than a third of all consumed toys (Tu et al. 2022). Toys are even a marketing strategy for other non-ludic spheres, such as fast food. In 2021, it was estimated that just McDonald's sold 1 billion toys in Happy Meals (Spectrum News 2021), and although some alternatives have been implemented, such as books instead of toys, the phase-out has only been tried in France,

Ireland, and United Kingdom, thus tangible results remain to be seen (McDonald 2022).

This means that some areas, like fast food consumption, could also be part of individual choices to decrease plastic per capita consumption. When it comes to toys, and beyond consumers' choices, eco-friendly designs are a crucial part of extending the life span of products thus reducing the rate at which they are both consumed and discarded (Karademir 2023).

9.2.2.5 Electronic and Electrical Items (e-Plastics)

One of the fastest-growing waste streams globally is e-waste (electronic and electrical waste) and it is expected to be as much as 74.7 million metric tons by 2030 (Shi et al. 2021). While electronic and electrical components are a major push for metal markets, plastic might constitute up to 20% of global waste in the e-waste stream (Shi et al. 2021). Most of the stream ends up in landfills all over the world and are part of intricate networks of plastic exports, unreported, unaccounted, and usually, illegal forms of waste flows as a manifestation of neocolonialism (Stoett 2024), where mostly e-waste is exported towards the economic south (Little 2022).

There are many drivers of the growth in the e-waste stream and, as previously stated, a lot of them are systemic (and by design), such as programmed obsolescence, non-existence of repair options, and lack of collection after the end of life; besides, that, generation is hardly uniform, being Europe the zone where most e-waste per capita is generated, albeit where also the most is recycled too (United Nations Institute for Training and Research (UNITAR) 2024).

Consumers hold an increasing influence on e-waste generation, driven by the demand for new electronics (Islam et al. 2021) and the main drivers on consumption seem to be accessibility to the Internet, ease of using and connecting devices among each other, increasing computing efficiency and safety (Statista 2024). Those areas are widely exploited by manufacturers reaching the end of the life of products as fast as possible, thus practicing a planned obsolescence loop (Statista 2024).

Apart from regulating planned obsolescence practices in markets with public policies, consumer behavior might be influenced when an item reaches the end of use (not life) (Ylä-Mella et al. 2022), thus extending further their life beyond waste. Increasing awareness of sustainable policies, certification, and practices might sway consumers towards brands that hold higher environmental standards, like demanding extended responsibility practices from producers, which makes them accountable for the environmental footprint of their products through all electronics' life stages (Li et al. 2023).

9.2.2.6 Packaging and Single Use Plastics, Including Food Packing

Food containers are a specific packaging category, given the associated properties deemed safe to contain food. They are defined, by The United States Environment Protection Agency (EPA), as products purchased not because of their properties, but their contents, and are "assumed to be discarded the same year the products they contain are purchased" (United States Environmental Protection Agency (EPA) 2023). Moreover, food containers, especially for their color and overall message, might play an important role whether a consumer makes a product purchase, and increasingly so, sustainability concerns and how they are being portrayed in packing plays a significant role too (Berthold et al. 2024). Although no specific data is available regarding purchase intentions solely related to sustainability claims, data suggest that the topic is becoming a variable that accounts for some consumer choices (Berthold et al. 2024).

Here, all types of single-use plastics (SUPs) must be included. It has been estimated that up to half of all produced plastics are made for single use and then rapidly discarded (World Economic Forum (WEF) 2024). The most common SUPs are bags, bottles, wraps, and straws, as a subcategory of food packaging (usually for to-go-food) and are also the most common littered plastics found in the environment (National Resources Defense Council (NRDC) 2024). Packaging is estimated to be around 40% of the year-round production (Pizzorno 2024) and has a short life span, with less than 6 months. This fact, that not all plastics would become waste rapidly, also determines the rates of waste production. Packaging has a shorter span compared to other products (see Fig. 9.2).

For some items, such as plastic bags and SUPs, bans have been or are being proposed. Straws and bags bans across the globe have been implemented, leading to widespread use of compostable alternatives (with different environmental implications), yet two main problems remain: alternatives might come with bigger environmental footprints, or they will not address root causes, such lack of proper waste management, lack of circular thinking and eco-friendly designs (Stanislaus 2018), and the throwaway culture (National Resources Defense Council (NRDC) 2024). Besides that, and specifically with straws, there is a need to account for accessibility for people with swallowing disabilities (Hemsley et al. 2023).

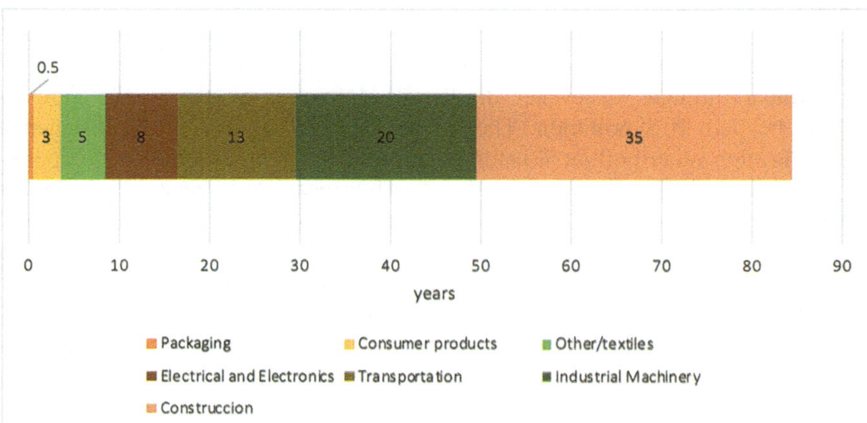

Fig. 9.2 Median life span of different plastics products in years. Elaboration based on Organisation for Economic Co-operation and Development (OECD) (2022)

9.2.2.7 E-Commerce and Delivery

It has been estimated that by 2027, at least 23% of all sales that occur in 1 year would be done virtually, through a variety of channels (Jose 2023). Although an upward trend was already in the making, the COVID-19 sanitary emergency sealed the fate of online shopping and delivery services, which, in turn, creates around 5 times more packaging waste than normal shopping, with most of it being plastic waste (Kim et al. 2022). In fact, the packaging sector consumes more than a third of produced plastic (Jose 2023) and now accounts for at least 10% of plastic waste in landfills, at least in developed countries (Geyer et al. 2017). At first, as a perceived safe and sanitary way to purchase and satisfy individual needs, online shopping has now become a trademark for new generations (Feng and Ivanov 2023). Most consumer drivers of e-commerce rest on the perception of usefulness and ease of use of new technological tools, which are perceived as convenient (Zerbini et al. 2022), while several online dynamics, like targeted adds, might also play a crucial role in purchase desire, intention, and final consumption.

Food delivery, although a subgenre of e-commerce, has its own drivers and the share market has grown almost three times the size it had in 2017 (Ahuja et al. 2022). While COVID19 also had an enormous impact on its dynamics, it has not stopped even though lockdowns have receded, mostly boosted by perceived convenience and time management issues in consumers. This delivery segment and its plastic waste footprint are beyond comprehension from a classical profit dynamic perspective, given that online platforms are still being operated with lower to zero profit margins (Zerbini et al. 2022) and yet, despite that, are a major source of plastic waste (Iimuro and Tabata 2024) that is usually very hard to recycle or even collected separately (Janairo 2021).

9.2.2.8 Clothing

Synthetic textile fibers are another source of MPs directly related to consumption patterns and individual choices, which, like microbeads, are introduced into the environment through wastewater discharges. Influence on the textile industry extends beyond MPs pollution. The textile industry has a cross-environmental impact through water pollution, landfill saturation, energy consumption, air pollution, and the overconsumption of resources (European Parliament 2024). Textile consumption has increased due to the exponential population growth and the emergence of new trends such as fast fashion; at the same time, synthetic fibers have replaced materials such as cotton, hemp, and linen (de Oliveira et al. 2023).

Synthetic fibers present a contamination problem related to the chemical additives that are incorporated during manufacturing, such as lubricants, flame retardants, antimicrobial products, halogenated compounds, plasticizing agents, formaldehyde-based substances, synthetic dyes, and pigments and stabilizers, among others, which can cause damage to ecosystems and to the organisms that ingest them (De Falco et al. 2018; de Oliveira et al. 2023).

Microfibers could be released from clothing due to the mechanical and chemical stresses that the fabrics undergo during the washing process, and it is estimated that in a typical 5 kg wash load of polyester fabrics, more than six million synthetic fibers could be released, a quantity that would be influenced by the type of detergent used, the water temperature in the wash cycle, the hardness of the water and the mechanical action of the washing center (De Falco et al. 2018). Even without considering the washing load, discarded textiles might account for at least 5% of all waste in a landfill, given that not even 1% of all the used clothes are recycled (European Parliament 2024).

Some actions that could reduce the release of microfibers include the use of fabric softener in each wash cycle, due to its ability to reduce friction between fibers, the use of special washing bags, and the coating of fabrics (De Falco et al. 2018; Napper et al. 2015). Unlike microbeads, it is more difficult to reduce or eliminate the use of synthetic fibers because the use of natural textile fibers is more expensive and represents a negative environmental impact derived from the extensive cultivation of cotton and linen (de Oliveira et al. 2023). Beyond that, natural fibers might also have their own environmental impacts on aquatic biota (Albarano et al. 2024). Thus, it can be argued that changing raw materials without changing the speed of consumption might not be the best approach.

Although comparable results are hard to obtain, a strong correlation between fibers detachment and the density of the garment has been suggested, which might inform consumer choice when acquiring products. Domestic discharges are a major source of MPs in wastewater streams, usually from washing clothes and the use of cosmetics and personal care products (Talukdar et al. 2024). This might be partially attended when wastewater comes into treatment plants, although the removal efficiency is highly dependent on the treatment stage, which might be 100% only when secondary treatment is available (Tang and Hadibarata 2021). And, when water is not treated, all contained MPs in the effluent would eventually reach water streams. This is dire because wastewater treatment varies greatly between high-income and low-income countries (74% to 4.3% respectively), with a global average of 52% of wastewater treated (Jones et al. 2021).

It has been estimated that, on average, a European citizen might consume around 15 kg of textiles (garments, household products, and shoes) in a year and that most of them would be used a few times before being discarded (European Environment Agency and European Environment Information and Observation Network 2024), while United States and China are the major consumers worldwide, being in United States (by 2018) where more new items per capita per year are consumed (Fashion United 2024). Moreover, most countries, regardless of developing status, have no formal, structured and institutionalized efforts to gather, funnels of recycled clothes and other textiles that, for the most part, end up unsorted in landfills (Sinha et al. 2022) and textiles end up in a variety of places, like the Atacama Desert (Bartlett 2024).

When it comes to specific measures on a domestic level, individuals can choose to wash clothes with cold or tempered water, given an increased shedding has been reported with water temperatures above 40 °C or choose the strict amount of detergent needed to clean clothes without overconsuming, given the used of detergent has been linked to increase in shedding, less so with powder; the use of softener seems to reduce the release of fibers (Zhang et al. 2021). Nevertheless, the cost and environmental implications of detergents and softeners should also be accounted for, mainly because both detergents and softeners come in plastic packaging.

Pressure to change the fast fashion trend that most textile firms had acquired (European Environment Agency and European Environment Information and Observation Network 2024) might be something individuals might take upon themselves, while also catering to local efforts to upcycle and recycle clothes. For instance, efforts like Wrap (2024), a recycling project located in the UK, Australia, and United States, or the second-hand store Humana Fundación Pueblo Para Pueblo (2024), present in some European countries, like Austria, Germany of Spain, might be a good first look at options post-consumption that consumers have to funnel their used clothes, although, as it can be seen, those efforts are quite regional.

9.2.2.9 Personal Care and Hygiene Products

Considering the origin of MPs, one of the major sources of generation is the high consumption of products that contain MPs or that are susceptible to being transformed into them, such as microbeads in Personal Care and Hygiene Products (PCP), including glittery makeup (Haque and Fan 2023).

Microbeads (usually made of polyethylene) are used in PCP as substitutes for minerals and natural ingredients whose function is to cleanse and exfoliate the skin (Haque and Fan 2023; Sun et al. 2020); the high demand for this type of product and the low availability of natural resources to be part of their components motivated the chemical industry to look for economical and easy-to-produce substitutes in massive quantities. An example of this are exfoliants, which in addition to microbeads, can also contain alumina and decahydrate particles of sodium tetraborate instead of keratinous materials such as pumice, oats, apricot, ground fruit cores or nutshells (Sun et al. 2020).

It is estimated that an exfoliant could release between 4594 and 94,500 microbeads in a sole use (Napper et al. 2015) and that, between 1970 and 2019, 3.0×10^5 tons of MPs derived from PCP accumulated in the environment, and that amount would only increase year by year (Sun et al. 2020). In developed countries such as the United States, the consumption of PCP and cosmetics containing microbeads caused an emission of 282 tons of MPs in 2015 and a per capita use of polyethylene microbeads of 2.4 mg/day (Sun et al. 2020).

Although the use of microbeads in PCP and cosmetics has been reported as an emerging environmental problem, a large part of the population continues to consume them because they are unaware of the negative impacts of the use of these components, believe that these elements are of natural origin (e.g., nuts, oats, sand, or sugar), or, simply, no much options might be available in local markets, such as in United States and Japan, where some larger companies drive the market (Rossi et al. 2007).

In the case of consumers who are aware of microbeads and their effects, they prefer to ignore them, arguing that other social, environmental, and economic problems with a greater impact on the world should be addressed before the problem caused by MPs (Napper et al. 2015). Considering the above, one of the greatest challenges facing the problem of MPs is the lack of public knowledge and understanding on the subject, since a large part of the population is still not aware of the extent of microplastic pollution and its influence on the environment and health (Borriello et al. 2022; Garcia-Vazquez and Garcia-Ael 2021).

To identify how this problem can be addressed, various citizen participation surveys have been carried out, in which it has been identified that the issue of MPs is not mentioned spontaneously, that the population is not aware that large plastic items are fragmented into smaller pieces in the environment, nor of how microplastics are transported to various ecosystems, nor are they aware of what microbeads are or their presence in PCP (Napper et al. 2015). However, there have also been positive responses from consumers, who in addition to knowing and informing themselves about this problem, have also mentioned their interest in regulating and/or eliminating the use of microbeads in PCP (Garcia-Vazquez and Garcia-Ael 2021; Napper et al. 2015).

To reduce the release of MPs in natural environments, different measures have been taken. One of the most important ones is regulating microbead usage in PCP, mainly by banning its addition. Some countries, such as the United States and the United Kingdom, have developed and implemented these kinds of legislation. For example, a legislative ban on the use of plastic microbeads in cosmetics and toiletries by the United Kingdom's Environmental Audit Committee in 2016 was approved (Borriello et al. 2022; Garcia-Vazquez and Garcia-Ael 2021; Napper et al. 2015) and the Microbead-Free Waters Act of 2015 that was gradually implemented in the United States between 2017 and 2018 (US Food and Drug Administration (USFDA) 2022).

Other strategies are aimed at awareness campaigns that inform consumers about plastic pollution and its consequences, and in which consumers are advised to check the ingredients of PCP and ensure that they do not contain microbeads, as well as to opt for natural exfoliants such as oats, ground nuts, and granulated sugar (Napper et al. 2015; Sun et al. 2020). Nevertheless, this area might prove to be quite complex. Consumer´s preferences, especially when it comes to PCP, are a function on a wide range of attributes like durability, strength, and brand-associated characteristics (Kamwendo and Maharaj 2022) and, although some studies have shown that environmental concerns might play a role when choosing a brand, the focus has been on intend rather than consumers purchase (Setiawan et al. 2023).

On the part of the industry, strategies are aimed at finding replacements for microbeads in PCP, such as organic waxes that can be manufactured in a suitable size and shape and are like plastics in terms of friction index, hardness, and temperature stability (Sun et al. 2020).

It is important to mention that, in addition to the lack of knowledge, the population could also continue consuming products containing microbeads because they are not included in the list of ingredients (Sun et al. 2020) or because they are cheaper and more accessible than products containing elements of natural origin, so regulations and actions would also have to be aimed at producing products with easily accessible natural ingredients. In fact, this is a key aspect of choice. Consumers have shown interest in choosing PCP that do not contain microbeads. Unfortunately, they have found few alternatives, either because they do not exist in the market, or when including natural ingredients, they are more expensive (Deng et al. 2020; Núñez-Cacho et al. 2020).

9.3 Freedom of Choice: How Much Can We Choose? A Brief Critical Overview of Personal Choices in Modern Markets

In this section, some critical ideas are discussed. Although it is important to acknowledge the role individual choices play in pollution generation and thus, pollution prevention, as seen previously, it is also important to recognize that, in a complex, modern, and market-based global economy with intertwined markets, individual choices have a limited impact when it comes to addressing planetary scale limits. This is particularly important to acknowledge, and address given that, by systemic design, profits are a function not only on productive and cutting cost, but overproduction (Pirgmaier 2020).

Moreover, behind life-saving plastic applications, even more disposable ones have been developed, and thus, our dependence on SUPs (Farrelly et al. 2021). Recognizing that societies are not monolithic (Farrelly et al. 2021) and some have more tools than others to actively choose to prevent and fight plastic pollution, while others end up at the end of the disposal cycle, rendering their citizen powerless to act. In fact, some links have been made between the local cleaning of the plastic environment and the possibility of consuming more plastic, when this pollution is being unfairly exported towards developing countries (Barnes 2019). Some argue that the main issue behind most environmental problems is over-consumption (Stafford and Jones 2019). However over-consumption is a byproduct of overproduction.

9.3.1 Transportation as an Example of Interactions in Personal Choice and Public Policies

One of the major sources of MPs in the oceans, even greater than rivers, might be airborne MPs generated as a byproduct of tire wear and brakes and the hotspots globally are tightly linked with highly urbanized areas, where road traffic and vehicle density is higher (Evangeliou et al. 2020). Thus, there is a direct relationship between the use of cars, from a private standpoint, and MPs emissions. Furthermore, it can be argued that appealing to a responsible car usage, the use of a private car can give multiple environmentally positive feedback, one of the being MPs. Nevertheless, although it might be seen, at first glance, that the use of cars in some areas is a personal choice, at least for people who own cars, this needs to be analyzed under the urban design landscape.

In areas where there is public transportation and bike paths, such as the Netherlands, at least 44% of travel is made either by foot or by bicycle (Ton et al.

2020), but in most urban areas, all urban design centers on individual car ownership (Glaeser and Kahn 2004) and most people travel by car (Ritcher 2022).

9.4 Public Perception and Personal Choices

9.4.1 External Measures to Promote Pro-environmental Behavior: From Nudges to Bans

Personal choices can be influenced by external incentives or even mandated through legal measures. In the former case, environmental nudges, subtle alterations to an individual's environment, can steer decision-making towards more sustainable options. These changes may involve design elements, social norms, or psychological principles that render the "pro-environmental" choice more appealing or convenient. The application of nudges is grounded in behavioral science, leveraging common human tendencies such as concern for self-image, conformity to social norms, and preference for default options (Schubert 2017).

Environmental nudges have been proposed as a strategy to reduce single-use plastic consumption and promote circularity. Suggested interventions include making disposable cups and cutlery available only upon request, increasing the ratio of recycling bins to general waste bins, and implementing a fee for the use of disposable cups (United Nations Environmental Programme (UNEP) 2020). While nudging policies have demonstrated the ability to influence pro-environmental decisions in waste production (Byerly et al. 2018), their effectiveness can be limited in certain sectors or contexts (Rivers et al. 2017).

Coercive measures have emerged to steer individual behavior towards environmentally preferable alternatives. Since the early 2000s, regulations limiting the production, commerce, use, and disposal of single-use plastics and microplastics have been adopted in over 127 countries (UNEP 2018). By 2020, the literature reported 67 international policies, 147 national policies, and 77 subnational policies (Diana et al. 2022), encompassing regulatory, economic, and informational measures.

In many cases, these regulations emerge as a response to pressure exerted by activists. Diverse groups worldwide are dedicated to combating plastic pollution through various efforts, including promoting recycling, organizing cleanups, advocating for informal waste workers, and lobbying for stricter regulations. This anti-plastics activism has yielded governance reforms, ranging from local bans on single-use plastics to national moratoriums on new incinerators (Dauvergne and Islam 2023).

Bans, the most common regulatory policies, frequently target specific products, such as disposable cutlery, straws, and packaging; plastic bags are the most regulated single-use application, typically through restrictions on their free retail distribution. Notably, these bans are often accompanied by exceptions, including biodegradable and biobased materials, products with recycled content, or those possessing specific physical characteristics. Another noteworthy ban, where social pressure played a significant role, is that of microbeads, which are intentionally added, manufactured particles in consumer products. Several countries, including Canada, the Republic of Korea, New Zealand, Sweden, the European Union, and the United Kingdom, have implemented legally binding bans on microbeads, primarily targeting their use in personal care products (UNEP 2018; Kramm 2023). Additionally, voluntary phase-outs and eco-labeling initiatives have been established through agreements between governments, NGOs, and industry.

9.4.2 Effectiveness of Bans

Within the regulatory landscape, there is a heterogeneity in the measures adopted, ranging from local to regional regulations (Diana et al. 2022). This intrinsic diversity hinders uniform application and generates considerable variability in implementation. Existing legal frameworks are mostly not designed to comprehensively address cross-cutting issues involving production, commercial, and waste management activities. The ambiguity in the delimitation of competencies between the different levels of authority has triggered discrepancies, contradictions, and even litigation, as documented in the United States (Wagner 2017).

The evaluation of the impact of these measures is hindered by the lack of robust monitoring mechanisms and the inherent complexity of the required analysis. According to UNEP data, in 50% of cases of bans or levies, there is insufficient information to determine their impact, 20% have had an insignificant impact, and only the remaining 30% have achieved an effective reduction in bag consumption and its consequent environmental impact (UNEP 2018). However, the effectiveness of these interventions can vary significantly depending on the specific product or material. In the case of plastic bags, for instance, studies have reported an average reduction of 66% in consumption following the implementation of regulatory and economic instruments (Schusser et al. 2013).

9.4.3 Towards Systemic Measures to Fight Plastic Pollution

While individual actions are relevant in preventing and reducing plastic pollution, it is evident that systemic conditions must be modified to achieve sustainable production, consumption, and management of plastics. Recognizing this situation, civil society organizations play an indispensable role in advocating for changes in public policies, starting at the local level and extending to global agreements. This has been acknowledged in the ongoing development of a global agreement against plastic pollution, led by the United Nations, which reflects the concern and interest of the global community regarding this issue (Edo et al. 2023).

9.5 Final Remarks: Individual Actions Matter, Just Not Alone

This chapter aimed to briefly present a global picture of sources and paths of MPs and how they intertwine with personal choices, essentially from a consumer's perspective. It also the interlinks between personal choices and systemic configuration that enhances MPs input into the environment. It covered specific sectors where it was considered that personal choice might have a greater influence and impact, provided awareness and willingness to adopt certain conducts, and avoid certain products.

In this sense, and for some relevant sectors, some discussed ideas are summarized in Table 9.1, with associated implications.

To showcase the interaction between public policies and personal choices (and thus systemic change), personal care products were chosen to illustrate the latter; Fig. 9.3 shows an example of how personal care product consumption might interact with systemic change.

Figure 9.3 illustrates feedback where awareness might have an impact on consumers' choices, which, in turn, might push companies and government in other directions, from public policies such as bans and waste management redesign, to companies adhering themselves to better standards manufacturing ecofriendly products and forwarding eco-designs.

All of the above points to a significant amount of plastic waste is related to products with shorter life spans, like packaging and consumer products; even textiles have an average life of just 5 years, be they degrade due to poor quality, or displacement trends This means that the need for interventions that extend the life span of products, change the collective mindset of their uses and increase productors' accountability towards recollection, repurpose, recycling and paying environmental costs of their products should be at the top of societal priorities when it comes to addressing plastics pollution.

In the end, raising awareness and engaging citizenship with active environmental, like participating in public consultation, protest, and education efforts, engaging in beaches and river cleaning ups, or simply separating and cleaning the generated waste, are small steps that contribute towards plastic pollution reduction. Nevertheless, actions that promote closing the tap on plastic production and urging the government to enact extended responsibility of producers, ban single-use plastics, and increase their public policies to implement circular economy models are a must.

In a few words, systemic actions that stem from public policies and profound changes in the current economic model that still holds up to the idea of infinite growth need to be questioned, tackle, and rethink if plastic waste is a serious fight being undertaken.

Table 9.1 Summarization of individual actions related to specific sectors

Sector	Individual actions	Are there already perceived outcomes?
Consumer plastics: Trash bags	In situ active separation at domestic level and the possible reduction of the amount and volume of waste discarded, might thus impact in direct consumption of trash bags	Some policies have been implemented to charge single-use plastic bags and implemented designated bags for municipal waste recollection (Nishijima and Nakatani 2024) Nevertheless, in many countries, where single used plastics bans have not been enacted or are partially endorsed, trash bag consumption is mixed with shopping bags, and the lines and efforts to reduce bags consumptions remains hinder by the bans´ own exceptions (Excell 2019)
Consumer plastics: Toys	It is highly possible that most toys will not be recycled, but they can be upcycled, repaired and, for electronic parts, they can enter e-waste streams. Consumer's might demand extended responsibility policies from big toy brands and prefer those that have circular economy and sustainability statements	Some small measures have been implemented by large chains to change plastic toys for textiles, with their own associated impacts
Consumer plastics: e-plastics from electrical and electronic devices	Choose to use an electronic device up to the end of life, and not change it because of fashion, trend or new features that do not considerably improve the use or performance or are in fact needed Look for recovery centers, especially for cellphones, printers, and batteries, that are common in electronical distributors, supermarkets, and convenience stores	Global initiatives call for small devices to be properly discarded or recycled, to recover and re-introduce valuable materials in the market, such as metals. Sadly, plastics are not considered when those efforts are being designed (WEEE Forum 2022)
Packaging and food delivery	Left default choices as not cutlery included and opt-out on delivery when possible. Dealing with packaging on deliveries might be more challenging but some sites also include options to opt out of extra packaging	Some apps include, by default, choices that reduce plastic usage, but it is mostly limited to cutlery, not food packaging, thus consumers should always choose not to have delivery with those SUPs. In fact, it has been probed that by making the "green choice" of no cutlery, up to 22 million of single-use sets of cutleries might be spared from ending in homes and eventually as waste (Ryan 2023)

(continued)

Table 9.1 (continued)

Sector	Individual actions	Are there already perceived outcomes?
E-commerce	Although there is no doubt about the convenience and practical matters regarding e-commerce, consumers might choose and plan their deliveries to maximize packaging and delivery efforts, use the opt-out option on packaging when existing and choose offline shopping when possible	Few measures have been undertaken by e-commerce platforms, like Amazon, although substitutions might come with their own environmental impacts, such as the use of paper fillers instead of plastic pillows (Amazon Sustainability 2023). In the long term extended producer responsibility and management programs should be part of the e-commerce endeavor itself
Clothing	Choose to repair, upcycle, reuse, and donate before throwing put, while also observing good practices when donating clothes, such as labeling and sorting, to prevent mismanagement of donated clothes (Wells 2024)	Although some big names in the apparel industry have reported that one-quarter of their textiles are being recycled (Richford 2024), the logic of fast fashion behind their brands is still at large (Robertson 2024)

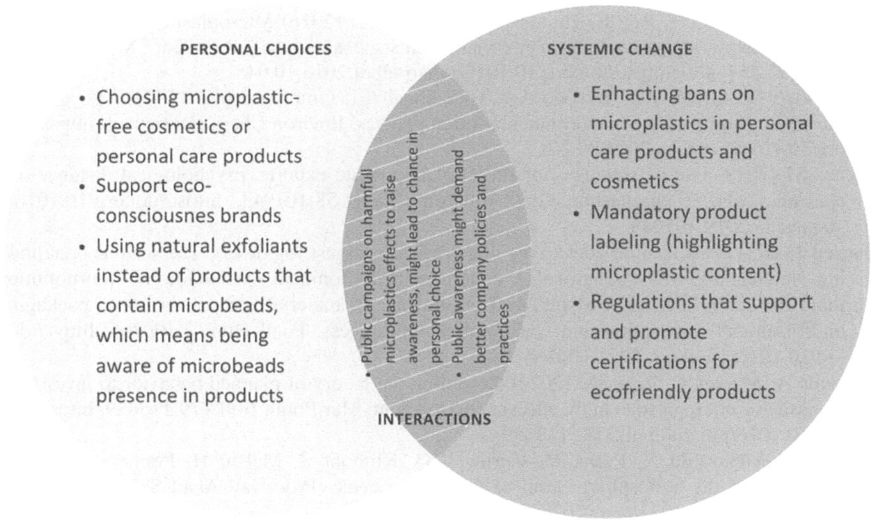

Fig. 9.3 Interaction between personal choices and systemic change when talking about personal care products. Own elaboration

Acknowledgments The authors would like to thank both El Colegio de la Frontera Sur and the Universidad Autónoma Metropolitana for their institutional support. This chapter has been made thanks to the funding provided by the Consejo Nacional de Humanidades, Ciencias y Tecnologías (CONAHCYT), in Mexico, for a doctoral grant from 2022 to 2026.

References

Acarer Arat S (2024) A review on cigarette butts: environmental abundance, characterization, and toxic pollutants released into water from cigarette butts. Sci Total Environ 928:172327. https://doi.org/10.1016/j.scitotenv.2024.172327

Ahuja K, Chandra V, Lord V, Peens C (2022) Ordering in: the rapid evolution of food delivery. In: Industries, technology, media and telecomunications. McKinsey, Chicago. https://www.mckinsey.com/industries/technology-media-and-telecommunications/our-insights/ordering-in-the-rapid-evolution-of-food-delivery

Albarano L, Maggio C, La Marca A, Iovine R, Lofrano G, Guida M, Vaiano V, Carotenuto M, Pedatella S, Spica VR, Libralato G (2024) Risk assessment of natural and synthetic fibers in aquatic environment: a critical review. Sci Total Environ 934:173398. https://doi.org/10.1016/j.scitotenv.2024.173398

Alimba CG, Faggio C, Sivanesan S, Ogunkanmi AL, Krishnamurthi K (2021) Micro(nano)-plastics in the environment and risk of carcinogenesis: insight into possible mechanisms. J Hazard Mater 416:126143. https://doi.org/10.1016/j.jhazmat.2021.126143

Allen S, Allen D, Phoenix VR, Gael LR, Jimenez PD, Simonneau A, Stephane B, Galop D (2019) Atmospheric transport and deposition of microplastics in a remote mountain catchment. Nat Geosci 12(5):339. https://doi.org/10.1038/s41561-019-0335-5

Amazon Sustainability (2023) Reducing waste & packaging. Waste. https://sustainability.aboutamazon.com/waste#:~:text=1%20of%202-,Packaging%20innovation,curbside%20recycling%20for%20our%20customers

Anderson AG, Grose J, Pahl S, Thompson RC, Wyles KJ (2016) Microplastics in personal care products: exploring perceptions of environmentalists, beauticians and students. Mar Pollut Bull 113(1-2):454–460. https://doi.org/10.1016/j.marpolbul.2016.10.048

Bamigboye O, Alfred MO, Bayode AA, Unuabonah EI, Omorogie MO (2024) The growing threats and mitigation of environmental microplastics. Environ Chem Ecotoxicol. https://doi.org/10.1016/j.enceco.2024.07.001

Barnes SJ (2019) Out of sight, out of mind: plastic waste exports, psychological distance and consumer plastic purchasing. Glob Environ Chang 58:101943. https://doi.org/10.1016/j.gloenvcha.2019.101943

Bartlett J (2024) Fast fashion goes to die in the world's largest fog desert. The scale is breathtaking. NatGeo. https://www.nationalgeographic.com/environment/article/chile-fashion-pollution

Berthold A, Guion S, Siegrist M (2024) The influence of material and color of food packaging on consumers' perception and consumption willingness. Food Hum 2:100265. https://doi.org/10.1016/j.foohum.2024.100265

Borriello A, Massey G, Rose JM (2022) Extending the theory of planned behavior to investigate the issue of microplastics in the marine environment. Mar Pollut Bull 179:113689. https://doi.org/10.1016/j.marpolbul.2022.113689

Brahney J, Mahowald N, Prank M, Cornwell G, Klimont Z, Matsui H, Prather KA (2021) Constraining the atmospheric limb of the plastic cycle. Proc Natl Acad Sci 118(16):1–10. https://doi.org/10.1073/pnas.2020719118

Byerly H et al (2018) Nudging pro-environmental behavior: evidence and opportunities. Front Ecol Environ 16(3):159–168. https://doi.org/10.1002/FEE.1777

Charitou A, Naasan Aga-Spyridopoulou R, Mylona Z, Beck R, McLellan F, Addamo AM (2021) Investigating the knowledge and attitude of the Greek public towards marine plastic pollution and the EU single-use plastics directive. Mar Pollut Bull 166:112182. https://doi.org/10.1016/j.marpolbul.2021.112182

Dauvergne P, Islam S (2023) The politics of anti-plastics activism in Indonesia and Malaysia. Cambridge Prisms Plast 1:e1. https://doi.org/10.1017/plc.2023.3

Davison SMC, White MP, Pahl S, Taylor T, Fielding K, Roberts BR, Economou T, McMeel O, Kellett P, Fleming LE (2021) Public concern about, and desire for research into, the human health effects of marine plastic pollution: results from a 15-country survey across Europe and Australia. Glob Environ Chang 69:102309. https://doi.org/10.1016/j.gloenvcha.2021.102309

De Falco F, Gullo MP, Gentile G, Di Pace E, Cocca M, Gelabert L, Brouta-Agnésa M, Rovira A, Escudero R, Villalba R, Mossotti R, Montarsolo A, Gavignano S, Tonin C, Avella M (2018) Evaluation of microplastic release caused by textile washing processes of synthetic fabrics. Environ Pollut 236:916–925. https://doi.org/10.1016/j.envpol.2017.10.057

de Oliveira CRS, da Silva Júnior AH, Mulinari J, Ferreira AJS, da Silva A (2023) Fibrous microplastics released from textiles: occurrence, fate, and remediation strategies. J Contam Hydrol 256:104169. https://doi.org/10.1016/j.jconhyd.2023.104169

Deng L, Cai L, Sun F, Li G, Che Y (2020) Public attitudes towards microplastics: perceptions, behaviors and policy implications. Resour Conserv Recycl 163:105096. https://doi.org/10.1016/j.resconrec.2020.105096

Diana Z et al (2022) The evolving global plastics policy landscape: an inventory and effectiveness review. Environ Sci Policy 134:34–45. https://doi.org/10.1016/j.envsci.2022.03.028

Dong S, Yu Z, Huang J, Gao B (2022) Fate and transport of microplastics in soils and groundwater. In: Emerging contaminants in soil and groundwater systems: occurrence, impact, fate and transport. Elsevier, Amsterdam, pp 301–329. https://doi.org/10.1016/B978-0-12-824088-5.00001-X

Dubey I, Khan S, Kushwaha S (2022) Developmental and reproductive toxic effects of exposure to microplastics: a review of associated signaling pathways. Front Toxicol 4:1–10. https://doi.org/10.3389/ftox.2022.901798

Ebrahimi P, Abbasi S, Pashaei R, Bogusz A, Oleszczuk P (2022) Investigating the impact of physicochemical properties of microplastics on human health: a short bibliometric analysis and review. Chemosphere 289:133146. https://doi.org/10.1016/j.chemosphere.2021.133146

Edo C, Fernández-Piñas F, Leganes F, Gómez M, Martínez I, Herrera A, Hernández-Sánchez C, González-Sálamo J, Borges JH, López-Castellanos J, Bayo J, Romera-Castillo C, Elustondo D, Santamaría C, Alonso R, García-Gómez H, Gonzalez-Cascon R, Martínez-Hernández V, Landaburu-Aguirre J, González-Pleiter M (2023) A nationwide monitoring of atmospheric microplastic deposition. Sci Total Environ 905:166923. https://doi.org/10.1016/j.scitotenv.2023.166923

Egbeocha CO, Malek S, Emenike CU, Milow P (2018) Feasting on microplastics: ingestion by and effects on marine organisms. Aquat Biol 27:93–106. https://doi.org/10.3354/ab00701

European Environment Agency & European Environment Information and Observation Network (2024) Textiles. In: Topics. EEA, Copenhagen, p 1

European Parliament (2024) The impact of textile production and waste on the environment (infographics). CircularEconomy. https://www.europarl.europa.eu/topics/en/article/20201208STO93327/the-impact-of-textile-production-and-waste-on-the-environment-infographics

Evangeliou N, Grythe H, Klimont Z, Heyes C, Eckhardt S, Lopez-Aparicio S, Stohl A (2020) Atmospheric transport is a major pathway of microplastics to remote regions. Nat Commun 11(1):3381. https://doi.org/10.1038/s41467-020-17201-9

Excell C (2019) 127 countries now regulate plastic bags. Why aren't we seeing less pollution? Insights World Resources Institute. https://www.wri.org/insights/127-countries-now-regulate-plastic-bags-why-arent-we-seeing-less-pollution#:~:text=These%20policies%20range%20from%20outright,in%20slowing%20down%20ocean%20pollution

Farrelly T, Taffel S, Shaw I (2021) Plastic legacies: pollution, persistence, and politics, vol 1, 1st edn. Athabasca University, Athabasca, pp 1–22. https://doi.org/10.15215/aupress/9781771993272.01

Fashion United (2024) Global fashion industry statistics. Fashion Web page

Feng R, Ivanov A (2023) Does a generational gap exist in online shopping attitudes? A comparison of Chinese consumer generations from the media system dependency perspective. Telemat Inf Rep 12:100106. https://doi.org/10.1016/j.teler.2023.100106

Garcia-Vazquez E, Garcia-Ael C (2021) The invisible enemy. Public knowledge of microplastics is needed to face the current microplastics crisis. Sustain Prod Consump 28:1076–1089. https://doi.org/10.1016/j.spc.2021.07.032

Gardner Business Media Inc (2024) Plastics use in the consumer products industry. Plastics Technology Topics. https://www.ptonline.com/topics/consumer-products#:~:text=Plastics%20in%20consumer%20products%20include,%3B%20garden%20hoses%3B%20outdoor%20furniture

Geyer R, Jambeck JR, Law KL (2017) Production, use, and fate of all plastics ever made. https://www.science.org

Ghosh S, Sinha JK, Ghosh S, Vashisth K, Han S, Bhaskar R (2023) Microplastics as an emerging threat to the global environment and human health. Sustain For 15(14):10821. https://doi.org/10.3390/su151410821

Glaeser EL, Kahn ME (2004) Chapter 56: sprawl and urban growth. In: Henderson JV, Thisse J-F (eds) Cities and geography, vol 4. Elsevier, Amsterdam, pp 2481–2527. https://doi.org/10.1016/S1574-0080(04)80013-0

Haque F, Fan C (2023) Fate and impacts of microplastics in the environment: hydrosphere, pedosphere, and atmosphere. Environments 10(5):70. https://doi.org/10.3390/environments10050070

Hemsley B, Darcy S, Given F, Murray BR, Balandin S (2023) Going thirsty for the turtles: plastic straw bans, people with swallowing disability, and sustainable development goal 14, life below water. Int J Speech Lang Pathol 25(1):15–19. https://doi.org/10.1080/17549507.2022.2127900

Henderson L, Green C (2020) Making sense of microplastics? Public understandings of plastic pollution. Mar Pollut Bull 152:110908. https://doi.org/10.1016/j.marpolbul.2020.110908

Hongsawat P, Thinjong W, Chouychai B, Punyapalakul P, Prarat P (2024) Microplastics in retail shellfish from a seafood market in eastern Thailand: occurrence and risks to human food safety. Mar Pollut Bull 201:116228. https://doi.org/10.1016/j.marpolbul.2024.116228

Humana Fundación Pueblo Para Pueblo (2024) Humana. HumanaWeb

Iimuro M, Tabata T (2024) Relationship between the demand for food delivery and takeaway services and the associated plastic packaging waste during the COVID-19 pandemic. J Mater Cycl Waste Manage 26(1):591–601. https://doi.org/10.1007/s10163-023-01861-y

International Union for Conservation of Nature (IUCN) (2024) Issues brief: plastic pollution. www.iucn.org/issues-briefs

Islam MT, Huda N, Baumber A, Shumon R, Zaman A, Ali F, Hossain R, Sahajwalla V (2021) A global review of consumer behavior towards e-waste and implications for the circular economy. J Clean Prod 316:128297. https://doi.org/10.1016/j.jclepro.2021.128297

Janairo JIB (2021) Unsustainable plastic consumption associated with online food delivery services in the new normal. Clean Responsible Consump 2:100014. https://doi.org/10.1016/j.clrc.2021.100014

Jones N (2023) Microplastics are filling the skies. Will they affect the climate? Yale Environ 1:e360

Jones ER, van Vliet MTH, Qadir M, Bierkens MFP (2021) Country-level and gridded estimates of wastewater production, collection, treatment and reuse. Earth Syst Sci Data 13:237–254

Jose R (2023) Addressing plastic packaging waste in E-commerce retail

Kamwendo AR, Maharaj M (2022) The preferences of consumers when selecting skin care products. J Contemp Manage 19(1):82–106. https://doi.org/10.35683/jcm20134.138

Karademir D (2023) Most materials are recyclable, so why can't children's toys be sustainable? Yale University, Yale

Kautish P, Sharma R, Mangla SK, Jabeen F, Awan U (2021) Understanding choice behavior towards plastic consumption: an emerging market investigation. Resour Conserv Recycl 174:105828. https://doi.org/10.1016/j.resconrec.2021.105828

Kida M, Koszelnik P (2021) Investigation of the presence and possible migration from microplastics of phthalic acid esters and polycyclic aromatic hydrocarbons. J Polym Environ 29(2):599–611. https://doi.org/10.1007/s10924-020-01899-1

Kim Y, Kang J, Chun H (2022) Is online shopping packaging waste a threat to the environment? Econ Lett 214:110398. https://doi.org/10.1016/j.econlet.2022.110398

Koelmans AA, Kooi M, Law KL, Van Sebille E (2017) All is not lost: deriving a top-down mass budget of plastic at sea. Environ Res Lett 12(11):9500. https://doi.org/10.1088/1748-9326/aa9500

Kooi M, van Nes EH, Scheffer M, Koelmans AA (2017) Ups and downs in the ocean: effects of biofouling on vertical transport of microplastics. Environ Sci Technol 51(14):7963–7971. https://doi.org/10.1021/acs.est.6b04702

Kramm J (2023) Agential cuts of regulatory science practices—the case of microplastics. Environ Plan E: Nat Space. https://doi.org/10.1177/25148486231221020

Law Insider (2024) Institutional product definition. Dictionary. https://www.lawinsider.com/dictionary/institutional-product

Leslie HA, van Velzen MJM, Brandsma SH, Vethaak AD, Garcia-Vallejo JJ, Lamoree MH (2022) Discovery and quantification of plastic particle pollution in human blood. Environ Int 167:107400. https://doi.org/10.1016/j.envint.2022.107400

Lett Z, Hall A, Skidmore S, Alves NJ (2021) Environmental microplastic and nanoplastic: exposure routes and effects on coagulation and the cardiovascular system. Environ Pollut 291:118190. https://doi.org/10.1016/j.envpol.2021.118190

Li H, Lu X, Wang S, Zheng B, Xu Y (2021) Vertical migration of microplastics along soil profile under different crop root systems. Environ Pollut 278:116833. https://doi.org/10.1016/j.envpol.2021.116833

Li K, Qin Y, Zhu D, Zhang S (2023) Upgrading waste electrical and electronic equipment recycling through extended producer responsibility: a case study. Circ Econ 2(1):100025. https://doi.org/10.1016/j.cec.2023.100025

Little PC (2022) Burning matters: life, labor and e-waste pyropolitics in Ghana, 1st edn. Oxford University Press, Oxford

Luo Y, Naidu R, Fang C (2024) Toy building bricks as a potential source of microplastics and nanoplastics. J Hazard Mater 471:134424. https://doi.org/10.1016/j.jhazmat.2024.134424

Lwanga EH, Beriot N, Corradini F, Silva V, Yang X, Baartman J, Rezaei M, van Schaik L, Riksen M, Geissen V (2022) Review of microplastic sources, transport pathways and correlations with other soil stressors: a journey from agricultural sites into the environment. In: Chemical and biological technologies in agriculture, vol 9. Springer, Cham. https://doi.org/10.1186/s40538-021-00278-9

McDonald (2022) Packaging, toys & waste. Corporate. https://corporate.mcdonalds.com/corpmcd/our-purpose-and-impact/our-planet/packaging-toys-and-waste.html#:~:text=Transitioning%20Away%20From%20Virgin%20Fossil%20Fuel%2DBased%20Plastics&text=Sourcing%20100%25%20of%20our%20primary%20guest%20packaging1%2C%202%20from,by%20the%20end%20of%202025

Moita Neto JM, Silva EA (2023) Sources of microplastic generation in the environment. Int J Environ Res Public Health 20(13):6202. https://doi.org/10.3390/ijerph20136202

Napper IE, Bakir A, Rowland SJ, Thompson RC (2015) Characterization, quantity and sorptive properties of microplastics extracted from cosmetics. Mar Pollut Bull 99(1–2):178–185. https://doi.org/10.1016/j.marpolbul.2015.07.029

National Resources Defense Council (NRDC) (2024) Single-use plastics 101. Stories. https://www.nrdc.org/stories/single-use-plastics-101#what

Ng E-L, Lwanga EH, Eldridge SM, Johnston P, Hu H-W, Geissen V, Chen D (2018) An overview of microplastic and nanoplastic pollution in agroecosystems. Sci Total Environ 627:1377–1388. https://doi.org/10.1016/j.scitotenv.2018.01.341

Nishijima A, Nakatani J (2024) Survey and analysis on the use and disposal of plastic shopping bags before and after the introduction of charges. J Mater Cycl Waste Manage 26(2):741–754. https://doi.org/10.1007/s10163-023-01856-9

Núñez-Cacho P, Leyva-Díaz JC, Sánchez-Molina J, van der Gun R (2020) Plastics and sustainable purchase decisions in a circular economy: the case of Dutch food industry. PLoS One 15(9):e239949. https://doi.org/10.1371/journal.pone.0239949

Okeke ES, Okoye CO, Atakpa EO, Ita RE, Nyaruaba R, Mgbechidinma CL, Akan OD (2022) Microplastics in agroecosystems impacts ecosystem functions and food chain. Resour Conserv Recycl 177:105961. https://doi.org/10.1016/j.resconrec.2021.105961

Organisation for Economic Cooperation and Development (OECD) (2022) Global plastics outlook: economic drivers, environmental impacts and policy options. OECD Publishing, Paris

Organisation for Economic Co-operation and Development (OECD) (2022) Plastics use by application. Dataset. https://data-explorer.oecd.org/vis?tenant=archive&df[ds]=DisseminateArchiveDMZ&df[id]=DF_PLASTIC_USE_10&df[ag]=OECD&pd=%2C&to[TIME_PERIOD]=false&vw=tb&lb=bt

Paruta P, Pucino M, Bouche J (2021) Plastic pains the environment: a global assessment of paint's contribution to plastic leakage to Land, Ocean and Waterways. www.e-a.earth/publications

Pelamatti T, Rios-Mendoza LM, Hoyos-Padilla EM, Galván-Magaña F, De Camillis R, Marmolejo-Rodríguez AJ, González-Armas R (2021) Contamination knows no borders: toxic organic compounds pollute plastics in the biodiversity hotspot of Revillagigedo archipelago National Park. Mexico Marine Pollut Bull 170:112623. https://doi.org/10.1016/j.marpolbul.2021.112623

Peries SD, Sewwandi M, Sandanayake S, Kwon HH, Vithanage M (2024) Airborne transboundary microplastics–a swirl around the globe. In: Environmental pollution, vol 353. Elsevier, Amsterdam. https://doi.org/10.1016/j.envpol.2024.124080

Petrovic M, Mihajlovic I, Tubic A, Novakovic M (2023) Microplastics in municipal solid waste landfills. Curr Opin Environ Sci Health 31:100428. https://doi.org/10.1016/j.coesh.2022.100428

Pirgmaier E (2020) Consumption corridors, capitalism and social change. Sustain Sci Pract Policy 16(1):274–285. https://doi.org/10.1080/15487733.2020.1829846

Pizzorno J (2024) Plastic food container safety the path ahead. Integr Med 23:6–10

Rasheed FN, Walraven G (2023) Cleaning up plastics in healthcare waste: the transformative potential of leadership. BMJ Innov 9(2):103–108. https://doi.org/10.1136/bmjinnov-2022-000986

Richford R (2024) H&M Group makes moves on recycled materials in annual sustainability report. WWD Sustainability. https://wwd.com/sustainability/business/hm-recycled-materials-interview-sustainability-leyla-ertur-1236288566/

Rios-Mendoza LM, Ontiveros-Cuadras JF, Leon-Vargas D, Ruiz-Fernández AC, Rangel-García M, Pérez-Bernal LH, Sanchez-Cabeza JA (2021) Microplastic contamination and fluxes in a touristic area at the SE gulf of California. Mar Pollut Bull 170:112638. https://doi.org/10.1016/j.marpolbul.2021.112638

Ritcher F (2022) Cars still dominate the American commute. Industries in depth world economic forum https://www.weforum.org/agenda/2022/05/commute-america-sustainability-cars/

Rivers N, Shenstone-Harris S, Young N (2017) Using nudges to reduce waste? The case of Toronto's plastic bag levy. J Environ Manag 188:9. https://doi.org/10.1016/j.jenvman.2016.12.009

Rizan C, Mortimer F, Stancliffe R, Bhutta MF (2020) Plastics in healthcare: time for a re-evaluation. J R Soc Med 113(2):49–53. https://doi.org/10.1177/0141076819890554

Robertson L (2024) How ethical is H&M? Good on you. https://goodonyou.eco/how-ethical-is-hm/

Rohais S, Armitage JJ, Romero-Sarmiento M-F, Pierson J-L, Teles V, Bauer D, Cassar C, Sebag D, Klopffer M-H, Pelerin M (2024) A source-to-sink perspective of an anthropogenic marker: a first assessment of microplastics concentration, pathways, and accumulation across the environment. Earth Sci Rev 254:104822. https://doi.org/10.1016/j.earscirev.2024.104822

Rossi E, Director M, Prlic A, Hoffman R (2007) A study of the European cosmetics industry final report prepared for: European Commission, Directorate General for Enterprise and Industry Contact Information

Ryan J (2023) Single-use plastic waste from food delivery apps could be reduced with one tweak. News Science

Ryan PG, Suaria G, Perold V, Pierucci A, Bornman TG, Aliani S (2020) Sampling microfibres at the sea surface: the effects of mesh size, sample volume and water depth. Environ Pollut 258:113413. https://doi.org/10.1016/j.envpol.2019.113413

Schreyers LJ, Van Emmerik THM, Huthoff F, Collas FPL, Wegman C, Vriend P, Boon A, De Winter W, Oswald SB, Schoor MM, Wallerstein N, Van Der Ploeg M, Uijlenhoet R (2024) River plastic transport and storage budget. Water Res 259:121786. https://doi.org/10.1016/j.watres

Schubert C (2017) Green nudges: do they work? Are they ethical? Ecol Econ 132:329–342. https://doi.org/10.1016/j.ecolecon.2016.11.009

Schusser S et al (2013) Degradation of thin poly(lactic acid) films: characterization by capacitance-voltage, atomic force microscopy, scanning electron microscopy and contact-angle measurements. Electrochim Acta 8:25. https://doi.org/10.1016/j.electacta.2013.08.025

Setiawan M, Widjojo R, Purnama Alamsyah D (2023) Consumers' reasonings for intention to purchase natural personal care products in Indonesian market. Cogent Business Manage 11(1):2289205. https://doi.org/10.1080/23311975.2023.2289205

Sewwandi M, Kumar A, Pallewatta S, Vithanage M (2024) Microplastics in urban stormwater sediments and runoff: an essential component in the microplastic cycle. TrAC Trends Anal Chem 178:117824. https://doi.org/10.1016/j.trac.2024.117824

Shelton J (2023) Sounding the alarm about disposable plastic in the health care industry. Yale News

Shi P, Wan Y, Grandjean A, Lee JM, Tay CY (2021) Clarifying the in-situ cytotoxic potential of electronic waste plastics. Chemosphere 269:128719. https://doi.org/10.1016/j.chemosphere.2020.128719

Singh A, Mishra B K (2023) Microbeads in personal care products: An overlooked environmental concern. Journal of Cleaner Production, 427:139082. https://doi.org/10.1016/j.jclepro.2023.139082

Sinha P, Dissanayke DGK, Abeysooriya RP, Bulathgama BHN (2022) Addressing post-consumer textile waste in developing economies. J Text Inst 113(9):1887–1907. https://doi.org/10.1080/00405000.2021.1954428

Spectrum News (2021) McDonald's to phase out plastic happy meal toys by 2025. All Boroughs News

Stafford R, Jones PJS (2019) Viewpoint—ocean plastic pollution: a convenient but distracting truth? Mar Policy 103:187–191. https://doi.org/10.1016/j.marpol.2019.02.003

Stanislaus M (2018) Banning straws and bags won't solve our plastic problem. https://www.wri.org/insights/banning-straws-and-bags-wont-solve-our-plastic-problem#:~:text=For%20example%2C%20Denmark's%20Ministry%20of,we%20must%20invest%20in%20alternatives

Statista (2024) Consumer electronics - worldwide. Statista. https://www.statista.com/study/55488/consumer-electronics-market-market-data-and-analysis/

Stefan C, Ansems N (2018) Web-based global inventory of managed aquifer recharge applications. Sustain Water Resour Manage 4(2):153–162. https://doi.org/10.1007/s40899-017-0212-6

Stoett P (2024) Plastic waste colonialism: a typology of global toxicity. In: Gündoğdu S (ed) Plastic waste trade: a new colonialist means of pollution transfer. Springer, Cham, pp 3–15. https://doi.org/10.1007/978-3-031-51358-9

Sun Q, Ren SY, Ni HG (2020) Incidence of microplastics in personal care products: an appreciable part of plastic pollution. Sci Total Environ 742:140218. https://doi.org/10.1016/j.scitotenv.2020.140218

Sunaga N, Okochi H, Niida Y, Miyazaki A (2024) Alkaline extraction yields a higher number of microplastics in forest canopy leaves: implication for microplastic storage. Environ Chem Lett 22(4):1599–1606. https://doi.org/10.1007/s10311-024-01725-3

Talukdar A, Kundu P, Bhattacharya S, Dutta N (2024) Microplastic contamination in wastewater: sources, distribution, detection and remediation through physical and chemical-biological methods. Sci Total Environ 916:170254. https://doi.org/10.1016/j.scitotenv.2024.170254

Tang KHD, Hadibarata T (2021) Microplastics removal through water treatment plants: its feasibility, efficiency, prospects and enhancement by proper waste management. In: Environmental challenges, vol 5. Elsevier, Amsterdam. https://doi.org/10.1016/j.envc.2021.100264

The Scientists' Coalition for an Effective Plastics Treaty (2024) The scientists' coalition for an effective plastics treaty. The Scientists' Coalition for an Effective Plastics Treaty. https://ikhapp.org/scientist-working-groups/

Ton D, Bekhor S, Cats O, Duives DC, Hoogendoorn-Lanser S, Hoogendoorn SP (2020) The experienced mode choice set and its determinants: commuting trips in The Netherlands. Transp Res A Policy Pract 132:744–758. https://doi.org/10.1016/j.tra.2019.12.027

Tu JC, Chu KH, Gao DZ, Yang C (2022) Analyzing decision-making factors of green design for kid's toys based on the concept of product lifecycle. PRO 10(8):1523. https://doi.org/10.3390/pr10081523

Umeh OR, Ophori DU, Ibo EM, Eke CI, Oyen TP (2024) Groundwater systems under siege: the silent invasion of microplastics and cock-tails worldwide. Environ Pollut 356:124305. https://doi.org/10.1016/j.envpol.2024.124305

UNEP (2018) Legal limits on single-use plastics and microplastics. https://wedocs.unep.org/bitstream/handle/20.500.11822/27113/plastics_limits.pdf?sequence=1&isAllowed=y

United Nations Environment Programme (UNEP) (2024) Our planet is choking on plastic. Beat Plastic Pollution https://www.unep.org/interactives/beat-plastic-pollution/

United Nations Environmental Programme (UNEP) (2020) The little book of green nudges. https://citaviprojectdata.blob.core.cloudapi.de/vdiypifpqwq6bxnhrrv3usa1yf7xgoxylm8t8uzk8ci5qseftq/80342cca-bfe9-4356-b395-0323261c1e98?sv=2018-03-28&sr=c&sig=vGklwAMQD85GKipyuw4VwI29At26HpdhYn7lDAPMhzE%3D&se=2020-01-04T16%3A45%3A38Z&sp=rl

United Nations Institute for Training and Research (UNITAR) (2024) Global e-waste monitor 2024. Global e-waste monitor 2024: electronic waste rising five times faster than documented e-waste recycling. https://unitar.org/about/news-stories/press/global-e-waste-monitor-2024-electronic-waste-rising-five-times-faster-documented-e-waste-recycling

United States Environmental Protection Agency (EPA) (2023) Containers and packaging: product-specific data. Facts and Figures

US Food and Drug Administration (USFDA) (2022) The microbead-free waters act: FAQs. Cosmetics. https://www.fda.gov/cosmetics/cosmetics-laws-regulations/microbead-free-waters-act-faqs#:~:text=What%20is%20the%20Microbead%2DFree,)%20drugs%2C%20such%20as%20toothpastes

van Emmerik T, Mellink Y, Hauk R, Waldschlager K, Schreyers L (2022) Rivers as plastic reservoirs. Front Water 3:786936. https://doi.org/10.3389/frwa.2021.786936

Van Rensburg ML, Nkomo SL, Dube T (2020) The 'plastic waste era'; social perceptions towards single-use plastic consumption and impacts on the marine environment in Durban, South Africa. Appl Geogr 114:102132. https://doi.org/10.1016/j.apgeog.2019.102132

Wagner TP (2017) Reducing single-use plastic shopping bags in the USA. Waste Manag 70:3–12. https://www.sciencedirect.com/science/article/pii/S0956053X17306335?dgcid=raven_sd_recommender_email. Accessed 22 April 2019

Wang J, Li Y, Lu L, Zheng M, Zhang X, Tian H, Wang W, Ru S (2019) Polystyrene microplastics cause tissue damages, sex-specific reproductive disruption and transgenerational effects in marine medaka (Oryzias melastigma). Environ Pollut 254:113024. https://doi.org/10.1016/j.envpol.2019.113024

WEEE Forum (2022) International e-waste day: of ~16 billion mobile phones possessed worldwide, ~5.3 billion will become waste in 2022. News. https://weee-forum.org/ws_news/of-16-billion-mobile-phones-possessed-worldwide-5-3-billion-will-become-waste-in-2022/

Wells K (2024) Purging old clothes? Here's how to responsibly donate and repurpose your castoffs. Wirecutter by The New York Times. https://www.nytimes.com/wirecutter/guides/how-to-get-rid-of-old-clothes/

World Economic Forum (WEF) (2024) Consumers beyond waste. Initiatives. https://initiatives.weforum.org/consumers-beyond-waste/home

Wrap (2024) Wrap. Textiles. https://www.wrap.ngo/taking-action/textiles#:~:text=The%20Textiles%20Action%20Network%20is,economy%20for%20fashion%20and%20textiles

Ylä-Mella J, Keiski RL, Pongrácz E (2022) End-of-use vs. end-of-life: when do consumer electronics become waste? Resources 11(2):18. https://doi.org/10.3390/resources11020018

Yuan Q, Sun R, Yu P, Cheng Y, Wu W, Bao J, Alvarez PJJ (2022) UV-aging of microplastics increases proximal ARG donor-recipient adsorption and leaching of chemicals that synergistically enhance antibiotic resistance propagation. J Hazard Mater 427:127895. https://doi.org/10.1016/j.jhazmat.2021.127895

Zerbini C, Bijmolt THA, Maestripieri S, Luceri B (2022) Drivers of consumer adoption of e-commerce: a meta-analysis. Int J Res Mark 39(4):1186–1208. https://doi.org/10.1016/j.ijresmar.2022.04.003

Zhang YQ, Lykaki M, Alrajoula MT, Markiewicz M, Kraas C, Kolbe S, Klinkhammer K, Rabe M, Klauer R, Bendt E, Stolte S (2021) Microplastics from textile origin-emission and reduction measures. Green Chem 23(15):5247–5271. https://doi.org/10.1039/d1gc01589c

Microplastic: Global Pollution Controlling Policy

10

Mohammad Oves, Djadjiti Namla, Majed Al-Shaeri, Amamad Samman, and Nasir Ahmed Al-Kinani

Abstract

Microplastics (MPs) pollution poses a significant and growing threat to the health of our oceans, marine life, and ultimately, human well-being. These minute plastic particles, measuring less than 5 mm in size, derive from a variety of sources including plastic debris breakdown, industrial processes, and microbeads in personal care products. One of the major challenges posed by the presence of microplastics in the environment is the difficulty in managing their emissions from sources such as textiles and tires. Marine organisms are more affected by MPs—by ingestion and entanglement, as well as their potential transfer through the food web. Additionally, MP pollution is associated with socio-economic ramifications because of cleanup efforts and the potential loss of ecosystem services. In response to the growing concern over microplastic pollution control, various mitigation actions have been identified to address their widespread distribution in the environment. The present chapter is mainly focused on the comprehensive global policy to address MP pollution in marine ecosystems such as national and regional policies targeting their proliferation. The chapter also presents a wide discussion of the various methods for quantitatively and qualitatively detecting MPs, management strategies, and emerging

M. Oves (✉)
Centre of Excellence in Environmental Studies, King Abdulaziz University, Jeddah, Kingdom of Saudi Arabia
e-mail: owais.micro@gmail.com

D. Namla · M. Al-Shaeri · N. A. Al-Kinani
Department of Biological Sciences, Faculty of Sciences, King Abdul Aziz University, Jeddah, Kingdom of Saudi Arabia

A. Samman
Department of Environmental Sciences, Faculty of Sciences, King Abdul Aziz University, Jeddah, Kingdom of Saudi Arabia

M. Oves et al. (eds.), *Marine Microplastics and the Quest for Remediation*,
https://doi.org/10.1007/978-981-96-8979-8_10

technologies for MPs reduction. Moreover, it examines the role of international collaboration in addressing this transboundary issue, emphasizing the importance of coordinated efforts and knowledge-sharing among stakeholders.

Keywords

Microplastics · Pollution · Policy · Marine · Ecosystems · Sustainability

10.1 Introduction

Microplastic pollution has become one of the most significant global environmental challenges in the twenty-first century, with widespread impacts on ecosystems, human health, and economic stability (Pachuau et al. 2024). Microplastics are plastic particles smaller than 5 mm in diameter, originating from a variety of sources such as the breakdown of larger plastic items, industrial activities, and personal care products containing microbeads. These small particles are present in both terrestrial and aquatic environments and have even been detected in some of the most remote areas of the planet (Pachuau et al. 2024; Ziani et al. 2023). Unlike larger plastic debris, microplastics are not easily visible to the naked eye, making them more difficult to detect, regulate, and remove. Consequently, these particles accumulate at alarming rates in marine and freshwater systems, soil, and even the atmosphere, posing a growing threat to global ecosystems (Ali et al. 2024).

The persistence of microplastics is primarily due to the inherent properties of plastic materials, which degrade slowly over time (Biswas and Pal 2024). Instead of breaking down completely, plastic particles fragment into smaller pieces, further increasing their impact on the environment (Gallo et al. 2018; Sutkar et al. 2023). The widespread use of plastics in modern society, coupled with inefficient waste management systems, has exacerbated the issue. Industries such as packaging, textiles, and automotive sectors are significant contributors to microplastic pollution through the improper disposal of plastic products and the release of synthetic fibers and particles into the environment (Qaiser et al. 2023). As a result, microplastic pollution has evolved into a complex, multifaceted issue that requires urgent attention from policymakers, researchers, and stakeholders worldwide (Ali et al. 2024; Qaiser et al. 2023).

Microplastics enter natural ecosystems through various pathways, including wastewater discharge, agricultural runoff, atmospheric deposition, and improper waste disposal. These particles tend to persist in the environment for extended periods, accumulating in water bodies, sediments, and even marine organisms (Duis and Coors 2016). Microplastics are easily carried by ocean currents and wind, allowing them to spread over long distances and making their pollution a transboundary issue that impacts countries and ecosystems worldwide (Emenike et al. 2023). The global nature of the problem highlights the need for international cooperation to tackle the growing crisis of microplastic pollution (Yang et al. 2021).

The impact of microplastics on marine ecosystems is both extensive and complex, affecting the physical environment and the organisms that inhabit it. Research has shown that microplastics have been found in the digestive systems of a wide range of marine species, from tiny plankton to large mammals (Marcharla et al. 2024). These particles can block the digestive tracts of marine animals, reduce their feeding efficiency, and cause internal injuries, all of which pose significant risks to marine biodiversity (Ali et al. 2024; Pothiraj et al. 2023). Furthermore, microplastics can act as carriers of harmful pollutants, such as persistent organic pollutants (POPs), that attach to the surface of the particles (Paul et al. 2024). These toxic substances can bioaccumulate and biomagnify through the food chain, eventually reaching humans who consume seafood, thereby posing serious health risks (Zolotova et al. 2022).

Humans are exposed to microplastics through various pathways, including the consumption of contaminated seafood, drinking water, and inhalation of airborne particles (Ali et al. 2024). Studies have shown that microplastics can accumulate in human tissues, which raises concerns about their long-term health effects (Ghosh et al. 2023; Tan et al. 2021; Lee et al. 2023). Potential health risks associated with microplastic exposure include inflammation, hormonal disruption, and an increased risk of cancer (Lee et al. 2023; Bora et al. 2024). While more research is needed to fully understand the health implications of microplastic exposure, the initial findings are alarming and indicate the need for immediate action to mitigate these risks. Figure 10.1 illustrates the health effects of microplastic exposures, derived from Lee et al. (2023). It highlights the various pathways via which microplastics enter the human body and their potential health consequences.

The economic consequences of microplastic pollution are equally significant. Industries that rely on marine resources, such as fisheries, aquaculture, and tourism, are particularly vulnerable to the negative impacts of microplastic contamination (Ziani et al. 2023; Ali et al. 2024). The degradation of marine ecosystems caused by

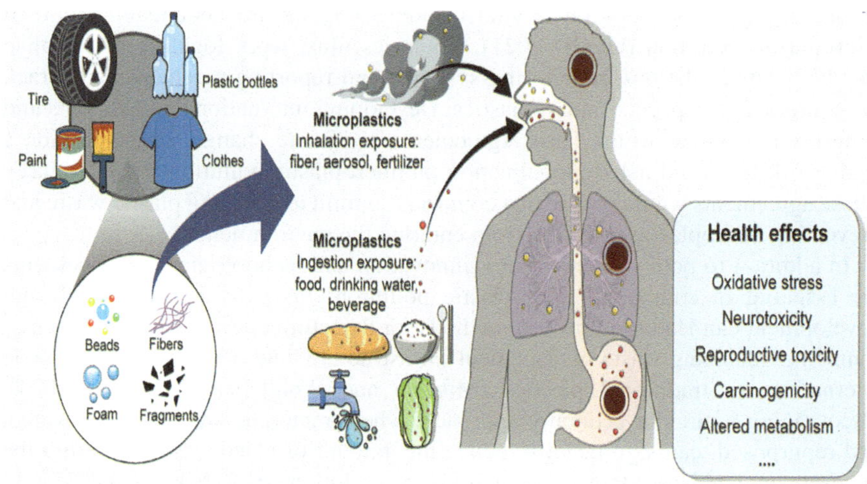

Fig. 10.1 Health effects of microplastic exposures, derived from Lee et al. (2023)

plastic pollution can reduce fish stocks, affect biodiversity, and disrupt the livelihoods of communities that depend on marine resources for their income (Ali et al. 2024). Additionally, the costs associated with cleaning up plastic pollution and addressing its effects on ecosystems place a significant financial burden on governments and municipalities (Anthony et al. 2024). As the issue continues to escalate, it is becoming increasingly evident that inaction will lead to long-term economic losses and environmental degradation (Kumar et al. 2021).

Given the complexity and global scale of the issue, there is an urgent need for effective policies to control and mitigate microplastic pollution. Despite growing awareness of the problem, existing policies and regulatory frameworks are often fragmented, inconsistent, and limited to national or regional levels (Rivera-Garibay et al. 2024). To address the root causes of microplastic pollution and promote sustainable practices, there is a critical need for coordinated, comprehensive global policies (Usman et al. 2022). Such policies should aim to reduce plastic production, improve waste management systems, and encourage the development of alternative materials that are biodegradable or recyclable (Yang et al. 2024).

Effective policies should also prioritize raising public awareness about the dangers of microplastic pollution and promoting responsible consumer behavior to reduce plastic waste at the source. Regulations should target industries that contribute significantly to microplastic pollution, such as the packaging, textile, and personal care sectors, by imposing stricter standards on the use of plastics and encouraging the adoption of sustainable alternatives. Public participation and collaboration between governments, industries, and communities are essential for the successful implementation of these policies. However, key policy approaches and frameworks that have been implemented globally to mitigate microplastic pollution are highlighted in Fig. 10.2, illustrating the international and regional governance strategies developed to combat plastic and microplastic pollution.

International collaboration is vital to addressing the transboundary nature of microplastic pollution (OECD 2021). Countries must work together to establish standardized regulations, monitoring systems, and reporting mechanisms to track the sources and impacts of microplastics. Developing international agreements and conventions, similar to the Paris Agreement on climate change, could provide a framework for coordinated global action on microplastic pollution (OECD 2021). These agreements would ensure that countries commit to reducing plastic waste and preventing microplastic pollution from entering the environment.

In addition to policy interventions, innovation and technological advancements are essential in combating microplastic pollution. Investments in research and development can lead to improved methods for detecting microplastics in the environment, enhancing waste management infrastructure, and developing sustainable alternatives to traditional plastics (Mitrano and Wohlleben 2020; Xu 2024). Encouraging a circular economy approach, where materials are reused, recycled, and repurposed, can significantly reduce the amount of plastic waste entering the environment and help address the microplastic pollution crisis (Nikpay and Toorchi Roodsari 2024).

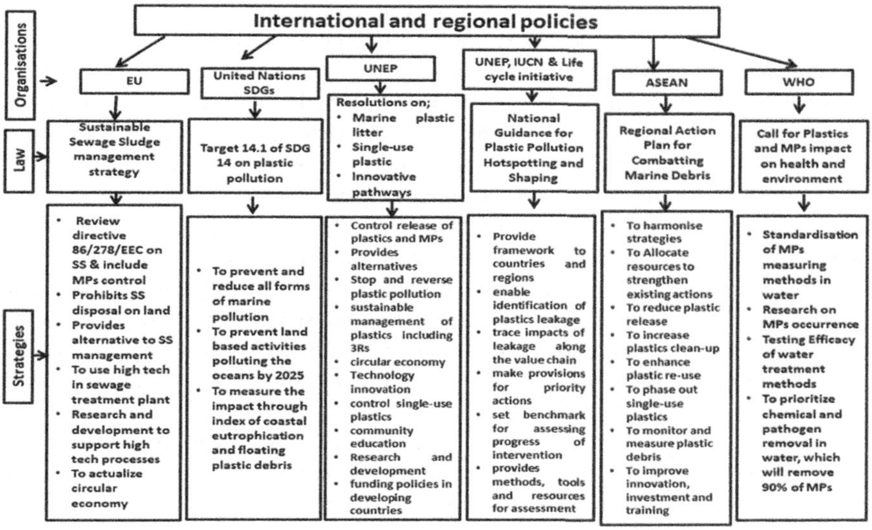

Fig. 10.2 Governance strategies by international and regional organizations to combat plastic and MPs pollution, derived from Usman et al. (2022)

Addressing this issue requires immediate action through the development and implementation of effective global policies, international collaboration, and sustainable practices. By addressing the root causes of microplastic pollution and promoting innovative solutions, we can mitigate its harmful effects and protect the health of both ecosystems and human populations. A comprehensive and coordinated approach is essential to tackle this complex environmental challenge and ensure a sustainable future for future generations.

10.2 Current Global Status of Microplastic Pollution

With the exponential rise in plastic production over the past few decades, microplastic particles have permeated nearly every part of the planet, from oceans and rivers to soil and even the air we breathe (Matavos-Aramyan 2024). The current global status of microplastic pollution underscores the urgent need for coordinated international policies and actions to address this pervasive environmental threat (Ghosh et al. 2023). According to recent statistics, over eight million metric tons of plastic waste enter the world's oceans every year, a significant portion of which breaks down into microplastic particles (Ghosh et al. 2023; OECD 2021; UNEP 2023). It is estimated that around 51 trillion microplastic particles currently float in the oceans, weighing up to 269,000 tons (UNEP 2023). These particles are found in every corner of the marine environment, from surface waters to deep-sea sediments. Alarmingly, microplastics have also been detected in freshwater bodies, soil, and atmospheric dust, indicating that no ecosystem is free from this form of pollution.

Scientific studies suggest that microplastics now outnumber marine organisms in certain areas of the ocean, with devastating consequences for biodiversity and ecosystem health (OECD 2021; UNEP 2023).

The production of plastics has skyrocketed globally, with industries producing over 400 million tons of plastic annually. Approximately 40% of all plastic produced is for single-use items, many of which end up as waste. A significant portion of this plastic waste breaks down into microplastic particles, which remain in the environment for hundreds of years (Kibria et al. 2023). One of the primary reasons microplastics are so prevalent is due to the widespread use of synthetic materials across industries, including packaging, textiles, cosmetics, and automotive sectors (OECD 2021; Šaravanja et al. 2022). The textile industry, for example, is one of the major contributors to microplastic pollution, as synthetic fibers shed from clothing during washing processes are discharged into water bodies through wastewater systems (Šaravanja et al. 2022; Akyildiz et al. 2024).

Certain regions and industries contribute disproportionately to the global microplastic pollution crisis. Developed countries, particularly in North America and Europe, are among the largest producers and consumers of plastic, generating significant amounts of plastic waste (Akyildiz et al. 2024). However, developing countries, especially in Asia, are often where the impacts of this pollution are most visible. Asia is considered a hotspot for plastic pollution due to inadequate waste management infrastructure, rapid urbanization, and high levels of plastic consumption. Countries such as China, Indonesia, the Philippines, and India are among the leading contributors to marine plastic pollution, largely due to the improper disposal of plastic waste (IUCN 2024). In Fig. 10.3, the life cycle assessment (LCA) framework for car tire production and microplastic release in real-world scenarios is described. The yellow box represents the stage where microplastics are actively produced and subsequently released through major pathways.

The packaging industry is the single largest contributor to plastic waste, accounting for nearly 50% of global plastic consumption. Disposable plastic items, such as bags, bottles, and food containers, often end up in the environment as litter or improperly managed waste (IUCN 2024). The personal care industry also plays a significant role, particularly through the use of microbeads in cosmetics, which are small plastic particles used as exfoliants in products such as facial scrubs and toothpaste (UNEP 2023). These microbeads pass through wastewater treatment plants and are released into water bodies, where they contribute to microplastic pollution (UNEP 2023; IUCN 2024).

Marine ecosystems are among the most heavily affected by microplastic pollution. Microplastic particles are now found in virtually all marine environments, from coastal waters to the deep sea (Ziani et al. 2023; Ali et al. 2024). Marine species at all trophic levels, from plankton to whales, are exposed to microplastics, often mistaking them for food (Kaliszewicz et al. 2023). This can lead to physical harm, such as blockages in digestive systems, and chemical harm, as microplastics often carry toxic pollutants like persistent organic pollutants (POPs). These toxins can bioaccumulate through the food chain, affecting predator species, including humans who consume seafood. Coral reefs, mangroves, and seagrass beds, which

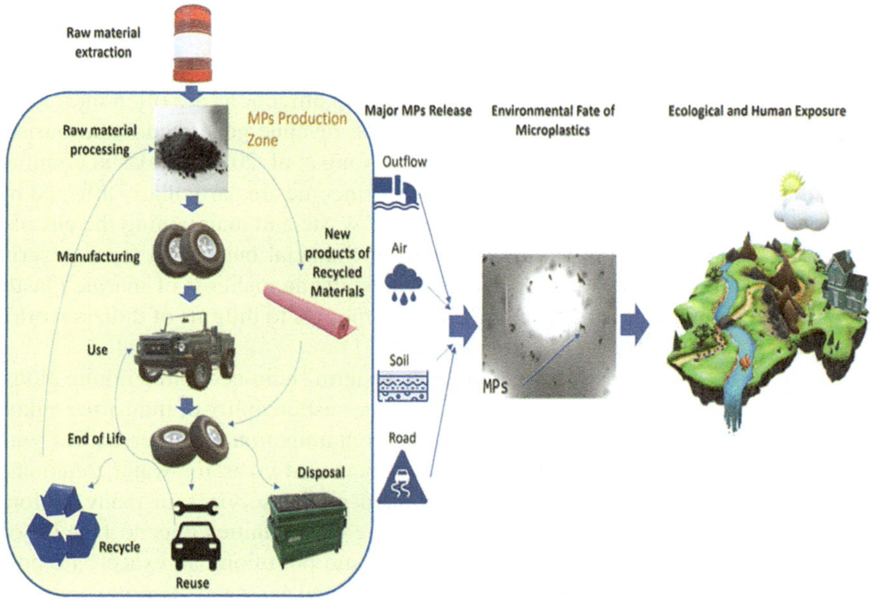

Fig. 10.3 Life cycle assessment framework for car tire production and microplastic release in real-world scenarios. The yellow box represents the stage where MPs are actively produced and subsequently released through major pathways, derived from the results presented by Nikpay and Toorchi-Roodsari (Nikpay and Toorchi Roodsari 2024)

are crucial for marine biodiversity, are also impacted by microplastic accumulation, leading to reduced habitat quality and overall ecosystem degradation (Ziani et al. 2023; Ali et al. 2024; Kaliszewicz et al. 2023).

Freshwater ecosystems, such as rivers and lakes, are similarly affected. Rivers act as conduits for microplastic pollution, carrying plastic waste from inland areas to the oceans. Studies have found significant microplastic contamination in major rivers worldwide, including the Yangtze River in China, the Ganges River in India, and the Mississippi River in the United States (Chen et al. 2024; Swain et al. 2025). These freshwater bodies serve as crucial water sources for millions of people and play vital roles in the agriculture and fishing industries. The presence of microplastics in these water bodies threatens both human health and the sustainability of aquatic ecosystems.

In addition to environmental impacts, microplastic pollution poses significant socio-economic risks to communities that depend on marine and freshwater resources. The fisheries and aquaculture industries, which provide food and livelihoods for millions of people, are particularly vulnerable to the negative effects of microplastic contamination (Matavos-Aramyan 2024; Wang et al. 2024a). Fish stocks are declining due to habitat degradation and pollution, reducing the income and food security of fishing communities, particularly in developing countries. The presence of microplastics in seafood also raises concerns about the safety and marketability of these products, which can impact global trade and economic stability (Landrigan et al. 2020; Wang et al. 2024b).

The tourism industry is another sector heavily impacted by microplastic pollution. Pristine beaches, coral reefs, and other natural attractions are often marred by plastic debris, deterring tourists and reducing the revenue generated from tourism activities (Chen et al. 2024; Swain et al. 2025; Wang et al. 2024a). Coastal communities that rely on tourism as a primary source of income are particularly affected by this issue (Raes et al. 2021). Cleaning up plastic waste and maintaining the cleanliness of tourist destinations places a significant financial burden on local governments and communities. According to estimates, the annual cost of marine plastic pollution cleanup and related economic losses amounts to billions of dollars worldwide (OECD 2021; UNEP 2023; Raes et al. 2021).

Furthermore, microplastic pollution has long-term socio-economic implications, as it threatens food security, public health, and the sustainability of industries reliant on natural resources (Raes et al. 2021). The contamination of water bodies with microplastics can lead to water scarcity and increased costs for water treatment, placing additional pressure on already strained public services in many regions (Iheanacho et al. 2023). In coastal areas, where communities depend heavily on marine resources for their livelihoods, microplastic pollution can exacerbate poverty and inequality, particularly in regions with limited access to alternative income sources (Fred-Ahmadu et al. 2024).

Certe, the current global status of microplastic pollution presents a dire environmental and socio-economic challenge. The pervasive nature of microplastic particles, their accumulation in key ecosystems, and their far-reaching impacts on human health and industries highlight the need for urgent and coordinated action. Addressing microplastic pollution requires not only effective waste management and reduction strategies but also global policies and international cooperation to tackle this growing crisis at its root causes. Without immediate and sustained efforts, the socio-economic and environmental costs of microplastic pollution will continue to rise, threatening ecosystems, livelihoods, and public health on a global scale.

10.3 Existing Policies and Regulations on Microplastic Pollution

The increase in research on the MP hazards, their presence, and potential risks to the environment and human health have drawn much attention. International treaties addressing different aspects of MP pollution often lack clarity and consistency in defining the target pollutants. National and local regulations on microplastic pollution are issued by different authorities, which sometimes leads to discrepancies (Abad López et al. 2023). Because current laws and regulations on MP pollution are not yet comprehensive, complete, and clear, most of them only focus on broad categories such as plastic pollution without taking the minature plastics into consideration (Usman et al. 2020, 2021). These policies and plans have different targets of microplastic pollution including cosmetic and solid waste, without coherence.

Ideally, a collective approach to control MP pollution could be made in a way that the policies from different levels of authority align with and support each other. Regulations should be recommended to comprehensively implement enforcement mechanisms and strengthen risk assessment and monitoring. Moreover, a collective approach to address the issue of global microplastic pollution is crucial.

Unarguably, plastic waste is difficult to degrade and is retained in the environment for a long time, breaking down into fragmentary plastics (particles less than 5 mm), so-called microplastics (Pothiraj et al. 2023; Yang et al. 2024). The main sources of microplastics in the environment are the primary (including eroded or discarded plastic products) and the secondary (friction and wear of larger plastics) microplastics. It is estimated that at least 5.25 trillion particles (weighing 268,940 tons) of marine plastic debris are currently floating at sea (UNEP 2023). They are both found in the water and the air, posing a serious threat to the ecological environment (Usman et al. 2021). There is still much uncertainty in the long-term behavior, transport, and ecological effects of microplastics in both solid and aquatic environments. Furthermore, they will also have serious impacts on human health and the sustainability of marine ecosystems as they enter the food chain (Usman et al. 2020).

10.3.1 Gaps and Challenges in Current Microplastic Control Policies

While some microplastic policies have been developed by many countries, there is an urgent need for concerted, uniform, and globally recognized policies. The absence of uniformity has led to inconsistencies in policy approaches, legislative frameworks, funding, enforcement mechanisms, reporting and documenting mechanisms, and multilevel institutional cooperation (Nikiema and Asiedu 2022). As a result, comprehensive and interdisciplinary global efforts are required to monitor, report, control, and inform on microplastic pollution at community, national, and transboundary levels (Nikiema and Asiedu 2022). However, there are significant financial gaps in the monitoring, sampling, extraction, quantification, categorization, and tracking microplastics (Antohi et al. 2023). The majority of communities cannot monitor or track adequately due to financial barriers, which complicate a rigorous understanding of its global distribution, abundance, polymers, shapes, sizes, sources, sinks, and subsequent overaccumulation in the environment (Yarahmadi et al. 2024).

The circularity in consumer behavior, including its expectations, reduction-reuse-recycle cultural enforcement, institutional compliance, and corporate responsibilities has been lacking (Hartmann et al. 2019). So, the implementation of various policies has resulted in a fragmented effort. Although several countries have banned single-use plastics, numerous exceptions for specific shapes, sizes, and polymer limitations have resulted in excessive and uncontrolled use (Cornago et al. 2021). Likewise, policy applications to certain sectors or countries have motivated circumvention strategies, including transshipment, smuggling, and unreported activities. Microplastics are also generated as secondary sources of larger plastic debris, such as shopping bags, films, foams, bottles, tire treads or threads, ropes, and kits (Kumar et al. 2021; Cornago et al. 2021). The control of secondary sources is more complicated than that of primary sources: (1) without proper care, utility, and education, large plastics deteriorate into vast microplastic granules; (2) the conversion mechanisms of atmospheric and aquatic environments are mechanical, thermal, oxidative, ultraviolet, and biodegradative; and (3) in granular form, it becomes particularly complex to prioritize and clean up (Pfohl et al. 2022). In spite of the undeniable challenges, global policy responses remain largely fragmented and do not give enough priority to the control of secondary plastic sources (Qaiser et al. 2023). Given the rapid diversification of plastic products, packaging, additives, composite agents and newly emerging pollutants, the current approach to developing global microplastic policies remains inadequate (OECD 2021; UNEP 2023).

Unambiguous policies and financial incentives should be developed and implemented to aid in both normalizing the transformation of all plastics into biodegradable, recyclable, and compostable materials, and then bioprocessing them. An additional dimension requiring urgent attention is appropriate global consideration for combating, mitigating, and adapting to the pervasive environmental and pollution effects of microplastics. Since the adverse effects of microplastics on humans, the food chain, ecosystems, biodiversity, and ecosystem resilience have been rather recently identified and reported, that concern is currently largely missing in the most of legal and control frameworks. Efforts to normalize, mainstream, and integrate the cost of environmental and disposal considerations in concert with emerging insight are to be commended (Lee et al. 2023). Despite various policies that have been designed to control and minimize plastic pollution-induced environmental degradation, they have led to the unintended consequence of exacerbating social and territorial inequality. Fundamentally deprived individuals and communities often bear the brunt of pollution wrongdoing. Policy transformations should be well-designed to ensure broad and inclusive engagement.

Moreover, justice and equity principles should not be an afterthought but rather should be a core component of the policy vision, objective, substance, focus, enactment, enforcement, and evaluation. Although environmental protection is critical to human well-being, it can not be achieved at the expense of other fundamental rights, including public health and social welfare. On the grounds of national security and public order, changes in law and policy should not contribute to a modus operandi of silencing, suspending, brutalizing critical voices, scholars, and dissenters.

10.3.2 Lack of Uniform Global Policies

There is no uniform global policy addressing microplastic pollution. When countries introduce a comprehensive regulation on plastic, which includes MP, the term might be defined as a low-confusing broad-to-narrow frame for clear management purposes or to comply with the nation's policy environment. The absence of a uniform global policy has led to variances in the availability of policy among countries and class-by-class regulatory disparities in the management of microplastics, the most recent emerging issue of the global plastic waste problem (Hartmann et al. 2019). Depending on how the country manages plastic issues, such as environmentally, socially, and economically whether the severe effects or the benefits have more weight, the policy responses might differ significantly. The disparities in national regulations will result in different environmental protections. Because of the complexity of MP sources and mass flows and different concerns, with many a severe environmental threat, countries cannot regulate every one of them simultaneously. Some govern water bodies focusing on the prevention of bioaccumulation and physical habitat damage and some substances including MP can affect animals via water vehicles, therefore, mainly govern product-specific regulations (Filiciotto and Rothenberg 2021). Apart from sectoral or substance governance, market-based measures in economic studies are also popular. However, the disparities in national regulations will result in differences in the content of national regulation type (Corinaldesi et al. 2021). One example is that, in those with high GDPs, regulatory approaches to voluntary initiatives like every other taking turns as the promoter of controls were taken because of market concerns (OECD 2021; UNEP 2023; Mattsson et al. 2022; Kaloyianni et al. 2021). Those differences make the participants in international trade face different regulations on microplastics (Aragaw 2021). Those impacts cannot be mitigated adequately with current international policy efforts and agreements. In addition, delegated regulations of a country mean an accountability issue. Poor national regulations might result in a lack of assurance of necessary protections (OECD 2021; UNEP 2023; Aragaw 2021). A strict federal regulation was formulated, but delegate creation was made to the states. However, not all 50 states need to manage MPs, and the government might not have enough capacity to examine the regulations of all delegates. Moreover, the understanding of MPs is still in its infancy, so prosecutions are even harder. Finally, different definitions of microplastics across jurisdictions would cause regulatory confusion like product life definitions. Harmonization of microplastics regulations is only now beginning to form the subject of international discussions. A UN-led review of global regulations with the involvement of G20 countries is ongoing, but it is said that it may take quite an extended period to reach a consensus (Aragaw 2021). Global policy efforts on micropollutants must be comprehensive and uniform (partial treatment cannot solve a global problem). The most recent and currently effective policy efforts taken in this regard made evidence of the exact opposite trend.

10.3.3 Challenges in Monitoring and Tracking Microplastic Pollution, Financial Constraints

There are multidimensional, technical, and operational challenges in the monitoring and tracking of microplastic pollution. This is largely due to the diversity and intrinsic nature of this pollutant as well as the fact that its presence and abundance over time, through various media, and across environments have yet to be properly established (Landrigan et al. 2023). The key concerns are related to the technical difficulties and challenges associated with the detection, analysis, and monitoring of microplastic particles in different ecosystems. Microplastic particles have a diverse range of sizes, morphologies, and densities that can complicate the isolation of these particles from a variety of environmental samples and can hinder their subsequent analysis. Progress in understanding and knowledge accumulation are the prerequisites to developing innovative and effective measures to confront this emerging pollutant (Zhang et al. 2018).

Despite the increasing attention given to microplastics via the publication of numerous research papers and reports, a knowledge base is still lacking to understand the key aspects of their environmental occurrence. Data on freshwater systems is fragmentary if not absent and this hampers the science-based assessment of the environmental risk of microplastics in these important ecosystems (Koelmans et al. 2019). There are calls for scientists to address the current knowledge gaps of microplastics in the freshwater, and to provide an adequate base for an evidence-based discussion. However, one of the main challenges is the substantial amount of funding that this will require. It is recognized that financial constraints on scientific programs and actions can result in sub-optimal results and the inability to draw meaningful conclusions (Wagner et al. 2014). Therefore, raising the issue of funding is necessary to complete or establish well-designed programs to address all facets of the microplastics problem with the best scientific and technical capacity. The need for significant financial investment is emphasized in academia, governmental institutions, and the general public (Wagner et al. 2014; Lusher et al. 2017). Finding, understanding and ultimately solving a problem of such a broad spectrum and global dimensions require a comprehensive approach and cannot be resolved by institutions or interdisciplinary scientists acting on their own. In this respect, mechanisms for collaborative funding between different sectors and actors could be applied to combine development, know-how, infrastructure, and human resources.

10.3.4 Consumer Behavior, Corporate Responsibility, and Fragmentation in Policy Implementation

Given the interplays between consumer behavior and corporate responsibility, consumer choice can directly impact overall plastic production and waste generation. China, being the largest importer of plastic waste for many years, banned the import of waste in early 2017, causing a shift of waste disposal strategies around the world (Bora 2020; Carr et al. 2019). The recyclers in many wealthier countries are now stockpiling or landfilling recovered plastics to respond to this change (Carr et al. 2019). From a consumer perspective, this means that even diligent efforts to recycle can be meaningless if the end-use country no longer accepts the imports, resulting in many wastes being incinerated or landfilled. While internationally there has been an emphasis on consumer ethics regarding disposal practices, the choice of what to purchase is ignored. Given that human civilization has only been producing plastics en masse for the last 70 years, the use and mismanagement of plastic waste is essentially a modifiable consumer behavior (Carr et al. 2019).

Currently, microplastic pollution control policies are fragmented and mostly focus on the downstream impacts, such as recycling origins and the pathways plastic waste undergoes in the environment, whereas other key areas, including environmental or health effects, are less addressed. However, there are still key players in the value chain that are often isolated from the policy efforts, namely, the corporations (Matavos-Aramyan 2024). Corporate strategies to address plastic production and waste generation are varied, and the results of these strategies have been relatively ineffective vis-à-vis the policy objectives, partially due to the complexity of global plastic value chains. As a result, policies implemented have been predominantly localized and focused on the least economically powerful polity, the consumers (Ghosh et al. 2023). It is clear that unless the upstream policy interventions are multipronged in nature, the efforts to control the microplastic issues will be inconsistent. Overall, the manipulative nature of plastic usage and its disposal suggests that promoting responsible practices can be very cost-effective and striking, and importantly the combined effect of educated consumers can partly drive the demand for sustainable products, spurring market innovation (OECD 2021). All that is needed is a greater alignment of consumer interests with reinforcing corporate responsibility, thereby fostering synchronous policy implementation across the global value chain.

10.3.5 Challenges in Secondary Microplastic Control

Plastic products are widely used for product packaging, industrial equipment, daily essentials, and other purposes (Williams and Rangel-Buitrago 2022). Due to their low price and durability, plastic items are used worldwide by a plethora of industries, economies, and consumers. However, plastic waste products have various negative ecological results, including pollution of aquatic environments, habitats, and ecosystems (UNEP 2023). Remaining plastic consumer goods in terrestrial, fluvial, and marine environments contribute directly or indirectly to the breakdown of enter-chain plastics in natural elements, eventually reaching natural resources, the atmosphere, and food chains (Landrigan et al. 2020). There are numerous harmful health effects of this ecological consumption, such as chronic fulfillment of contaminated foodstuffs and drinks, following ingestion, respiration, or dermal exposure to micro- and macro-plastic goods (Hartmann et al. 2019).

Macroplastics initially decay into smaller-sized plastic constituents, fairly dependent on environmental conditions permitting degradation processes. Degradation comprises a sequence of photochemical alterations resulting in the inception of microscopic plastic particles; conversely, mechanical procedures typically damage or divide the polymer into smaller untouched-ish polymers with really constant monomer species (Kumar et al. 2021; UNEP 2023; Landrigan et al. 2020; Hartmann et al. 2019). After a few decades, natural disasters, wear, and tear, or operating conditions make plastic objects more brittle, like grains smaller than 5 mm, breaking up or shaping microplastic powders (Williams and Rangel-Buitrago 2022).

10.3.6 Limited Integration of Emerging Technologies

Plastic pollution events and the subsequent global investigations have increased attention to microplastic pollution and facilitated the understanding of microplastic pollution sources, distribution characteristics, potential risks, and pollutant exposure pathways (Williams and Rangel-Buitrago 2022; Campanale et al. 2020). The distribution and variation of these pollutants depend on the local environment and human activities, so they are greatly related to geography and human activities. To better support scientific policymaking, it is necessary to have a more comprehensive understanding of the spatial distribution and change of microplastic pollution (Yahaya et al. 2024). With the development of human knowledge and technology globally, the level of monitoring and the accuracy of the data are continuously improved. Advanced monitoring tools such as remote sensing, drones, sensors, electronic noses and eyes, satellite imagery, monitors, and artificial intelligence can more accurately show the distribution of pollutants (Yarahmadi et al. 2024). However, for most existing frameworks, these technologies are insufficiently utilized. Understanding how these technologies can be used and aggregated under the framework requires specific schools of wet-laboratory studies.

Rising knowledge and technology understanding enable more flexible and adaptable policies. One solution to the plastic crisis involves the creation of innovative technologies. Such clean-up technologies have the potential to remove some of the plastic waste now circulating in the environment, protecting marine life and benefitting the health of ecosystems and local communities (Matavos-Aramyan 2024; Hartmann et al. 2019). Additionally, the involvement of clean-up technologies has the potential to create good environmental adaptation and plastic waste removal (Alaghemandi 2024). Despite recent studies and field investigations concentrating on this emerging type of waste, microplastic pollution remains a huge challenge for scientists and policymakers due to the limited knowledge of various systems (Nikiema and Asiedu 2022). To tackle this complex and neglected issue, innovative approaches and technologies are essential (Mathiarasan and Hüls 2021; Kassab et al. 2023). Such a situation requires researchers, policymakers, and businesses to work together and innovate. Citizens, businesses, and governments, especially those of developed countries, need to work in conjunction to ensure all plastic is collected and sent for recycling. If better collaboration across activities is reached but in the wider and deeper implementation of the 3Rs into daily practice due to effective policy measures, many of the adverse impacts of current linear plastic-related socioeconomic systems or behavior patterns on the environment, and human health would probably be mitigated at the global scale.

10.3.7 Socioeconomic Disparities

Socioeconomic disparity can be a significant hurdle in the comprehensive management of microplastic pollution. Marginalized communities often bear the brunt of environmental pollution, yet have fewer resources to mitigate and prevent pollution

(Mathiarasan and Hüls 2021). People living in poverty or economic hardship are often negatively impacted by pollution due to their limited access to sustainable lifestyle practices. A mutual interaction can be established between economic status and sustainable and/or unsustainable environmental practices. As economic status increases, an individual has more opportunity for sustaining a lifestyle with less negative impact on the environment. Wealthier individuals tend to be more informed about environmental issues and obtain resources for dealing with them more effectively, creating a cycle that neglects poorer communities (Phelan et al. 2020). Consequently, the economic status can be an additional contributor to the increase in disparities regarding polluted areas.

In order to achieve equity in the comprehensive management of microplastic pollution, it is necessary to adopt an inclusive approach considering different socioeconomic contexts. Conducting and evaluating community-led initiatives is an effective way to reduce environmental injustice and promote equity. Providing funding or other necessary support for community-based organizations (CBOs) creating or maintaining microplastic pollution management projects is important for an inclusive approach. Additionally, in order to effectively control the impacts of microplastic pollution on safety and public health, there is an urgent need to establish integrated environmental policies and strategies for water, soil, and sediments for developing countries. Policy challenges include determining an adequate level of funding for microplastics control technology and the frequency at which site location, sampling, testing, and analysis will be conducted (Felipe-Rodriguez et al. 2022). Addressing social disparities will be important as well. In many regions, vulnerable populations (e.g., indigenous communities and migrants) live along rivers, estuaries, and coasts. As a result, these communities are disproportionately exposed to pollutants, including plastics, via water.

10.4 Proposed Framework for a Global Microplastic Pollution Control Policy

Existing studies have not provided a comprehensive overview of current legislation on plastic pollution, particularly regarding microplastics. Thus, it is crucial to design a structured, rigorous, and detailed summary of all existing policies addressing microplastics. A novel review framework is developed to consider the origin and effect of pollution in determining what constitutes control and also utilizes global institutional categories described and maintained by large data-driven social science databases (Lampitt et al. 2023). By sharing responsibilities and capabilities, all nations can contribute to the protection and sustainable use of the oceans, their resources, and to preserve their inherent capacities for adaptation and self-renewal. The new platform for ocean governance has been prepared for the development of a global, comprehensive, and coordinated response to the challenge of preventing, reducing, and controlling microplastic pollution. It is recognized that such a response must be based on the best available scientific evidence and take into account variations where appropriate.

Hence, a significant emphasis has been placed on the goals of the current study, which are to describe actions and strategies in a structured, comparatist manner in order to address the global nature of the phenomenon of microplastic pollution, arising from its shared nature. Due to the current state of knowledge and relatively limited data availability, uncertainties, gaps, and challenges tend to be addressed (OECD 2021). But in recognizing potential imperfections, it underscored the necessity to adopt both stringent precautionary measures and a more harmonized policy. A call is also made for the development and/or enhancement of current monitoring networks and for the sharing of data, knowledge, and best practices to support informed decisions.

10.5 Public Awareness and Stakeholder Involvement

It is widely accepted that greater public awareness of microplastic pollution is critical for adopting protective practices and supporting regulatory measures. Microplastics have frequently been identified even in everyday consumer products, such as seafood, honey, sugar, and drinking water (Raab and Bogner 2021). In particular, consumers with limited knowledge or awareness of microplastics are unable to evaluate or avoid hazardous products (Willis and Fytianos 2022; Deme et al. 2022; Prata et al. 2019). It became clear that microplastics are a ubiquitous contaminant of the environment and the food chain that is potentially harmful to human health. However, the lack of robust and up-to-date knowledge still creates skepticism.

It is necessary to better inform the wider public about the sources and risks of microplastics through initiatives. Action such as community meetings, social media education, and the development of educational materials would help increase the awareness of the public (Vincent and Hoellein 2021; Gouin et al. 2019; Coffin et al. 2021; Seeley and Lynch 2023). Furthermore, it is important to involve a wide range of stakeholders, including communities, industries, and policymakers. Although some regulations and policies have been established in some regions and countries, the impact of these on microplastic pollution is still inadequate. A joint and multidisciplinary effort is needed, including awareness-raising activities and the establishment of new policies involving all the main actors.

Greater transparency in the shaping of regulations and policies would increase understanding and improve confidence among stakeholders. This may be achieved by maintaining and enhancing open and inclusive procedures for regulatory development. Relevant stakeholders would be regularly consulted. Transparency in the development, and subsequent monitoring and evaluation, of all relevant measures to address microplastic contamination would promote confidence and cooperation within all sectors. Additionally, effective science reporting in newspapers, television, radio, or other media can have a noticeably larger effect on public perceptions and, ultimately, molding opinions and attitudes. To further fuel public exposure, cooperation with schools or non-governmental organizations to enable visits or lectures would be beneficial. Information about European initiatives should be announced on the web and social media pages of the institution to promote events such as workshops, communications, and campaigns and increase interest. These initiatives are often picked up by the media and reach a broader audience. In fact, research-driven policies and collaborative strategies across different sectors play a crucial role in tackling microplastic pollution. Figure 10.4, derived from Deme et al. (2022), illustrates a multi-sectoral approach to microplastic pollution reduction with national policies informed by research.

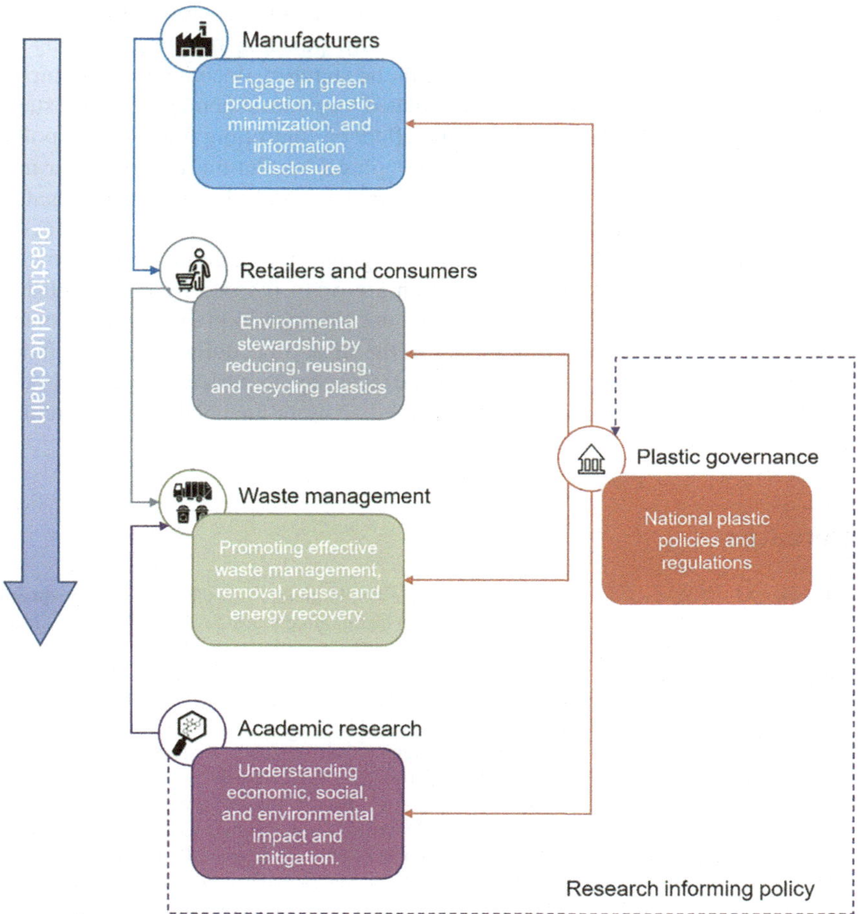

Fig. 10.4 An overview of a multi-sectoral approach in microplastic pollution reduction with national policies informed by research, derived from Deme et al. (2022)

10.6 Policy Recommendations and Action Plans

Since the early 2000s, the world has been paying increasing attention to microplastic pollution. In 2008, a conference proposed for the first time to address the issue of marine microplastic debris, and it was proposed to prevent and reduce the impact of microplastic pollution starting from the source of the microplastic debris. Many relevant studies have emerged from 2008 to the present, making the research on microplastic pollution a hotspot in the field of marine environmental protection. Microplastics are widely distributed in the environment and have a long accumulation period, so it is a necessity and practical significance to give timely attention and support to the management and control of this problem. The analysis shows the commitment and responsibilities of various countries and regions in controlling global microplastic pollution. However, it is still puzzling that target setting on microplastic pollution management is rare in the statutory documents of countries. Although some countries also have time-related target arrangements in their statutory documents on waste management, the effect of controlling microplastic pollution is unclear. In comparison, the microplastic pollution control measures established by most countries are to adopt strategies and plans to avoid negative social or environmental impacts rather than to take measures to reduce human activities that generate microplastics. In terms of regulatory requirements, measures to reduce sources of microplastic consumption are often ignored by the established measures, and no relevant instructions are set up, although food safety requirements can be found in most regulatory measures. At the same time, dollar figures are memorably lacking in most of the waste statutes enacted by countries or regions. In this context, a growing concern about how to mobilize more financial capital to reduce microplastic pollution.

References

Abad López AP, Trilleras J, Arana VA, Garcia-Alzate LS, Grande-Tovar CD (2023) Atmospheric microplastics: exposure, toxicity, and detrimental health effects. RSC Adv 13(11):7468–7489. https://doi.org/10.1039/d2ra07098g

Akyildiz SH, Fiore S, Bruno M, Sezgin H, Yalcin-Enis I, Yalcin B, Bellopede R (2024) Release of microplastic fibers from synthetic textiles during household washing. Environ Pollut 357:124455. https://doi.org/10.1016/j.envpol.2024.124455

Alaghemandi M (2024) Sustainable solutions through innovative plastic waste recycling technologies. Sustain For 16(23):10401. https://doi.org/10.3390/su162310401

Ali N, Khan MH, Ali M, Sidra A, Khan A, Nabi G, Ali F, Bououdina M, Kyzas GZ (2024) Insight into microplastics in the aquatic ecosystem: properties, sources, threats and mitigation strategies. Sci Total Environ 913:169489. https://doi.org/10.1016/j.scitotenv.2023.169489

Anthony J, Varalakshmi S, Kumar Sekar A, Thalavai Sivasankarasubbiah K, Harikrishnan T, Rangamaran VR, Gopal D, Ramalingam K (2024) Microplastics pollution in Indian marine environment: sources, effects and solutions. Front Mar Sci 11:1512802. https://doi.org/10.3389/fmars.2024.1512802

Antohi VM, Ionescu RV, Zlati ML, Iticescu C, Georgescu PL, Calmuc M (2023) Regional regression correlation model of microplastic water pollution control using circular economy tools. Int J Environ Res Public Health 20(5):4014. https://doi.org/10.3390/ijerph20054014

Aragaw TA (2021) Microplastic pollution in African countries' water systems: a review on findings, applied methods, characteristics, impacts, and managements. SN Appl Sci 3(6):629. https://doi.org/10.1007/s42452-021-04619-z

Biswas T, Pal SC (2024) Emerging threats of microplastics on marine environment: a critical review of toxicity measurement, policy practice gap and future research direction. J Clean Prod 434:139941. https://doi.org/10.1016/j.jclepro.2023.139941

Bora DK (2020) Rise of the sustainable circular economy platform from waste plastics: a biotechnological perspective. MRS Energy Sustain 7:E28. https://doi.org/10.1557/mre.2020.28

Bora SS, Gogoi R, Sharma MR, Anshu B, Deka P, Bora J, Naorem RS, Das J, Teli AB (2024) Microplastics and human health: unveiling the gut microbiome disruption and chronic disease risks. Front Cell Infect Microbiol 14:1492759. https://doi.org/10.3389/fcimb.2024.1492759

Campanale C, Massarelli C, Savino I, Locaputo V, Uricchio VF (2020) A detailed review study on potential effects of microplastics and additives of concern on human health. Int J Environ Res Public Health 17(4):1212. https://doi.org/10.3390/ijerph17041212

Carr A, Fetherston É, Makled T, Meyer L (2019) Towards a circular plastics economy: closing the loop on plastics. University of Michigan, Ann Arbor. https://deepblue.lib.umich.edu/bitstream/handle/2027.42/148831/Towards?sequence=1

Chen H, Wang T, Ding Y, Yuan F, Zhang H, Wang C, Wang Y, Wang Y, Song Y, Fu G, Zou X (2024) A catchment-wide microplastic pollution investigation of the Yangtze River: the pollution and ecological risk of tributaries are non-negligible. J Hazard Mater 466:133544. https://doi.org/10.1016/j.jhazmat.2024.133544

Coffin S, Wyer H, Leapman JC (2021) Addressing the environmental and health impacts of microplastics requires open collaboration between diverse sectors. PLoS Biol 19(3):e3000932. https://doi.org/10.1371/journal.pbio.3000932

Corinaldesi C, Canensi S, Dell'Anno A, Tangherlini M, Di Capua I, Varrella S, Willis TJ, Cerrano C, Danovaro R (2021) Multiple impacts of microplastics can threaten marine habitat-forming species. Commun Biol 4(1):431. https://doi.org/10.1038/s42003-021-01961-1

Cornago E, Börkey P, Brown A (2021) Preventing single-use plastic waste: implications of different policy approaches (OECD environment working paper no. 14). Organisation for economic co-operation and development (OECD), Paris. https://one.oecd.org/document/ENV/WKP(2021)14/en/pdf

Deme GG, Ewusi-Mensah D, Olagbaju OA, Okeke ES, Okoye CO, Odii EC, Ejeromedoghene O, Igun E, Onyekwere JO, Oderinde OK, Sanganyado E (2022) Macro problems from microplastics: toward a sustainable policy framework for managing microplastic waste in Africa. Sci Total Environ 804:150170. https://doi.org/10.1016/j.scitotenv.2021.150170

Duis K, Coors A (2016) Microplastics in the aquatic and terrestrial environment: sources (with a specific focus on personal care products), fate and effects. Environ Sci Eur 28(1):2. https://doi.org/10.1186/s12302-015-0069-y

Emenike EC, Okorie CJ, Ojeyemi T, Egbemhenghe A, Iwuozor KO, Saliu OD, Okoro HK, Adeniyi AG (2023) From oceans to dinner plates: the impact of microplastics on human health. Heliyon 9(10):e20440. https://doi.org/10.1016/j.heliyon.2023.e20440

Felipe-Rodriguez M, Böhm G, Doran R (2022) What does the public think about microplastics? Insights from an empirical analysis of mental models elicited through free associations. Front Psychol 13:920454. https://doi.org/10.3389/fpsyg.2022.920454

Filiciotto L, Rothenberg G (2021) Biodegradable plastics: standards, policies, and impacts. ChemSusChem 14(1):56–72. https://doi.org/10.1002/cssc.202002044

Fred-Ahmadu OH, Ahmadu FO, Adedapo AE, Oghenovo I, Ogunmodede OT, Benson NU (2024) Microplastics and chemical contamination in aquaculture ecosystems: the role of climate change and implications for food safety—a review. Environ Sci Eur 36:181. https://doi.org/10.1186/s12302-024-00995-6

Gallo F, Fossi C, Weber R, Santillo D, Sousa J, Ingram I, Nadal A, Romano D (2018) Marine litter plastics and microplastics and their toxic chemicals components: the need for urgent preventive measures. Environ Sci Eur 30(1):13. https://doi.org/10.1186/s12302-018-0139-z

Ghosh S, Sinha JK, Ghosh S, Vashisth K, Han S, Bhaskar R (2023) Microplastics as an emerging threat to the global environment and human health. Sustain For 15(14):10821. https://doi.org/10.3390/su151410821
Gouin T, Becker RA, Collot AG, Davis JW, Howard B, Inawaka K, Lampi M, Ramon BS, Shi J, Hopp PW (2019) Toward the development and application of an environmental risk assessment framework for microplastic. Environ Toxicol Chem 38(10):2087–2100. https://doi.org/10.1002/etc.4529
Hartmann NB, Hüffer T, Thompson RC, Hassellöv M, Verschoor A, Daugaard AE, Rist S, Karlsson T, Brennholt N, Cole M, Herrling MP, Hess MC, Ivleva NP, Lusher AL, Wagner M (2019) Are we speaking the same language? Recommendations for a definition and categorization framework for plastic debris. Environ Sci Technol 53(3):1039–1047. https://doi.org/10.1021/acs.est.8b05297
Iheanacho S, Ogbu M, Bhuyan MS, Ogunji J (2023) Microplastic pollution: an emerging contaminant in aquaculture. Aquac Fish 8(6):603–616. https://doi.org/10.1016/j.aaf.2023.01.007
IUCN (2024) Plastic pollution. https://iucn.org/resources/issues-brief/plastic-pollution
Kaliszewicz A, Panteleeva N, Karaban K, Runka T, Winczek M, Beck E, Poniatowska A, Olejniczak I, Boniecki P, Golovanova EV, Romanowski J (2023) First evidence of microplastic occurrence in the marine and freshwater environments in a remote polar region of the Kola peninsula and a correlation with human presence. Biology 12(2):259. https://doi.org/10.3390/biology12020259
Kaloyianni M, Bobori DC, Xanthopoulou D, Malioufa G, Sampsonidis I, Kalogiannis S, Feidantsis K, Kastrinaki G, Dimitriadi A, Koumoundouros G, Lambropoulou DA, Kyzas GZ, Bikiaris DN (2021) Toxicity and functional tissue responses of two freshwater fish after exposure to polystyrene microplastics. Toxics 9(11):289. https://doi.org/10.3390/toxics9110289
Kassab A, Al Nabhani D, Mohanty P, Pannier C, Ayoub GY (2023) Advancing plastic recycling: challenges and opportunities in the integration of 3D printing and distributed recycling for a circular economy. Polymers 15(19):3881. https://doi.org/10.3390/polym15193881
Kibria MG, Masuk NI, Safayet R, Nguyen HQ, Mourshed M (2023) Plastic waste: challenges and opportunities to mitigate pollution and effective management. Int J Environ Res 17(1):20. https://doi.org/10.1007/s41742-023-00507-z
Koelmans AA, Mohamed Nor NH, Hermsen E, Kooi M, Mintenig SM, De France J (2019) Microplastics in freshwaters and drinking water: critical review and assessment of data quality. Water Res 155:410–422. https://doi.org/10.1016/j.watres.2019.02.054
Kumar R, Verma A, Shome A, Sinha R, Sinha S, Jha PK, Kumar R, Kumar P, Shubham D (2021) Impacts of plastic pollution on ecosystem services, sustainable development goals, and need to focus on circular economy and policy interventions. Sustain For 13(17):9963. https://doi.org/10.3390/su13179963
Lampitt RS, Fletcher S, Cole M, Kloker A, Krause S, O'Hara F, Ryde P, Saha M, Voronkova A, Whyle A (2023) Stakeholder alliances are essential to reduce the scourge of plastic pollution. Nat Commun 14(1):2849. https://doi.org/10.1038/s41467-023-38613-3
Landrigan PJ, Stegeman JJ, Fleming LE, Allemand D, Anderson DM, Backer LC, Brucker-Davis F, Chevalier N, Corra L, Czerucka D, Bottein MD, Demeneix B, Depledge M, Deheyn DD, Dorman CJ, Fénichel P, Fisher S, Gaill F, Galgani F, Gaze WH, Rampal P (2020) Human health and ocean pollution. Ann Glob Health 86(1):151. https://doi.org/10.5334/aogh.2831
Landrigan PJ, Raps H, Cropper M, Bald C, Brunner M, Canonizado EM, Charles D, Chiles TC, Donohue MJ, Enck J, Fenichel P, Fleming LE, Ferrier-Pages C, Fordham R, Gozt A, Griffin C, Hahn ME, Haryanto B, Hixson R, Ianelli H, Dunlop S (2023) The Minderoo-Monaco commission on plastics and human health. Ann Glob Health 89(1):23. https://doi.org/10.5334/aogh.4056
Lee Y, Cho J, Sohn J, Kim C (2023) Health effects of microplastic exposures: current issues and perspectives in South Korea. Yonsei Med J 64(5):301–308. https://doi.org/10.3349/ymj.2023.0048
Lusher AL, Hollman PCH, Mendoza-Hill JJ (2017) Microplastics in fisheries and aquaculture: status of knowledge on their occurrence and implications for aquatic organisms and food safety

(FAO fisheries and aquaculture technical paper No. 615). Food and Agriculture Organization of the United Nations, Rome. https://openknowledge.fao.org/server/api/core/bitstreams/a9a298e0-9db6-4769-beac-37325be3e280/content

Marcharla E, Vinayagam S, Gnanasekaran L, Soto-Moscoso M, Chen W-H, Thanigaivel S, Ganesan S (2024) Microplastics in marine ecosystems: a comprehensive review of biological and ecological implications and its mitigation approach using nanotechnology for the sustainable environment. Environ Res 256:119181. https://doi.org/10.1016/j.envres.2024.119181

Matavos-Aramyan S (2024) Addressing the microplastic crisis: a multifaceted approach to removal and regulation. Environmental Advances 17:100579. https://doi.org/10.1016/j.envadv.2024.100579

Mathiarasan S, Hüls A (2021) Impact of environmental injustice on children's health-interaction between air pollution and socioeconomic status. Int J Environ Res Public Health 18(2):795. https://doi.org/10.3390/ijerph18020795

Mattsson K, Ekstrand E, Granberg M, Hassellöv M, Magnusson K (2022) Comparison of pre-treatment methods and heavy density liquids to optimize microplastic extraction from natural marine sediments. Sci Rep 12(1):15459. https://doi.org/10.1038/s41598-022-19623-5

Mitrano DM, Wohlleben W (2020) Microplastic regulation should be more precise to incentivize both innovation and environmental safety. Nat Commun 11(1):5324. https://doi.org/10.1038/s41467-020-19069-1

Nikiema J, Asiedu Z (2022) A review of the cost and effectiveness of solutions to address plastic pollution. Environ Sci Pollut Res Int 29(17):24547–24573. https://doi.org/10.1007/s11356-021-18038-5

Nikpay M, Toorchi Roodsari S (2024) Crafting a scientific framework to mitigate microplastic impact on ecosystems. Microplastics 3(1):165–183. https://doi.org/10.3390/microplastics3010010

OECD (2021) Policies to reduce microplastics pollution in water: focus on textiles and tyres. OECD Publishing, Paris. https://doi.org/10.1787/7ec7e5ef-en

Pachuau L, Rebecca V, Khawlhring L, Punuri JB (2024) Microplastics: their effects on the environment, human health, and plant ecosystems. Environ Pollut Manage 1(1):248–259. https://doi.org/10.1016/j.epm.2024.11.004

Paul S, Nath S, Bhattacharjee S, Mukherjee S (2024) Unveiling the effects of microplastics pollution on marine fauna. Blue Biotechnol 1:6. https://doi.org/10.1186/s44315-024-00006-6

Pfohl P, Wagner M, Meyer L, Domercq P, Praetorius A, Hüffer T, Hofmann T, Wohlleben W (2022) Environmental degradation of microplastics: how to measure fragmentation rates to secondary micro- and nanoplastic fragments and dissociation into dissolved organics. Environ Sci Technol 56(16):11323–11334. https://doi.org/10.1021/acs.est.2c01228

Phelan AA, Ross H, Setianto NA, Fielding K, Pradipta L (2020) Ocean plastic crisis-mental models of plastic pollution from remote Indonesian coastal communities. PLoS One 15(7):e0236149. https://doi.org/10.1371/journal.pone.0236149

Pothiraj C, Amutha Gokul T, Ramesh Kumar K, Ramasubramanian A, Palanichamy A, Venkatachalam K, Pastorino P, Barcelò D, Balaji P, Faggio C (2023) Vulnerability of microplastics on marine environment: a review. Ecol Indic 155:111058. https://doi.org/10.1016/j.ecolind.2023.111058

Prata JC, Silva ALP, da Costa JP, Mouneyrac C, Walker TR, Duarte AC, Rocha-Santos T (2019) Solutions and integrated strategies for the control and mitigation of plastic and microplastic pollution. Int J Environ Res Public Health 16(13):2411. https://doi.org/10.3390/ijerph16132411

Qaiser Z, Aqeel M, Sarfraz W, Rizvi ZF, Noman A, Naeem S, Khalid N (2023) Microplastics in wastewaters and their potential effects on aquatic and terrestrial biota. Case Stud Chem Environ Eng 8:100536. https://doi.org/10.1016/j.cscee.2023.100536

Raab P, Bogner FX (2021) Conceptions of university students on microplastics in Germany. PLoS One 16(9):e0257734. https://doi.org/10.1371/journal.pone.0257734

Raes L, Litulo C, Pereira MAM, Manyara P (2021) Marine plastics, fisheries, and livelihoods in Mozambique. IUCN, Cham. www.iucn.org/theme/marine-and-polar/our-work/close-plastic-tap-programme/reports

Rivera-Garibay O, Méndez-López ME, Torres-Irineo E, Rivas M, Santillo D, Álvarez-Filip L (2024) Presence of microplastic in target species of small-scale fisheries and possible social implications on the local communities. Mar Biol 171:78. https://doi.org/10.1007/s00227-024-04399-1

Šaravanja A, Pušić T, Dekanić T (2022) Microplastics in wastewater by washing polyester fabrics. Materials 15(7):2683. https://doi.org/10.3390/ma15072683

Seeley ME, Lynch JM (2023) Previous successes and untapped potential of pyrolysis-GC/MS for the analysis of plastic pollution. Anal Bioanal Chem 415(15):2873–2890. https://doi.org/10.1007/s00216-023-04671-1

Sutkar PR, Gadewar RD, Dhulap VP (2023) Recent trends in degradation of microplastics in the environment: a state-of-the-art review. J Hazard Mater Adv 11:100343. https://doi.org/10.1016/j.hazadv.2023.100343

Swain PR, Parida PK, Majhi PJ, Behera BK, Das BK (2025) Microplastics as emerging contaminants: challenges in inland aquatic food web. Water 17(2):201. https://doi.org/10.3390/w17020201

Tan SMA, Wan Mohd Khalik WMA, Ong MC, Shao YT, Pan H-J, Bhubalan K (2021) Marine microplastics as vectors of major ocean pollutants and its hazards to the marine ecosystem and humans. Prog Earth Planet Sci 8:12. https://doi.org/10.1186/s40645-020-00405-4

UNEP (2023) Our planet is choking on plastic. https://www.unep.org/interactives/beat-plastic-pollution/

Usman S, Abdull Razis AF, Shaari K, Amal MNA, Saad MZ, Mat Isa N, Nazarudin MF, Zulkifli SZ, Sutra J, Ibrahim MA (2020) Microplastics pollution as an invisible potential threat to food safety and security, policy challenges and the way forward. Int J Environ Res Public Health 17(24):9591. https://doi.org/10.3390/ijerph17249591

Usman S, Abdull Razis AF, Shaari K, Amal MNA, Saad MZ, Mat Isa N, Nazarudin MF (2021) Polystyrene microplastics exposure: an insight into multiple organ histological alterations, oxidative stress and neurotoxicity in Javanese Medaka fish (Oryzias javanicus Bleeker, 1854). Int J Environ Res Public Health 18(18):9449. https://doi.org/10.3390/ijerph18189449

Usman S, Abdull Razis AF, Shaari K, Azmai MNA, Saad MZ, Mat Isa N, Nazarudin MF (2022) The burden of microplastics pollution and contending policies and regulations. Int J Environ Res Public Health 19(11):6773. https://doi.org/10.3390/ijerph19116773

Vincent AES, Hoellein TJ (2021) Distribution and transport of microplastic and fine particulate organic matter in urban streams. Ecol Appl 31(8):e02429. https://doi.org/10.1002/eap.2429

Wagner M, Scherer C, Alvarez-Muñoz D, Brennholt N, Bourrain X, Buchinger S, Fries E, Grosbois C, Klasmeier J, Marti T, Rodriguez-Mozaz S, Urbatzka R, Vethaak AD, Winther-Nielsen M, Reifferscheid G (2014) Microplastics in freshwater ecosystems: what we know and what we need to know. Environ Sci Eur 26:12. https://doi.org/10.1186/s12302-014-0012-7

Wang L, Huang J, Chen M, Jin H, Wu Y, Chen X (2024a) Investigation of microplastics in urban rivers of eastern China in summer: abundance, characteristics and ecological risk assessment. Environ Sci Process Impacts 26(7):1245–1256. https://doi.org/10.1039/d4em00153b

Wang C, Liu X, Ma Q, Xing S, Yuan L, Ma Y (2024b) Distribution and effects of microplastics as carriers of heavy metals in river surface sediments. J Contam Hydrol 266:104396. https://doi.org/10.1016/j.jconhyd.2024.104396

Williams AT, Rangel-Buitrago N (2022) The past, present, and future of plastic pollution. Mar Pollut Bull 176:113429. https://doi.org/10.1016/j.marpolbul.2022.113429

Willis B, Fytianos G (2022) Towards microplastic reduction within institutions. Water Air Soil Pollut 233(8):337. https://doi.org/10.1007/s11270-022-05790-8

Xu S (2024) Environmental legislation analysis improvement approach of global marine plastic pollution from the perspective of holistic system view. Front Mar Sci 11:1481635. https://doi.org/10.3389/fmars.2024.1481635

Yahaya T, Adewale MK, Ibrahim AB, Abdulkadir B, Emmanuela CC, Fari AZ, Attahiru AK, Wanda JD (2024) Abundance, characterization, and health risk evaluation of microplastics in borehole water in Birnin Kebbi, Nigeria. Environ Anal Health Toxicol 39(2):e2024017. https://doi.org/10.5620/eaht.2024017

Yang H, Chen G, Wang J (2021) Microplastics in the marine environment: sources, fates, impacts and microbial degradation. Toxics 9(2):41. https://doi.org/10.3390/toxics9020041

Yang X, Zhou Y, Li Y, Wang S (2024) International law and regulation of marine microplastics: current situation, problems, and development. Sustain For 16(21):9337. https://doi.org/10.3390/su16219337

Yarahmadi A, Heidari S, Sepahvand P, Afkhami H, Kheradjoo H (2024) Microplastics and environmental effects: investigating the effects of microplastics on aquatic habitats and their impact on human health. Front Public Health 12:1411389. https://doi.org/10.3389/fpubh.2024.1411389

Zhang K, Shi H, Peng J, Wang Y, Xiong X, Wu C, Lam PKS (2018) Microplastic pollution in China's inland water systems: a review of findings, methods, characteristics, effects, and management. Sci Total Environ 630:1641–1653. https://doi.org/10.1016/j.scitotenv.2018.02.300

Ziani K, Ioniță-Mîndrican CB, Mititelu M, Neacşu SM, Negrei C, Moroşan E, Drăgănescu D, Preda OT (2023) Microplastics: a real global threat for environment and food safety: a state of the art review. Nutrients 15(3):617. https://doi.org/10.3390/nu15030617

Zolotova N, Kosyreva A, Dzhalilova D, Fokichev N, Makarova O (2022) Harmful effects of the microplastic pollution on animal health: a literature review. PeerJ 10:e13503. https://doi.org/10.7717/peerj.13503

A Greener Blueprint: Bioplastics and Sustainable Alternatives to Control Microplastics in the Environment

11

Syeda Tahmida Mutahara Abdal, Mohammad Toha, Sadia Sikder, and Md. Mostafizur Rahman

Abstract

The widespread presence of MPs in the environment has raised concerns and encouraged research. Multiple studies have identified MPs across various environmental compartments. There is an increasing global concern regarding the environmental impact and accumulation associated with MPs. Bioplastics are currently being advocated as an alternative to conventional plastics. However, investigations focused on solutions for MP contamination are limited. Effective management of MP contamination requires understanding the distinctions between sustainable and conventional solutions. Hence, it is necessary to understand the sustainable solutions to MP pollution in freshwater and terrestrial ecosystems. Moreover, assessing the capacity of bioplastics to mitigate MP pollution in natural ecosystems through a review of existing research and innovative approaches is crucial for developing nations. Hence, this chapter will identify the shortcomings of conventional plastics and their contribution to MP pollution and highlight the necessity for sustainable alternatives. Lastly, this chapter discusses

S. T. M. Abdal · M. M. Rahman (✉)
Laboratory of Environmental Health and Ecotoxicology, Department of Environmental Sciences, Jahangirnagar University, Dhaka, Bangladesh

Department of Environmental Science, Bangladesh University of Professionals (BUP), Dhaka, Bangladesh
e-mail: rahmanmm@juniv.edu

M. Toha · S. Sikder
Laboratory of Environmental Health and Ecotoxicology, Department of Environmental Sciences, Jahangirnagar University, Dhaka, Bangladesh

Department of Environmental Science, Bangladesh University of Professionals (BUP), Dhaka, Bangladesh

Department of Environmental Science and Disaster Management, Daffodil International University, Dhaka, Bangladesh

M. Oves et al. (eds.), *Marine Microplastics and the Quest for Remediation*,
https://doi.org/10.1007/978-981-96-8979-8_11

bio-based materials, biodegradable polymers, and other sustainable alternatives to enhance the environmental friendliness of plastic manufacturing and usage. This will assist policymakers and decision-makers in implementing appropriate measures to ensure sustainable solutions for MP pollution.

Keywords

Microplastics · Bioplastics · Sustainable solutions · Green technology · Eco-friendly

11.1 Introduction

In recent years, there has been growing concern over small polymer aggregates (microplastics) in environmental matrices due to their detrimental impact on human health and ecosystems (Aysha et al. 2024; Geyer et al. 2017). These MPs are generally produced from two distinct sources. The primary sources of plastic fragments are (1) cosmetic microbeads (<5 mm) and (2) industrial pellets (<5 mm), with a significant portion of plastic fragments also resulting from (3) the secondary degradation of larger plastic debris (Rahman et al. 2024). MPs are challenging to environmental sustainability and public health due to their persistence, mobility, and ability to adsorb hazardous substances (Rochman et al. 2013; Aysha et al. 2024). However, their relevance cannot be sidelined due to the rising status of their pervasiveness (Toha et al. 2024c). However, the widespread distribution of MPs across all environmental compartments, such as air, water, and soil, has been detected (Toha et al. 2024b; Rahman et al. 2024). Most atmospheric MP contamination is caused by urban activities, industrial emissions, and vehicular abrasion (Dris et al. 2016a, b). Airborne MPs have been confirmed in urban, suburban, and remote areas and are suggested to be transported over long distances. Humans and animals can also inhale these particles, possibly resulting in respiratory distress, oxidative stress, and inflammation (Prata 2018). MP pollution acts as the major sink in such water bodies, with MPs detected in oceans, rivers, lakes, and drinking water sources (Li et al. 2018). They occur in aquatic systems when municipal wastewater discharges, industrial effluents, or surface runoff is present. MPs are an ingestion risk for marine and freshwater organisms and act as vectors for heavy metals, POPs, and pathogenic microbes (Koelmans et al. 2016a, b). The bioaccumulation and biomagnification of MPs within the food web raise concerns regarding human exposure through seafood consumption (Barboza et al. 2018; Gautam et al. 2022; Toha et al. 2024a). Agricultural practices, improper plastic waste management, and atmospheric fallout are reasons terrestrial ecosystems are currently being recognized as important reservoirs of MPs (Rillig 2012). Microbial communities are influenced by MPs in soil, which change soil physicochemical properties, and agronomic products decrease crop productivity. Broad adverse effects of MPs are observed on ecosystems and human health. These can be assumed as one affecting the physical, chemical, and biological effects (Wright and Kelly 2017). Ingestion of MPs by aquatic and terrestrial organisms can interfere with digestion, leading to gastrointestinal tract blockage, reduced nutrient absorption, and impaired energy metabolism (Foley et al. 2018). Malnutrition and decreased survival rates are frequent in zooplankton, fish, and birds due to mistaking MPs for food. These MPs can adsorb and concentrate toxic pollutants from their environment (polycyclic aromatic hydrocarbons (PAHs), bisphenol A (BPA), heavy metals, etc.) and act as carriers of such pollutants (Teuten et al. 2009). After ingestion, these contaminants can leach into organisms, and such endocrine disruption, genotoxicity, and metabolic disorders may cause. MPs have been demonstrated to change the proportion of microbial populations and help antibiotic-resistant genes grow (Lu et al. 2019). Currently, numerous treatment technologies have been employed to tackle MP pollution, and each technique has specific advantages and disadvantages (Rajala et al. 2020). Filtration and separation techniques remove MPs from wastewater and drinking water sources. MPs are degraded into smaller fragments or mineralized through advanced oxidation processes (AOPs) techniques (Zhang et al. 2020). However, during this process, incomplete degradation could lead to the emergence of nanoplastics. It is important to highlight that certain microbes can decompose MPs (Danso et al. 2019; Toha et al. 2024c). While mechanical and thermal treatments can mitigate MP pollution, these methods are costly and generate greenhouse gases (Zhao et al. 2018; Toha et al. 2024b). Conversely, certain green plastics, known as bioplastics, are derived from renewable sources and hold significant promise for diminishing MPs (Niaounakis 2013a, b). Unlike petroleum-based plastics, bioplastics are partly or completely biodegradable and are less persistent in the environment. PHAs, PLA, and polybutylene succinate (PBS) are designed to degrade under specific environmental conditions such as humidity, composting, or soil application (Tokiwa et al. 2009). Some of them are non-biodegradable bioplastic such as bio-based polyethylene (Bio-PE) and bio-based polyethylene terephthalate (Bio-PET) that have properties similar to conventional plastics but from renewable sources (Siracusa and Blanco 2020). Bioplastics have many advantages as they reduce reliance on fossil fuels, lower carbon footprint, and have the potential for biodegradability. Moreover, it can work with existing plastic manufacturing processes. However, it has a much higher production cost, and the industrial composting facilities are minimal worldwide. Unfortunately, the variation of degradation rate and environmental factors in the degradation facilities is a huge drawback to using bioplastics. Besides bioplastics, MP pollution could be controlled by several sustainable strategies and promote a circular economy (Jiang et al. 2020). New legislations include bans on single-use plastics, EPR programs, and very tight regulations on microplastic emissions (Giacovelli 2018). The establishment of plant-based fibers, mycelium-based packaging, and algae-derived polymers provide alternatives to current plastic technology (Toha et al. 2024c). Individuals can be stimulated to switch to eco-friendly product alternatives and adopt practices that reduce plastic waste generation through public education campaigns, incentives for sustainable modes of consumption, and corporate responsibility initiatives. Some advanced techniques can be used to reduce MP pollution in various life cycle stages, such as nanocellulose-based filters, bio-inspired plastic degrading enzymes, and MPs-trapping textiles (Xiang et al. 2021).

Moreover, it is necessary to explore bioplastics and sustainable alternatives for better outcomes. Hence, this chapter will discuss bioplastics and sustainable alternatives to controlling MPs in the environment.

11.2 Sources of MPs

According to multiple studies, microplastics (MPs) generally come from two significant sources such as primary and secondary sources (Andrady 2011; Geyer et al. 2017; Toha et al. 2024b). The description of these sources is given below:

11.2.1 Primary Microplastics

Primary microplastics (MPs) are found in a wide range of items, including toothpaste, body wash, scrubs, and cosmetics which directly enter the environment as raw materials of products (Rayhan et al. 2024).

11.2.1.1 Microbeads in Personal Care Products

Generally, the microbeads consist of polyethylene (PE) or polypropylene (PP) and are commonly used in exfoliating scrubs, toothpaste, and cosmetic formulations. Napper et al. (2015) mentioned that these particles are too small to be captured by conventional wastewater treatment plants and must be discharged directly into aquatic environments.

11.2.1.2 Industrial and Commercial Sources

Nurdles are microplastic pellets used as raw materials in plastic manufacturing. Accidental spills during transportation and handling contribute to environmental contamination. Moreover, MPs in paints, coatings, and adhesives are released into the environment due to the plastic-based coatings in them that degrade over time (Gewert et al. 2015; Rahman et al. 2024).

11.2.1.3 Synthetic Fibers from Textiles

Polyester, nylon, and acrylic are examples of synthetic textiles which shed microplastic fibers when washing. These fibers enter wastewater treatment systems, where only incomplete filtration allows their release into the water bodies (Browne et al. 2011; Aysha et al. 2024). Their distribution in urban and remote areas is also due to atmospheric fallout (Dris et al. 2016a, b).

11.2.2 Secondary Microplastics

Secondary microplastics come from the fragmentation of larger plastics caused by environmental degradation processes (photodegradation, mechanical abrasion, and chemical weathering) (Barnes et al. 2009; Toha et al. 2024a).

11.2.2.1 Degradation of Plastic Waste

UV exposure and mechanical forces break plastics like plastic bags, bottles, and packaging materials into smaller microplastic fragments over time. This issue is further worsened by improper disposal and waste management in marine and terrain systems (Thompson et al. 2004).

11.2.2.2 Tire Wear and Road Runoff

Synthetic rubber and additives comprise microplastic particles derived from abrasion of vehicle tires. Urban MP pollution in water bodies occurs due to the accumulation of these particles on the road surface and transportation via stormwater runoff (Kole et al. 2017).

11.2.2.3 Agricultural Practices

The application of sewage sludge, plastic mulch films, and irrigation with dirty water supply introduce microplastics into agricultural soils. As time passes, these particles accumulate in soil ecosystems and impact microbial communities and soil health (de Souza Machado et al. 2018).

11.2.2.4 Fishing Gear and Marine Debris

Fishing nets, ropes, and buoys are lost or discarded to become marine microplastic pollution. Consequently, marine life ingests these MPs and faces detrimental impacts (Lusher et al. 2017).

Lastly, a large number of MPs are continuously derived from various sources. Hence, it is crucial to implement specific policies, advanced technology, and public awareness to reduce MP emissions and their environmental impacts.

11.3 Fate and Transport of Microplastics in Air, Water, and Soil Compartments

According to the research studies, MPs are widely distributed in every environmental compartment (Zhang et al. 2020; Li et al. 2018). Hence, researchers and policymakers need precise and specific information to reduce its environmental and human health effects.

11.3.1 Fate and Transport in Air

Although airborne microplastics (MPs) primarily originate from various sources, industrial activities and tire emissions are most responsible for the spread of airborne MPs (Dris et al. 2016a, b). Notably, atmospheric currents and travel distances contribute to the ultimate deposition of particles in polar and mountainous regions (Bergmann et al. 2019a, b). MPs travel through gravitational settling into dry deposition sites and impaction (Allen et al. 2019). The attributes of MPs in the air are directly related to the shapes, dimensions, and density of MP. The duration of suspended motion in the atmosphere is prolonged for irregularly shaped lighter particles relative to dense, compact particles (Lu et al. 2019). Introducing microplastics into the atmosphere risks human safety, as pollutants adhere to these particles (Wright and Kelly 2017).

11.3.2 Fate and Transport in Water

According to Andrady (2017a, b), the aquatic environment attracts MPs through surface runoff, wastewater effluents, and maritime operations. Water particle movements rely on three hydrodynamic elements, which include water flow, turbulent conditions, and density fluctuations (Koelmans et al. 2016a, b; Rahman et al. 2024). Surface water contains buoyant polyethylene and polypropylene suspended before polyvinyl chloride and polystyrene materials sink towards sedimentation (Kowalski et al. 2016). Water-borne MP transport is influenced by biofouling because MP-surface-integrated microorganisms change density and buoyancy properties (Reisser et al. 2014; Sikder et al. 2024). Marine and freshwater bodies deploy MPs to disperse heavy metals together with persistent organic pollutants. Aquatic organisms that uptake MPs develop food chain accumulation, which results in trophic transfer (Setälä et al. 2014).

11.3.3 Fate and Transport in Soil

Soil becomes contaminated with MPs through the combination of agricultural work along with atmospheric loss, wastewater drainage, and natural plastic degradation (Rillig 2012). Soil microplastic movement depends upon several factors, including plastic dimensions mat, aerial types, and natural conditions such as the soil permeability rates, water content, and organic matter degeneration rates (Zhang et al. 2020). Microscopic plastic particles use various soil depths before they enter groundwater systems (Horton et al. 2017). The environment triggers two significant factors: UV radiation with mechanical stress and microbial activity, which causes soil fragmentation where nanoplastic formation occurs (Gigault et al. 2018). Soil biota and environmental factors produce MP redistribution and degradation by interacting with the pollutants (Kwak et al. 2021). Soil structure and functional processes change in response to MPs, leading to adverse effects on soil fertility alongside plant development (de Souza Machado et al. 2018).

11.4 Environmental and Human Health Effects of Microplastics

Microplastics (MPs) are widespread pollutants that negatively affect both environmental ecosystems and human health systems. As persistent and small, MPs cause complex reactions in ecosystems, resulting in environmental organism effects and potential health threats to humans. The study explores how MPs generate environmental effects on natural systems and human well-being through toxicological, physiological, and ecological results.

11.4.1 Environmental Effects

The environmental systems receive direct impacts from MPs through failures in water quality, damage to soil, and deterioration of air quality. Aquatic species consume MPs, which create physical and chemical stress for the organisms that ingest them (Toha et al. 2024a; Rochman et al. 2013). Zooplankton species that filter feed obtain MP particles by mistake, lowering their nutritional value to the point of causing them to starve (Setälä et al. 2014). Fish and shellfish that consume MPs face impaired reproduction, growth problems, and gastrointestinal complications (Cole et al. 2011). Through their lifecycle, MPs serve as organic pollutants and heavy metal passageways that boost their accumulation levels through biological food chain networks (Teuten et al. 2009). The toxic effects from MP exposure accumulate from lower to higher trophic levels, harming predator species, including human beings (Gouin et al. 2011). MPs inside soil ecosystems bring detrimental changes to microbial communities and soil structure, which impedes plant health and reduces soil fertility (Rillig 2012). The burrowing behavior of earthworms becomes inhibited when they encounter MPs in the environment, which negatively impacts soil oxygen circulation and soil nutrient sustenance (Huerta Lwanga et al. 2016).

11.4.2 Human Health Effects

The main ways humans become exposed to MPs include swallowing these particles, breathing them into their lungs, and absorbing them through the skin (Wright and Kelly 2017). The ingestion of seafood and the consumption of water and processed foods derived from the ocean results in dietary exposure (2014). Microplastics entering the gastrointestinal tract can trigger inflammatory reactions that disturb essential gut microbes and lead to metabolic disorder development (Lehner et al. 2019). The intake of airborne MPs in urban areas causes health problems for our respiratory system. The respiratory tissue of the lungs allows MPs to move deeply, which initiates oxidative stress and inflammation, just like other fine airborne pollutants (Prata 2018). MPs, BPA, and phthalates act as toxic additives with endocrine-disrupting characteristics (Rochman et al. 2013). Hormonal regulation faces disruption when exposed to these chemicals and such exposure leads to reproductive and developmental problems.

11.4.3 Toxicological Implications

Cellular and molecular biological systems connect with MPs, leading to oxidative stress damage and genotoxicity with immunotoxic effects (Bhattacharya et al. 2010). The study of MPs under laboratory conditions shows that these microplastics cause aquatic organisms to die by apoptosis and disrupt their gene expression patterns (Foley et al. 2018). Researchers using human cell lines have demonstrated that MP exposure leads to inflammatory cell reactions besides causing functional impairment of cells (Hussain et al. 2020). Long-term contact with MPs intensifies health problems, specifically within susceptible segments of the population, which includes young persons alongside those who confront respiratory diseases (Wright and Kelly 2017). To completely comprehend the combined health effects of MPs on humans, researchers need to conduct extensive toxicology studies across prolonged periods. As MPs exist throughout environmental areas, ecological systems, and human populations experience critical dangers. MPs create dual threats in aquatic environments and on land since they influence the biological functions of living species while also transferring harmful pollutants. Human contact with MPs through oral consumption and breathing can trigger metabolic problems, lung diseases, and disruption of the endocrine system. Studies with regulatory actions and public education efforts must be performed to fight the adverse consequences of MP pollution.

11.5 Existing Treatment Techniques

Due to their small size and the vast volumes they occupy in aquatic ecosystems, microplastics have recently been singled out as one of several important aspects of the marine pollution problem. Microplastics can be found in diverse and far-reaching marine environments; they are present in organisms at all levels of trophic strata and now even in the tissues of humans (Miller et al. 2022). Environmental organization's sheer volume of this kind of pollution has allowed substantial research into how to clean it up. Microplastics are not unlike other pollutants in the environment. They also have various treatment methods to remediate the pollution levels. The treatment methods are applied based on necessity and functionality.

11.5.1 Classification of the Treatment Techniques

Removal techniques of microplastics are similar to wastewater treatment plants since microplastics are heavily present in water (Koelmans et al. 2019). Khan et al. (2022) categorized microplastic treatment methods based on their removal efficacy: primary, secondary, tertiary, and advanced techniques. The nature and function of microplastic dictate what treatment technique needs to be applied. However, another emerging microplastic treatment method has been capturing attention since it may help us achieve an ecologically sustainable solution to this problem (Lee et al. 2022). That method is called the "Biomimetics Method." The categories are subdivided into various treatment strategies, explained later in this chapter. Microplastic removal techniques are assessed by calculating efficiency (Xu et al. 2021). It is done by the following equation

$$\eta = \left(\frac{C_I - C_E}{C_I} \right) \times 100$$

Here, η (%) is the removal efficiency, and C_I (n L^{-1}) and C_E (n L^{-1}) are the microplastic abundances in the influent and effluent treatment water respectively. This chapter categorizes the treatment methods into three main categories. A brief description of these categories is given below.

11.6 Primary Techniques

The primary techniques are classified into three types based on mechanisms. They are (a) grit chamber, (b) primary sedimentation, and (c) coagulation-flocculation techniques.

11.6.1 Secondary and Tertiary Techniques

Secondary and tertiary treatments reduce contaminants, such as dissolved organic matter and suspended sediments. The techniques are subcategorized based on the type of reaction that takes place. They are (a) biological techniques (membrane bioreactors and biofilters) and (b) chemical techniques (reverse osmosis and electrocoagulation).

11.6.2 Advanced Methods

Chemical (O_3, H_2O_2, or O_3/H_2O_2) oxidation, electrochemical, photocatalysis, plasma, and Fenton and its variants are some of the advanced treatments (Table 11.1). Moreover, these techniques are frequently necessary due to the complexity of the effluents regarding the type, composition, and form of the microplastics present in them (Kim et al. 2022; Rodríguez-Narvaez et al. 2021).

11.6.2.1 Primary Treatment Techniques

Primary treatment involves methods that remove microplastics at a primary level. It is used to remove larger microplastics. MP at this stage is removed by sedimentation, coagulation-flocculation, and flotation techniques. These techniques are done by instruments such as desanders, flotation tanks, and primary settlers (Khan et al. 2022). The functionality of this process depends on the weight of MPs. Designers and primary settlers precipitate and remove the heavier MPs in flocs. The flotation process removes the lighter ones. After the pretreatment stage, about 40.7%–97.9%

Table 11.1 Classification of the treatment techniques and their efficiency

Technique	Type	Advantages	Disadvantages	Microplastic removal efficiency	References
Grit chamber	Primary	Simple, low-cost, effective for large particles	Not effective for small microplastics	Low (~20–40%)	Khan et al. (2022)
Primary sedimentation		Removes larger sediments and microplastics	Limited efficiency for small microplastics and dissolved contaminants	Moderate (~40–60%)	Khan et al. (2022)
Coagulation-flocculation		Enhances particle aggregation for easier removal	Requires chemical additives, can generate sludge	High (~70–90%)	Khan et al. (2022)
Membrane bioreactors (MBR)	Secondary—biological	High efficiency in removing microplastics can be combined with other methods	Expensive, requires high maintenance and energy	Very high (~90–99%)	Kim et al. (2022)
Biofilters		Cost-effective, utilizes natural filtration mechanisms	Limited efficiency for very small particles	Moderate (~50–80%)	Kim et al. (2022)
Biomimetic techniques		Mimics natural processes for effective filtration	Still under research, high cost	High (~80–95%)	Kim et al. (2022)
Reverse osmosis (RO)	Secondary—chemical	Very high removal efficiency, removes nano-sized plastics	High energy consumption, membrane fouling	Very high (~95–99%)	Kim et al. (2022)
Electrocoagulation		Effective for charged microplastics, reduces coagulant use	Produces sludge, high operational costs	High (~80–95%)	Kim et al. (2022)

Chemical oxidation (O_3, H_2O_2, O_3/H_2O_2)	Advanced	Strong oxidation power, breaks down microplastics chemically	Chemical costs, potential by-products	High (~85–98%)	Rodríguez-Narvaez et al. (2021)
Electrochemical methods		Efficient and can be applied to various contaminants	Electrode degradation, high energy use	High (~85–97%)	Rodríguez-Narvaez et al. (2021)
Photocatalysis		Highly reactive, uses light-driven degradation	Requires specific catalysts, expensive	High (~85–98%)	Rodríguez-Narvaez et al. (2021)
Plasma treatment		Breaks down microplastics at molecular level	High energy consumption, complex setup	Very high (~90–99%)	Rodríguez-Narvaez et al. (2021)
Fenton and variants		Highly efficient, effective against a variety of microplastics	Requires chemical handling, generates secondary pollutants	Very high (~95–99%)	Rodríguez-Narvaez et al. (2021)

of MPs are removed (Cheng et al. 2021). The physical treatment techniques that involve the removal techniques of microplastics on a primary level are:

Grit Chamber Grit chambers are used for removing sand, including large particles. Depending on their designs and operations, grit chambers can also remove microplastics, oils, grease, and suspended solids. Initial treatments can potentially eliminate as much as 59% of the microplastics found in wastewater (Zhang et al. 2020).

Primary Sedimentation This is usually done after the preliminary process of microplastic removal is completed. The removal efficacy is up to 98%, the highest among all other primary treatment techniques. Since the method is highly dependent on gravity, the removal efficacy can differ occasionally depending on the size and type of microplastic removed. Gravity acts as the key actor in the settlement process by removing solids from liquids with its force. It removes suspended solids, biosolids, organic matter, and other particulates. After the screening and grit removal, the process is completed to remove microplastics from the water further. The settlement tank is designed so that the contaminated water flows in and the clean water goes out as outflow. As the wastewater progresses through the tank, solids sediment to the bottom due to gravitational forces. This results in a residual solid substance known as primary sludge, which is collected and transported for additional treatment or disposal.

Coagulation–Flocculation Sedimentation (CFS)
The procedure involves the introduction of a coagulant to counteract the charge of microplastics, thereby diminishing their stability. The microplastics undergo destabilization and aggregation, forming flocs that settle as sediments at the bottom (Girish et al. 2023; Khan et al. 2022). Coagulation is extensively utilized in wastewater treatment and has effectively eliminated microplastics (Xu et al. 2021). The polyethylene, polystyrene, and polyester analysis revealed a remarkable removal efficiency, achieving 97% for polyester fibers and 99% for polyethylene microfibers (Lapointe et al. 2020). A coagulant, like aluminum sulfate or ferric chloride, is introduced to the water. It acts as a chemical trigger that destabilizes negatively charged microplastic particles by neutralizing their surface charge, allowing them to clump together.

11.6.2.2 Secondary and Tertiary Treatment Techniques

Chemical Treatment Techniques

Reverse Osmosis Reverse osmosis is a pressure-driven membrane separation process that separates ions, molecules, and larger particles from water using a semipermeable membrane. Pure water is forced under high pressure through a membrane in RO, opposite to the sole natural osmotic pressure that typically prohibits water flow from a concentrated solution to a dilute one. The membrane's pore size is typically 0.0001 microns, and water molecules can pass and reject dissolved salts, heavy metals, and other impurities. RO is the most widely used process in desalination and removing various contaminants. Moreover, it is favored for municipal water supply and industrial processes. Although the rejections can be high, the process has high energy input for holding the required pressure and frequently suffers from membrane fouling and buildup of contaminants on the membrane surface (Shannon et al. 2008). Fouling and extending the membrane life are frequently achieved by pretreatment processes, whereas post-treatment may be necessary to adjust the water chemistry to the appropriate standards.

Electrocoagulation Electrocoagulation is an electrochemical process destabilizing water contaminants by in situ generation of coagulants from sacrificial anode electrodes. Typically, they use electrodes of aluminum or iron, but when an electric current is applied, the electrodes 'dissolve' into the water and release metal ions. Thus, these ions hydrolyze to give metal hydroxides that behave as coagulants because they neutralize the charge on colloidal particles, thus forming larger flocs (Toha et al. 2024c). After aggregation of the contaminants, they can be removed by subsequent sedimentation, flotation, or filtration. For removing heavy metals, oils, and emulsified contaminants, electrocoagulation has proven effective in treating industrial wastewater (Rahman et al. 2024). This method has one great advantage in reducing the amount of additional chemical coagulants, which often lead to the generation of secondary waste streams. However, performance is optimized by carefully controlling operational parameters such as current density, electrode spacing, and pH. The energy requirements and electrode consumption affect total operating costs.

Biological Treatment Techniques

Membrane Bioreactor (MBR) The activated sludge process, a conventional biological treatment, can be combined with membrane filtration in a membrane bioreactor (Judd 2011). In an MBR, microorganisms biologically accomplish the degradation of organic contaminants in a bioreactor. After treating the water, biomass is physically separated from treated water by passing the treated water through a membrane, usually microfiltration or ultrafiltration. The hybrid system has several advantages. The delicate pores of the membrane allow virtually all the biomass and colloidal particles to be retained, giving rise to a high-quality effluent that is often reusable. Second, membrane integration enables the system to operate at higher biomass concentration than conventional systems, increasing the degradation rate of organic pollutants and reducing the reactor volume (Meng et al. 2009). However, membrane fouling remains an issue of concern in the case of MBR applications. Various strategies, such as periodic back washing, chemical cleaning, and optimization of aeration, mitigate fouling and maintain system performance.

Biofilter Biofilter is one of the most accepted methods for microplastic removal for its eco-friendly processes. In addition to media filtration, biological processes can be employed to improve the effectiveness of depth filtration for pollutant removal (Thuptimdang et al. 2020). It converts environmental pollutants through attached microbial cells (biofilms) on filter media while physically separating particles within the filter bed, typically composed of sand, anthracite, and granular activated carbon (GAC) (Pujol et al. 1994; Basu et al. 2015). Engineering systems use a packed bed of media colonized by microorganisms (biofilters) where contaminants are removed from water (or air). In water treatment applications, the biofilters are designed to let the contaminated water flow through the media, and microbial communities are added to the surface to degrade pollutants (Vymazal 2010). An important part of the design of the media is to maintain a large media surface area in which microbial attachment and growth can occur. Since biofilters are very effective in removing organic compounds and other nutrients by biological processes, they are particularly suited for ammonia removal. Moreover, as they have low energy requirements and simple operation, they are also attractive for use in decentralized wastewater treatment systems and constructed wetlands (Tchobanoglous et al. 2014). These techniques are sensitive to environmental conditions, and temperature, pH, and contaminant concentration changes may affect microbial activity and treatment efficiency. To guarantee consistent performance, maintaining optimal operating conditions and periodically monitoring the process is required to adjust the operating parameters accordingly.

Comparative Analysis and Integrated Approaches

Chemical treatment methods, including reverse osmosis (RO) and electrocoagulation (EC), effectively eliminate dissolved salts, heavy metals, and colloidal particles. The significance of reverse osmosis in contexts requiring exceptionally high water purity is paramount, mainly when it achieves high rejection rates. Electrocoagulation presents environmentally beneficial advantages compared to the extensive use of chemical additives for particle destabilization and aggregation (Chen 2009). However, biological treatment techniques such as MBRs and biofilters are designed to break down organic contaminants. When integrated into MBRs, membranes promote biological degradation, offer physical (separation) components, and yield high-quality effluent with low turbidity and pathogen content. They are more straightforward in design and rely upon microorganisms' natural metabolic capabilities, making them highly suitable for decentralized treatment applications. Chemical and biological techniques are increasingly combined in integrated treatment systems. An example would be that the treatment train starts with electrocoagulation to take care of solids and decrease the contaminant load and finishes with an MBR to frost organic material and furnish physical differentiation of biomass. Such synergistic systems enhance efficiency and sustainability while minimizing energy usage and chemical use. Overall, they optimize treatment performance. Over the past decades, the water treatment sector has advanced due to technological developments in materials and process engineering. Research in reverse osmosis aims to improve membrane materials with increased permeability and reduced fouling resistance. New electrode configurations are also being developed to increase the coagulant produced and decrease the energy required. For the biological side, the cultivation of such specialized microbial consortia is attempted to develop robustness under variable influent conditions, thereby raising the softness of MBRs and biofilters (Meng et al. 2009; Tchobanoglous et al. 2014).

Biomimetics: The Greener Approach to Microplastic Treatment Techniques
Biomimetics is a relatively new treatment method for microplastic removal. Biomimetics emulates the principles of natural processes, systems, and elements to solve anthropogenic problems (Rashid et al. 2021). Biomimetics is the clever idea of replicating natural processes, which are proven to be the fittest and best because they survived the evolution processes of the Earth (Rashid et al. 2021). Instead of playing defense, one viable strategy might be looking for biomimetic methods for removing microplastics. Biomimetics, the practice of copying nature's designs and processes, has proven to be a powerful tool against environmental problems. Although existing techniques for eliminating microplastics have not been satisfactory, innovative green technologies based on natural systems could point toward a more comprehensive solution.

11.7 Sustainable Alternatives to Plastic Use

The use of plastics has numerous alternatives. Plastics that have degraded and broken down into microscopic forms are known as microplastics. Therefore, substituting eco-friendly products with plastic is the only approach to lower microplastic contamination globally. Renewability, minimal ecological impact, and environmental degradability should be the top priorities for sustainable plastic substitutes. A product is "sustainable" if it comes from a source that can be used again. Generally, bioplastics are formulated using bio-based polymers, including polylactic acid (PLA), polyhydroxyalkanoates (PHA), and starch-derived materials (Harnkarnsujarit et al. 2021). Their production involves incorporating natural polymers, reactive additives, and plasticizers to improve mechanical strength (Costa et al. 2023). The specific formulations containing agarose and silicon dioxide ensure complete biodegradability without emitting harmful gases (Rosenboom et al. 2022). These sustainable types of plastics can be categorized into three major types based on their source and degradability (European Bioplastics 2022): Bio-based and Biodegradable Plastics degrade naturally through microbial activity without any external artificial intervention. Bio-based but Non-Biodegradable Plastics are made from renewable resources but do not decompose as fast as biodegradable plastics. Fossil-based but biodegradable plastics are derived from petrochemicals or non-renewable sources and engineered to break down biologically and naturally.

11.7.1 Plastics Derived from Biodegradable and Compostable Materials

Polylactic Acid (PLA) The renewable resource-based thermoplastic polylactic acid (PLA) is developed from cornstarch, sugar beet, and other renewable sources (Dehnou et al. 2023). The existence of PLA as a linear aliphatic polyester has gained notable interest since the discovery of its environmentally compatible nature together with its easy processing characteristics and biocompatible properties (Dehnou et al. 2023; Ilyas et al. 2022). The biopolymer possesses excellent potential as a sustainable replacement for conventional petroleum-based plastic materials and simultaneously resolves environmental issues and the requirement for environmentally friendly materials (Ilyas et al. 2022; Wu et al. 2023).

Polyhydroxyalkanoates (PHA) Natural polyesters, called polyhydroxyalkanoates (PHAs), are synthesized by numerous microbial species during periods of unbalanced growth (Samui and Kanai 2019). Biopolymers are increasingly attracting attention from researchers seeking alternatives to conventional plastics due to their notable ecological properties, biocompatibility, and natural degradation abilities (Bano et al. 2024). The material properties of PHAs align with those of petrochemical plastics, characterized by hydrophobicity, thermoplasticity, piezoelectricity, and stereospecificity (Yogeswar et al. 2023). The applications of these polymers span various industrial sectors, including medical applications, environmental remediation, industrial manufacturing, and agriculture (Bano et al. 2024).

Starch-Based Plastics and Cellulose-Based Films Starch, a cost-effective biopolymer, is frequently modified with cellulose to enhance its mechanical and barrier properties, which are generally insufficient in pure starch films (Kour et al. 2024; Wahab et al. 2024). Research demonstrates that the inclusion of cellulose, primarily sourced from agricultural waste such as banana peels and midribs, markedly improves tensile strength and decreases water absorption, thereby rendering these bioplastics appropriate for uses including food packaging (Izzati et al. 2024) (Wahab et al. 2024).

11.7.2 Bio-Based but Non-Biodegradable Plastics

Bio-Polyethylene (Bio-PE) Experts have developed Bio-PE as a renewable plastic product by utilising ethanol derived from sugarcane instead of traditional petroleum-based raw materials (Pinto et al. 2016). Bio-PE maintains recyclability while producing lower greenhouse gas emissions than primary PE production (Van den Oever et al. 2017). Furthermore, it is not subject to biodegradation. The substance is available in bottles, plastic sacks, and packaging films.

Bio-polyethylene Terephthalate (Bio-PET) The enhancement of sustainability in Bio-PET is attributed to ethylene glycol derived from renewable sources, while its terephthalic acid component is sourced from petroleum. The plastic properties of conventional PET remain unchanged in Bio-PET for bottles and textile products (Mohanraj et al. 2020).

Bio-Polyamide (Bio-PA) Bio-PA obtains its name from bio-nylon and represents a material that derives its sources from castor oil together with alternative renewable materials (Carole et al. 2004). Bio-PA establishes its crucial position in automotive manufacturing and the textile sector because of its remarkable strength and heat tolerance. According to Meereboer et al. (2020), Bio-PA enables sustainability because it helps minimize the use of fossil fuels.

11.7.3 Fossil-Based but Biodegradable Plastics

Polybutylene Adipate Terephthalate (PBAT) represents a synthetic biodegradable plastic that emerges from fossil fuel materials. However, manufacturers typically combine it with starch-based bioplastics to boost flexibility and decay rates. Bio-PA appears in compostable bags and agricultural films. Moreover, it is also used in food packaging (Chinaglia et al. 2018). The natural decomposition rate of PBAT exceeds the rate of PLA, representing itself as a practical choice to reduce plastic waste (Song et al. 2009). Polycaprolactone (PCL) represents a second fossil-based bioplastic, whose composting conditions allow it to break down. The material finds its main applications for biomedical purposes since it meets biocompatibility requirements and has adjustable degradation behaviors. The applications of bioplastics are given in Fig. 11.1.

11.7.4 Applications of Bioplastics

11.7.4.1 Packaging Industry

The packaging industry is one of the most extensive applications of bioplastics. Food packaging of biodegradable and bio-based plastics is used in disposable cutlery and shopping bags. This sector is one of the most common uses of bioplastics, namely polyhydroxyalkanoates (PHA) and polylactic acid (PLA). Packaging types include: Bioplastics (such as PLA) are extensively used for packaging perishable goods as they match well with compostability constraints and food safety requirements (Avérous 2008). Conversely, restaurants and fast food chains have switched to using bioplastic plates and cutlery to cut plastic waste (Ashter 2016). Starch-based bioplastics are an eco-friendly material for use in places other than ordinary plastics, such as shopping and garbage bags, in that they decompose into composting conditions (Müller et al. 2012).

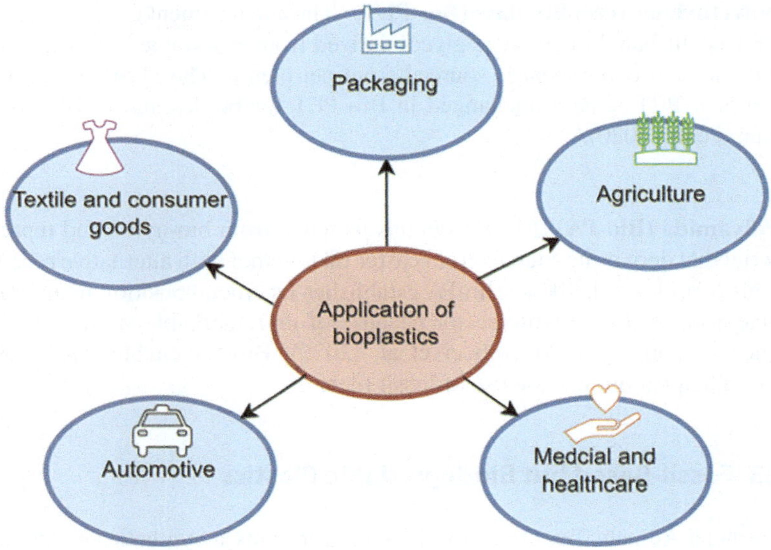

Fig. 11.1 Application of bioplastics

11.7.4.2 Agricultural Applications

Biodegradable mulching films of PHA or starch-based polymer are being used in agricultural fields to improve soil health and minimize plastic pollution (Thakur et al. 2014). Bioplastic-based plant containers naturally degrade with time so there is no worry about waste disposal (Reddy et al. 2013). These are also used to control the release of fertilizers and improve efficiency (Koller et al. 2017).

11.7.4.3 Medical and Healthcare Applications

Polycaprolactone (PCL) and PLA-based sutures and implants degrade over time in the body, thereby needing to be surgically removed; although 'degrading over time in the body' and 'surgically removed' are slightly similar in the meaning, the process is likely to be different. For controlled-release applications, biodegradable polymers like PHA and PLA are used as drug carriers (Chen 2009). Sterilizable and biodegradable bioplastics can assist in the reduction of hospital waste and promote more sustainable medical practices (Meereboer et al. 2020).

11.7.4.4 Automotive and Electronics Industry

Dashboards, upholstery, and trim parts in vehicles are made with polyamide (Bio-PA) and polyethylene (Bio-PE) (Carole et al. 2004). Biodegradable and recyclable bioplastics are increasingly used to make electronic housings.

11.7.4.5 Textile and Consumer Goods

Bio-based polyesters and PLA fibers are also used in textiles, replacing petroleum-based fabric, which is renewable (Shen et al. 2009). Shoe manufacturers increasingly use polymers made from bio-based materials in soles and uppers, reverting away from synthetic materials (Mekonnen et al. 2013).

11.7.5 Bioplastics and the Environment

Bioplastics have great potential to reduce microplastic pollution (Mekonnen et al. 2013). Hence, future research should focus on improving the biodegradability of non-degradable bioplastics and developing bioplastics from non-food biomass sources to enhance sustainability.

11.7.6 Life Cycle Analysis of Bioplastics

The environmental effects of bioplastics undergo lifecycle analysis, beginning with the farming of raw materials and continuing through the entire product's life duration until disposal. Bioplastics provide sustainability as renewable resource-derived materials serve as a pollution-reducing replacement for petrochemical plastics that chiefly cause environmental impact. The production stages of bioplastics come with unique environmental effects that run through their complete lifecycle. The analysis includes stages explained according to the following details:

11.7.6.1 Raw Material Cultivation

The production process of bioplastics starts with bio-based polymers that need sustainable agricultural methods to minimize environmental impacts. Extreme pesticide applications and forest clearing result in soil structure degradation, water contamination, and more excellent carbon dioxide production (Yadav and Nikalje 2024).

11.7.6.2 Manufacturing

The manufacturing operations for bioplastics allow continuous improvement to minimize their energy requirements, water consumption, and chemical usage. According to Yadav and Nikalje (2024), the production of bioplastics receives improved environmental results when manufacturers adopt renewable resources in conjunction with advanced extraction methods.

11.7.6.3 Environmental Impact

Bioplastics show reduced carbon emissions during their life cycle since they use bio-derived carbon instead of traditional fossil carbon sources (Mohapatra et al. 2024). According to Park and Won (2024), bioplastic polyhydroxy butyrate (PHB) and polylactic acid (PLA) possess global warming potential (GWP) levels exceeding those of conventional fossil-based plastics at the production stage.

11.7.6.4 End-of-Life Management

Biodegradable plastics are well-suited for anaerobic digestion and incineration treatments to decrease their total GWP as part of controlled waste management programs (Park and Won 2024). The proper disposal system and recycling programs decrease dependence on fresh resources while decreasing plastic environmental damage (Yadav and Nikalje 2024). The study of bioplastics as an alternative to standard plastics demonstrates multiple advantages and disadvantages throughout their existence.

11.7.7 Degradation Mechanism of Bioplastics

11.7.7.1 Mechanisms of Bioplastic Degradation

Bioplastic degradation is governed by several processes, including hydrolysis, enzymatic breakdown, and microbial assimilation. The polymer's composition, environmental circumstances, and microbial population activity dictate the degree and speed of breakdown (Kale et al. 2015).

Hydrolytic Degradation The degradation of bioplastics, including polylactic acid (PLA) and polycaprolactone (PCL), relies on hydrolysis. Water molecules hydrolyze the ester bonds in the polymer chain, producing smaller oligomers and monomers (Niaounakis 2013a, b). The hydrolytic degradation rate is influenced by temperature, pH, and moisture levels (Rudnik 2019). However, enzymatic and microbial degradation are also involved in the mechanism of bioplastic degradation (Tokiwa et al. 2009; Shah et al. 2008).

11.7.7.2 Biodegradation Mechanism of Plastics

Biodeterioration, fragmentation, assimilation, and mineralization are among the biodegradations that occur during the decomposition of plastic. These actions are crucial to halting the accumulation of plastic garbage in the environment (Narancic and O'Connor 2019).

Biodeterioration

According to Lucas et al. (2008), the first stage happens when environmental elements including UV rays, mechanical pressures, and temperature fluctuations degrade the polymer structure, enabling microbes to break it down.

Bio-fragmentation

During fragmentation, microorganisms release enzymes that split polymer chains into smaller pieces. According to Gewert et al. (2015), this process is dependent on the breakdown of extremely resistant plastics, including polyolefin plastics.

Assimilation

After being fragmented, monomers and oligomers are taken up by microbial cells and assimilated by their metabolic pathways. This converts the plastic-derived carbon to biomass, energy, and metabolic by-products (Song et al. 2009).

Mineralization

The last process in biodegradation includes the total breakdown of plastic portions into carbon dioxide, water, and biomass. Under aerobic conditions, CO_2 and water will be produced, while under anaerobic conditions, the possibility of methane production exists (Kyrikou and Briassoulis 2007).

11.7.7.3 Factors Affecting Plastic Biodegradation

The rate of plastics' biodegradation is affected by more than one factor, including polymer characteristics, environmental conditions, and microbial communities.

Plastic Susceptibility to Biodegradation Depends on the Polymer Characteristics

It highly depends on the crystallinity, molecular weight, and polymer composition (Albertsson et al. 2018). Temperature, humidity, oxygen levels, and pH significantly influence the degradation rates (Artham and Doble 2009; Davey 2007). Plastic-degrading microorganisms are present, increasing the biodegradation process in composting and marine environments (Jacquin et al. 2019). The following table compares the advantages and disadvantages of bioplastic use and plastic use (Table 11.2).

11.7.8 Advantages of Bioplastics

11.7.8.1 Environmental Benefits

Among the bioplastics, the ones based on plant origin have the least carbon footprint compared to petroleum-based plastics. Bioplastics are different from traditional plastic as they absorb CO_2 during plant growth, while in the production and disposal phase, they off CO_2 (Shen et al. 2009). Bio-production, such as bio-polyethylene (Bio PE) and polylactic acid (PLA) production, significantly reduces greenhouse gas emissions (Pinto et al. 2016). For example, polyhydroxyalkanoates (PHA) and starch-based polymers are biodegradable and naturally decompose in soil and marine environments (Chen 2009). As a result, plastic accumulation in landfills and oceans is reduced, which combats the severe problem of plastic pollution. Moreover, compostable bioplastics help from a waste management perspective by turning organic waste into a compostable product (Tokiwa et al. 2009). Plastics degrade into microplastics, which persist in the ecosystems and threaten marine and terrestrial life. On the other hand, biodegradable bioplastics decompose into non-harmful ingredients, minimizing the persistent effect of plastic contamination (Koller et al. 2017).

Table 11.2 Comparative analysis of the advantages and disadvantages of bioplastic use and plastic use

Criteria	Bioplastics	Regular plastics
Biodegradability	Biodegradable under specific conditions (e.g., industrial composting). Some types, like PLA, take longer to degrade in natural environments (Brodin et al. 2017)	Non-biodegradable; persists in the environment for hundreds of years. Degrades into microplastics (Andrady 2017a, b)
Advantages	• Derived from renewable resources (corn starch, sugarcane, etc.) • Lower carbon footprint • Some types are compostable, reducing landfill waste (Emadian et al. 2017)	• Durable, flexible, and cost-effective • Widely available with established production infrastructure • Resistant to degradation and chemical interactions
Disadvantages	• Requires specific composting facilities for degradation • Higher production costs • May compete with food production for raw materials (Khoo et al. 2011)	• Derived from fossil fuels, contributing to greenhouse gas emissions • Causes long-term environmental pollution and marine plastic contamination • Recycling rates are low (Geyer et al. 2017)

11.7.8.2 Economic Benefits

Bioplastic production from renewable feedstocks such as corn, sugarcane, or algae reduces dependence on nonrenewable products like petroleum, a finite resource with an uneven price (Ashter 2016). This improves energy security and maintains the raw material supply line. The global bioplastics market is growing, creating new economic opportunities. On the other hand, bioplastic research and production are sources of job creation for industries that invest in the area, including agriculture, biotechnology, and manufacturing (Meereboer et al. 2020). Furthermore, the governments of different countries are offering incentives for the production of bioplastics to accelerate the pace of industry growth (Mekonnen et al. 2013). However, the fabrication of bio-based polymers has enhanced due to the bio-based advancement in the synthesis of PHA and PLA by improvisation in the fermentation technique, decreasing the cost of bioplastics and gaining the competitiveness to discuss (Mohanraj et al. 2020).

11.8 Conclusions and Future Directions

Using bioplastics from natural resources reduces the exposure of consumers to toxic chemicals. Moreover, it is a safer consumer product and packaging. Thus, developed countries are creating policies to phase out single-use plastics and use sustainable alternatives. As for regulations, the European Union's Single-Use Plastics

Directive helps encourage bioplastic adoption. Moreover, in many countries, prohibition on non-biodegradable plastic bags also exhibits a growing inclination toward eco–friendly alternatives. A growing consumer awareness based on environmental issues and high demand for sustainable products has sprung up in recent years. The businesses that take bioplastics for their use in packaging and its products are the ones that follow corporate social responsibility (CSR) initiatives and are targeted at eco-conscious consumers. Although bioplastics are a sustainable alternative form of plastics, many restrictions are associated with their widespread use. One major challenge is that they are biodegradable, and this is conditioned on specific environmental conditions. The limitation of this degradation is that it interferes with the biodegradation of PLA and PHA-based bioplastics in freshwater and marine settings. Moreover, the bioplastics' lifecycle, including production and disposal, incurs cost competitiveness and recyclability issues. Bioplastics can also be integrated into existing waste management systems with technological innovations and policy changes to increase their economic feasibility and environmental benefits. These also exhibit mechanical and thermal properties inferior to those of conventional plastics. Accordingly, additives and complex processing procedures need to be used for improvement, which leads to an increase in production cost. Some factors restrict the market for bioplastics, namely available biomass sources and subsequent sustainable production methods, which would diminish the scalability and commercial viability of the product. However, public awareness and media attention to the benefits of bioplastics and these limitations make it important to continue research and development to solve the technical, economic, and environmental burdens of using bioplastics.

References

Albertsson AC, Hakkarainen M, Karlsson S (2018) Degradation of polymeric materials. In: Handbook of environmental degradation of materials. Elsevier, Amsterdam, pp 19–42

Allen S, Allen D, Phoenix VR, Le Roux G, Durantez Jiménez P, Simonneau A, Galop D (2019) Atmospheric transport and deposition of microplastics in a remote mountain catchment. Nat Geosci 12(5):339–344. https://doi.org/10.1038/s41561-019-0335-5

Andrady AL (2011) Microplastics in the marine environment. Marine Pollution Bulletin 62(8):1596–1605. https://doi.org/10.1016/j.marpolbul.2011.05.030

Andrady (2017a) Plastics and the environment. Wiley, Hoboken, pp 433–450

Andrady AL (2017b) The plastic in microplastics: a review. Mar Pollut Bull 119(1):12–22. https://doi.org/10.1016/j.marpolbul.2017.01.082

Artham T, Doble M (2009) Biodegradation of aliphatic and aromatic polycarbonates. Int Biodeterior Biodegradation 63(7):870–874

Ashter SA (2016) Introduction to bioplastics engineering. Elsevier, Amsterdam

Avérous L (2008) Chapter 21 - polylactic acid: synthesis, properties and applications. In: Monomers, polymers and composites from renewable resources. Elsevier, Amsterdam, pp 433–450. https://doi.org/10.1016/b978-0-08-045316-3.00021-1

Aysha S, Sultan MB, Bhuiyan MAR, Toha M, Ria RT (2024) Do we unknowingly eat breakfast that contains microplastics? Unveiling the microplastic contamination in commercial flour. Food Hum 3:100333

Bano S, Aslam AA, Khan A, Shabbir A, Qayyum F, Wahab N, Jabar A, Ul Islam I, Ng SL (2024) A mini-review on polyhydroxyalkanoates: synthesis, extraction, characterization, and applications. Process Biochem 146:250–261. https://doi.org/10.1016/j.procbio.2024.07.033

Barboza LGA, Dick Vethaak A, Lavorante BRBO, Lundebye AK, Guilhermino L (2018) Marine microplastic debris: an emerging issue for food security, food safety and human health. Mar Pollut Bull 133:336–348. https://doi.org/10.1016/j.marpolbul.2018.05.047

Barnes DKA, Galgani F, Thompson RC, Barlaz M (2009) Accumulation and fragmentation of plastic debris in global environments. Philos Trans R Soc B Biol Sci 364(1526):1985–1998. https://doi.org/10.1098/rstb.2008.0205

Basu OD, Dhawan S, Black K (2015) Applications of biofiltration in drinking water treatment - a review. J Chem Technol Biotechnol 91(3):585–595. https://doi.org/10.1002/jctb.4860

Bergmann M, Gutow L, Klages M (2019a) Marine anthropogenic litter. Springer, Cham

Bergmann M, Mützel S, Primpke S, Tekman MB, Trachsel J, Gerdts G (2019b) White and wonderful? Microplastics prevail in snow from the Alps to the Arctic. Sci Adv 5(8):1157. https://doi.org/10.1126/sciadv.aax1157

Bhattacharya P, Lin S, Turner JP, Ke PC (2010) Physical adsorption of charged plastic nanoparticles affects algal photosynthesis. J Phys Chem C 114(39):16556–16561. https://doi.org/10.1021/jp1054759

Brodin M, Klintström E, Boldizar A (2017) Plastics recycling: Challenges and opportunities. Clean Technologies and Environmental Policy 19:423–429. https://doi.org/10.1007/s10098-016-1220-7

Browne MA, Crump P, Niven SJ, Teuten E, Tonkin A, Galloway T, Thompson R (2011) Accumulation of microplastic on shorelines worldwide: sources and sinks. Environ Sci Technol 45(21):9175–9179. https://doi.org/10.1021/es201811s

Carole TM, Pellegrino J, Paster MD (2004) Opportunities in the industrial biobased products industry. Appl Biochem Biotechnol 113(1-3):871–885

Chen GQ (2009) A microbial polyhydroxyalkanoates (PHA) based bio- and materials industry. Chem Soc Rev 38(8):2434–2446

Cheng YL, Kim J-G, Kim H-B, Choi JH, Fai Tsang Y, Baek K (2021) Occurrence and removal of microplastics in wastewater treatment plants and drinking water purification facilities: a review. Chem Eng J 410:128381. https://doi.org/10.1016/j.cej.2020.128381

Chinaglia S, Tosin M, Degli-Innocenti F (2018) Biodegradation rate of biodegradable plastics at molecular level. Polymer Degradation and Stability 150:104–113. https://doi.org/10.1016/j.polymdegradstab.2018.02.003

Cole M, Lindeque P, Halsband C, Galloway TS (2011) Microplastics as contaminants in the marine environment: a review. Mar Pollut Bull 62(12):2588–2597. https://doi.org/10.1016/j.marpolbul.2011.09.025

Costa A, Encarnação T, Tavares R, Todo Bom T, Mateus A (2023) Bioplastics: innovation for green transition. Polymers 15(3):517. https://doi.org/10.3390/polym15030517

Danso D, Schmeisser C, Chow J, Zimmermann W, Wei R, Leggewie C (2019) New insights into the function and global distribution of polyethylene terephthalate (PET) degrading bacteria and enzymes. Eng Life Sci 19(5):256–265. https://doi.org/10.1002/elsc.201900033

Davey RJ (2007) The environmental impact of plastic waste. Environmental Science and Technology 41(10):3289–3290

de Souza Machado AA, Kloas W, Zarfl C, Hempel S, Rillig MC (2018) Microplastics as an emerging threat to terrestrial ecosystems. Glob Chang Biol 24(4):1405–1416. https://doi.org/10.1111/gcb.14020

Dehnou KH, Majidipour M, Norouzi GS (2023) A review: studying the effect of graphene nanoparticles on mechanical, physical, and thermal properties of polylactic acid polymer. RSC Adv 13(6):3976–4006. https://doi.org/10.1039/d2ra07011a

Dris R, Gasperi J, Rocher V, Saad M, Renault N, Tassin B (2016a) Microplastic contamination in an urban area: a case study in greater Paris. Environ Chem 13(6):592–598. https://doi.org/10.1071/EN14167

Dris R, Gasperi J, Saad M, Mirande C, Tassin B (2016b) Synthetic fibers in atmospheric fallout: a source of microplastics in the environment? Mar Pollut Bull 104(1–2):290–293. https://doi.org/10.1016/j.marpolbul.2016.01.006

Emadian SM, Onay TT, Demirel B (2017) Biodegradation of bioplastics in natural environments. Waste Management 59:526–536. https://doi.org/10.1016/j.wasman.2016.10.006

European Bioplastics (2022) What are bioplastics? https://www.european-bioplastics.org/bioplastics/

Foley CJ, Feiner ZS, Malinich TD, Höök TO (2018) A meta-analysis of the effects of exposure to microplastics on fish and aquatic invertebrates. Sci Total Environ 631-632:550–559. https://doi.org/10.1016/j.scitotenv.2018.03.046

Gautam R, Bhattacharya J, Yadav A (2022) Emerging contaminant of concern: Microplastics in aquatic environment. Environmental Research 204:111984. https://doi.org/10.1016/j.envres.2021.111984

Gewert B, Plassmann MM, MacLeod M (2015) Pathways for degradation of plastic polymers floating in the marine environment. Environ Sci Processes Impacts 17(9):1513–1521. https://doi.org/10.1039/c5em00207a

Geyer R, Jambeck JR, Law KL (2017) Production, use, and fate of all plastics ever made. Sci Adv 3(7):e1700782. https://doi.org/10.1126/sciadv.1700782

Giacovelli C (2018) Single-use plastics: a roadmap for sustainability. United Nations Environment Programme, Nairobi

Gigault J, Ter Halle A, Baudrimont M, Pascal PY, Gauffre F, Phi TL, Reynaud S (2018) Current opinion: what is a nanoplastic? Environ Pollut 235:1030–1034. https://doi.org/10.1016/j.envpol.2018.01.024

Girish N, Parashar N, Hait S (2023) Coagulative removal of microplastics from aqueous matrices: recent progresses and future perspectives. Sci Total Environ 899:165723. https://doi.org/10.1016/j.scitotenv.2023.165723

Gouin T, Roche N, Lohmann R, Hodges G (2011) A thermodynamic approach for assessing the environmental exposure of chemicals absorbed to microplastic. Environ Sci Technol 45(4):1466–1472. https://doi.org/10.1021/es1032025

Harnkarnsujarit N, Wongphan P, Chatkitanan T, Laorenza Y, Srisa A (2021) Bioplastic for sustainable food packaging. In: Sustainable food processing and engineering challenges. Elsevier, Amsterdam, pp 203–277. https://doi.org/10.1016/b978-0-12-822714-5.00007-3

Horton AA, Vijver MG, Lahive E, Spurgeon DJ, Svendsen C (2017) Acute toxicity of microplastics to fish and freshwater macroinvertebrates. Environ Toxicol Chem 36(11):2821–2829. https://doi.org/10.1002/etc.3825

Huerta Lwanga E, Mendoza Vega J, Ku Quej V, Chi J, Sanchez del Cid L, Chi C, Geissen V (2016) Field evidence for transfer of plastic debris along a terrestrial food chain. Sci Rep 6(1):21481. https://doi.org/10.1038/srep21481

Hussain N, Sarfraz M, Nadeem M, Bilal M (2020) Microplastics as emerging pollutants: Detection techniques, environmental fate, and impact. Environmental Chemistry Letters, 18(5):1563–1575. https://doi.org/10.1007/s10311-020-01001-5

Ilyas R, Aisyah H, Sharma S, Syafri E, Sari N, Bangar S, Zainudin E, Faudzi A, Zuhri M, Jumaidin R, Abral H, Asrofi M, Asyraf M, Sapuan S, Hassan S, Nawab Y (2022) Natural fiber-reinforced polylactic acid, polylactic acid blends and their composites for advanced applications. Polymers 14(1):202. https://doi.org/10.3390/polym14010202

Izzati NAR, Madihah HAM, Nurazzi MN, Haafiz MMK (2024) The effect of cellulose loadings on the properties of tapioca/banana peels starch bioplastics. Res Sq 1:22. https://doi.org/10.21203/rs.3.rs-4210538/v1

Jacquin J, Cheng J, Odobel C, Pandin C, Conan P, Pujo-Pay M, Barbe V (2019) Microbial ecotoxicology of marine plastic debris: a review on colonization and biodegradation by the "plastisphere". Front Microbiol 10:865

Jiang J, Wang Z, Hu J (2020) Turning old clothes into new fibers: closing the loop for textiles. Environ Sci Technol 54(24):15353–15361. https://doi.org/10.1021/acs.est.0c05680

Judd S (2011) The MBR Book: Principles and Applications of Membrane Bioreactors for Water and Wastewater Treatment (2nd ed.). Butterworth-Heinemann https://doi.org/10.1016/C2009-0-21588-7

Kale G, Auras R, Singh SP, Narayan R (2015) Biodegradability of polylactide bottles in real and simulated composting conditions. Polym Test 32(6):1001–1011

Khan NA, Khan AH, López-Maldonado EA, Alam SS, López López JR, Méndez Herrera PF, Mohamed BA, Mahmoud AED, Abutaleb A, Singh L (2022) Microplastics: occurrences, treatment methods, regulations and foreseen environmental impacts. Environ Res 215:114224. https://doi.org/10.1016/j.envres.2022.114224

Khoo HH, Tan RBH, Chng KWL (2011) Environmental impacts of conventional plastic and biodegradable plastic bags. International Journal of Life Cycle Assessment 16:231–242. https://doi.org/10.1007/s11367-011-0254-9

Kim S, Sin A, Nam H, Park Y, Lee H, Han C (2022) Advanced oxidation processes for microplastics degradation: a recent trend. Chem Eng J Adv 9:100213. https://doi.org/10.1016/j.ceja.2021.100213

Koelmans AA, Bakir A, Burton GA, Janssen CR (2016a) Microplastic as a vector for chemicals in the aquatic environment: critical review and model-supported reinterpretation of empirical studies. Environ Sci Technol 50(7):3315–3326. https://doi.org/10.1021/acs.est.5b06069

Koelmans AA, Besseling E, Shim WJ (2016b) Nanoplastics in the aquatic environment: critical review. In: Marine anthropogenic litter. Springer, Cham, pp 325–340. https://doi.org/10.1007/978-3-319-16510-3_12

Koelmans AA, Mohamed Nor NH, Hermsen E, Kooi M, Mintenig SM, De France J (2019) Microplastics in freshwaters and drinking water: critical review and assessment of data quality. Water Res 155(1):410–422. https://doi.org/10.1016/j.watres.2019.02.054

Kole PJ, Löhr AJ, Van Belleghem FGAJ, Ragas AMJ (2017) Wear and tear of tyres: a stealthy source of microplastics in the environment. Int J Environ Res Public Health 14(10):1265. https://doi.org/10.3390/ijerph14101265

Koller M, Salerno A, Dias M, Reiterer A, Braunegg G (2017) Modern biotechnological polymer synthesis: a review. Food Technol Biotechnol 55(2):187–210

Kour M, Kumar R, Chaudhary S (2024) Innovations in starch based bioplastics: a comprehensive review of methods and applications. Chem Sci Adv 1(2):47–66. https://doi.org/10.69626/csa.2024.0047

Kowalski N, Reichardt AM, Waniek JJ (2016) Sinking rates of microplastics and potential implications of their alteration by physical, biological, and chemical factors. Mar Pollut Bull 109(1):310–319. https://doi.org/10.1016/j.marpolbul.2016.05.064

Kyrikou I, Briassoulis D (2007) Biodegradation of agricultural plastic films: a critical review. J Polym Environ 15(3):125–150

Lapointe M, Farner JM, Hernandez LM, Tufenkji N (2020) Understanding and improving microplastic removal during water treatment: impact of coagulation and flocculation. Environ Sci Technol 54(14):8719–8727. https://doi.org/10.1021/acs.est.0c00712

Lee H, Shim JE, Park IH, Choo KS, Yeo M-K (2022) Physical and biomimetic treatment methods to reduce microplastic waste accumulation. Mol Cell Toxicol 19(1):13–25. https://doi.org/10.1007/s13273-022-00289-z

Lehner R, Weder C, Petri-Fink A, Rothen-Rutishauser B (2019) Emergence of nanoplastic in the environment and possible impact on human health. Environ Sci Technol 53(4):1748–1765. https://doi.org/10.1021/acs.est.8b05512

Li J, Liu H, Chen JP (2018) Microplastics in freshwater systems: a review on occurrence, environmental effects, and methods for microplastics detection. Water Res 137:362–374. https://doi.org/10.1016/j.watres.2017.12.056

Lu L, Luo T, Zhao Y, Cai C, Fu Z (2019) Interaction between microplastics and microorganisms as well as gut microbiota: a consideration on environmental animal and human health. Sci Total Environ 667:94–100. https://doi.org/10.1016/j.scitotenv.2019.02.380

Lucas N, Bienaime C, Belloy C, Queneudec M, Silvestre F, Nava-Saucedo JE (2008) Polymer biodegradation: mechanisms and estimation techniques. Chemosphere 73(4):429–442

Lusher AL, Hollman PCH, Mendoza-Hill JJ (2017) Microplastics in fisheries and aquaculture. FAO Fish Aquac Techn Pap 615:1–122
Meereboer KW, Misra M, Mohanty AK (2020) Review of recent advances in the sustainable production of bio-based polyamides. ACS Sustain Chem Eng 8(24):8881–8905
Meng F, Zhang H, Yang F, Liu L (2009) Identification of activated sludge characteristics affecting membrane fouling in submerged membrane bioreactors. Separation and Purification Technology 51(1):95–103. https://doi.org/10.1016/j.seppur.2006.01.017
Mekonnen T, Mussone P, Khalil H, Bressler D (2013) Progress in bio-based plastics and plasticizing modifications. J Mater Chem A 1(43):13379–13398
Miller EA, Yamahara KM, French C, Spingarn NE, Birch JM, Van KS (2022) A Raman spectral reference library of potential anthropogenic and biological ocean polymers. Sci Data 9(1):980. https://doi.org/10.1038/s41597-022-01883-5
Mohanraj R, Sethuraman MG, Mukherjee A (2020) Biodegradable polymer-based nanocomposites for advanced packaging applications. J Appl Polym Sci 137(27):48858
Mohapatra S, Behera S, Das S, Giri J, Dash M, Palai SP, Senapati S, Pritam P, Rath P, Bastia TK, Panda AK (2024) Life cycle analyses and carbon footprint of bioplastics. Elsevier, Amsterdam, pp 355–369. https://doi.org/10.1016/b978-0-323-95199-9.00007-x
Müller RJ, Kleeberg I, Deckwer WD (2012) Biodegradation of polyesters containing aromatic constituents. J Biotechnol 86(2):87–95
Napper IE, Bakir A, Rowland SJ, Thompson RC (2015) Characterization, quantity and sorptive properties of microplastics extracted from cosmetics. Mar Pollut Bull 99(1–2):178–185. https://doi.org/10.1016/j.marpolbul.2015.07.029
Narancic T, O'Connor KE (2019) Plastic waste as a global challenge: are biodegradable plastics the answer to the plastic waste problem? Microbiology 165(2):129–137
Niaounakis M (2013a) Biopolymers: applications and trends. William Andrew, Norwich
Niaounakis M (2013b) Biopolymers: processing and products. Elsevier, Amsterdam
Park D, Won W (2024) Unveiling the environmental gains of biodegradable plastics in the waste treatment phase: a cradle-to-grave life cycle assessment. Chem Eng J. https://doi.org/10.1016/j.cej.2024.150540
Pinto T, Dias MOS, Bonomi A (2016) Sugarcane-based biorefineries for the production of bio-PE: a sustainability assessment. Biofuels Bioprod Biorefin 10(2):210–223
Prata JC (2018) Airborne microplastics: consequences to human health? Environ Pollut 234:115–126. https://doi.org/10.1016/j.envpol.2017.11.043
Pujol R, Vázquez M, Cortina M, Olivella MÁ (1994) Pre-treatment of industrial wastewater before membrane separation: Optimization of flocculation conditions. Desalination 95(1):77–90. https://doi.org/10.1016/0011-9164(94)00016-0
Rahman MM, Toha M, Sikder S (2024) Micro/nanoplastics pollution from landfill sites: a comprehensive review on the formation, distribution, effects and potential mitigation. In: Technical landfills and waste management: volume 1: landfill impacts, characterization and valorisation. Springer, Cham, pp 247–265
Rajala K, Heiskanen A, Illikainen M (2020) Current trends in microplastic removal methods in wastewater treatment: a review. J Environ Chem Eng 8(5):104376. https://doi.org/10.1016/j.jece.2020.104376
Rashid M, Roni MA, Rahman M (2021) Clinical status of bioinspired and biomimetic materials. Elsevier, Amsterdam, pp 277–294. https://doi.org/10.1016/b978-0-12-821352-0.00009-5
Rayhan MRI, Toha M, Rahman MM (2024) Microplastics in air and their health hazard. In: Microplastics pollution and its remediation. Springer, Singapore, pp 189–214
Reddy MM, Vivekanandhan S, Misra M, Bhatia SK, Mohanty AK (2013) Biobased plastics and bionanocomposites: current status and future opportunities. Prog Polym Sci 38(10–11):1653–1689
Reisser J, Shaw J, Hallegraeff G, Proietti M, Barnes DKA, Thums M, Pattiaratchi C (2014) Millimeter-sized marine plastics: A new pelagic habitat for microorganisms and invertebrates. PLOS ONE, 9(6):e100289. https://doi.org/10.1371/journal.pone.0100289
Rillig MC (2012) Microplastic in terrestrial ecosystems and the soil? Environ Sci Technol 46(12):6453–6454. https://doi.org/10.1021/es302011r

Rochman CM, Browne MA, Halpern BS, Hentschel BT, Hoh E, Karapanagioti HK, Thompson RC (2013) Policy: classify plastic waste as hazardous. Nature 494(7436):169–171. https://doi.org/10.1038/494169a

Rodríguez-Narvaez OM, Goonetilleke A, Perez L, Bandala ER (2021) Engineered technologies for the separation and degradation of microplastics in water: a review. Chem Eng J 414:128692. https://doi.org/10.1016/j.cej.2021.128692

Rosenboom J-G, Langer R, Traverso G (2022) Bioplastics for a circular economy. Nat Rev Mater 7:117–137. https://doi.org/10.1038/s41578-021-00407-8

Rudnik E (2019) Compostable polymer materials. Elsevier, Amsterdam

Samui AB, Kanai T (2019) Polyhydroxyalkanoates based copolymers. Int J Biol Macromol 140:522–537. https://doi.org/10.1016/j.ijbiomac.2019.08.147

Setälä O, Fleming-Lehtinen V, Lehtiniemi M (2014) Ingestion and transfer of microplastics in the planktonic food web. Environmental Pollution 185:77–83. https://doi.org/10.1016/j.envpol.2013.10.013

Shah AA, Hasan F, Hameed A, Ahmed S (2008) Biological degradation of plastics: a comprehensive review. Biotechnol Adv 26(3):246–265

Shannon MA, Bohn PW, Elimelech M, Georgiadis JG, Marinas BJ, Mayes AM (2008) Science and technology for water purification in the coming decades. Nature 452(7185):301–310. https://doi.org/10.1038/nature06599

Shen L, Worrell E, Patel MK (2009) Present and future development in plastics from biomass. Biofuels Bioprod Biorefin 4(1):25–40

Sikder S, Toha M, Rahman MM (2024) An overview on the current state of heavy metal pollution and recent remediation approaches. In: Heavy metal toxicity: human health impact and mitigation strategies. Springer, Cham, pp 217–263

Siracusa V, Blanco I (2020) Bio-based biodegradable polymers for sustainable packaging applications: A review. Polymers 12(7):1647. https://doi.org/10.3390/polym12071647

Song JH, Murphy RJ, Narayan R, Davies GB (2009) Biodegradable plastics and composting. Philos Trans R Soc B Biol Sci 364(1526):2127–2139

Tchobanoglous G, Stensel HD, Tsuchihashi R, Burton FL (2014) Wastewater engineering: Treatment and resource recovery (5th ed.). McGraw-Hill Education

Teuten EL, Saquing JM, Knappe DRU, Barlaz MA, Jonsson S, Björn A, Takada H (2009) Transport and release of chemicals from plastics to the environment and to wildlife. Philos Trans R Soc B Biol Sci 364(1526):2027–2045. https://doi.org/10.1098/rstb.2008.0284

Thakur VK, Thakur MK, Raghavan P (2014) Progress in green polymer composites from lignin for multifunctional applications: A review. ACS Sustainable Chemistry & Engineering 2(5):1072–1092. https://doi.org/10.1021/sc500087z

Thompson RC, Olsen Y, Mitchell RP, Davis A, Rowland SJ, John AWG, Russell AE (2004) Lost at sea: where is all the plastic? Science 304(5672):838. https://doi.org/10.1126/science.1094559

Thuptimdang P, Siripattanakul-Ratpukdi S, Ratpukdi T, Youngwilai A, Khan E (2020) Biofiltration for treatment of recent emerging contaminants in water: current and future perspectives. Water Environ Res 93(7):972–992. https://doi.org/10.1002/wer.1493

Toha M, Sikder S, Rahman MM, Muhib MI (2024a) Microplastic occurrences in freshwater fish of Bangladesh. IntechOpen, London. https://doi.org/10.5772/intechopen.114897

Toha M, Rahman RR, Sikder S, Akbor MA, Rahman MM (2024b) Tracking of microplastics distribution patterns and their characterisation in deposited road dust from Dhaka city, Bangladesh. Emerg Contamin 10:100381

Toha M, Rayhan MRI, Rahman MM (2024c) Microplastics in terrestrial ecosystems. In: Microplastics pollution and its remediation. Springer, Singapore, pp 85–114

Tokiwa Y, Calabia BP, Ugwu CU, Aiba S (2009) Biodegradability of plastics. Int J Mol Sci 10(9):3722–3742

Van den Oever M, Molenveld K, van der Zee M, Bos H (2017) Bio-based and biodegradable plastics: Facts and figures. Wageningen Food & Biobased Research, Report 1722. https://doi.org/10.18174/408350

Vymazal J (2010) Constructed wetlands for wastewater treatment. Water 2(3):530–549. https://doi.org/10.3390/w2030530

Wahab DN, Siddique BM, Khairuddin N, Chew JJ, Su HT (2024) Mechanical, structural and barrier properties of starch-based film reinforced with cellulose microfibres extracted from midribs of Musa Saba. Food Res 8(3):117–123. https://doi.org/10.26656/fr.2017.8(s3).12

Wright SL, Kelly FJ (2017) Plastic and human health: a micro issue? Environ Sci Technol 51(12):6634–6647. https://doi.org/10.1021/acs.est.7b00423

Wu Y, Nguyen TT, Gao X, Wang Y, Zhou T, Wu J (2023) Biodegradable polylactic acid and its composites: characteristics, processing, and sustainable applications in sports. Polymers 15(14):3096. https://doi.org/10.3390/polym15143096

Xiang Y, Jiang L, Zhou Y, Luo Z, Zhi R (2021) Innovative solutions for mitigating microplastic pollution: technologies and sustainability perspectives. J Clean Prod 296:126546. https://doi.org/10.1016/j.jclepro.2021.126546

Xu Q, Huang Q-S, Luo T-Y, Wu R-L, Wei W, Ni B-J (2021) Coagulation removal and photocatalytic degradation of microplastics in urban waters. Chem Eng J 416(4):129123. https://doi.org/10.1016/j.cej.2021.129123

Yadav K, Nikalje GC (2024) Comprehensive analysis of bioplastics: life cycle assessment, waste management, biodiversity impact, and sustainable mitigation strategies. PeerJ 12:e18013. https://doi.org/10.7717/peerj.18013

Yogeswar M, Valle N, Nagarajan A (2023) Cyanobacteria as a source of biodegradable plastics. IntechOpen, London. https://doi.org/10.5772/intechopen.110376

Zhang C, Chen X, Wang J, Tan L (2020) Advanced oxidation processes for microplastic degradation. Chem Eng J 397:125424. https://doi.org/10.1016/j.cej.2020.125424

Zhao X, Cornish K, Vodovotz Y (2018) Narrowing the gap for bioplastic use in food packaging: an update. Environ Sci Technol 52(5):3128–3144. https://doi.org/10.1021/acs.est.7b05582

The manufacturer's authorised representative in the EU is Springer Nature Customer Service Centre GmbH, Europaplatz 3, 69115 Heidelberg, Germany. If you have any concerns regarding our products, please contact ProductSafety@springernature.com

Printed and bound by CPI Group (UK) Ltd, Croydon, CR0 4YY
30/09/2026
02225300-0013